Geotechnical Engineering and Earth Science

Geotechnical Engineering and Earth Science

Edited by **Agnes Nolan**

SYRAWOOD
PUBLISHING HOUSE

New York

Published by Syrawood Publishing House,
750 Third Avenue, 9th Floor,
New York, NY 10017, USA
www.syrawoodpublishinghouse.com

Geotechnical Engineering and Earth Science
Edited by Agnes Nolan

International Standard Book Number: 978-1-68286-091-5 (Hardback)

The publisher's policy is to use permanent paper from mills that operate a sustainable forestry policy. Furthermore, the publisher ensures that the text paper and cover boards used have met acceptable environmental accreditation standards.

Trademark Notice: Registered trademark of products or corporate names are used only for explanation and identification without intent to infringe.

Printed in the United States of America.

Contents

Permissions

List of Contributors

Preface

Geotechnical Engineering is a significant field of study in the discipline of civil engineering which primarily aims to understand the characteristics and composition of different soils, rocks, etc. This book aims to outline different data measurement systems and instrument used for assessment and evaluation in geotechnical engineering and related fields. The topics encompassed in this book elucidate some of the most significant techniques and instruments like magnetometer, data calibration mechanisms, remote sensing, etc. Researches and case studies included in this book are contributed by some of the most eminent experts and scientists in the field. Students and researchers actively engaged in the field will find this book full of crucial and unexplored concepts.

Significant researches are present in this book. Intensive efforts have been employed by authors to make this book an outstanding discourse. This book contains the enlightening chapters which have been written on the basis of significant researches done by the experts.

Finally, I would also like to thank all the members involved in this book for being a team and meeting all the deadlines for the submission of their respective works. I would also like to thank my friends and family for being supportive in my efforts.

Editor

Results from the intercalibration of optical low light calibration sources 2011

B. U. E. Brändström[1], C.-F. Enell[2], O. Widell[3], T. Hansson[3], D. Whiter[4], S. Mäkinen[4], D. Mikhaylova[1], K. Axelsson[1], F. Sigernes[5], N. Gulbrandsen[6], N. M. Schlatter[7], A. G. Gjendem[8], L. Cai[9], J. P. Reistad[10], M. Daae[8], T. D. Demissie[8], Y. L. Andalsvik[11], O. Roberts[12], S. Poluyanov[13], and S. Chernouss[14]

[1]Swedish Institute of Space Physics, Kiruna, Sweden

[2]Sodankylä Geophysical Observatory, University of Oulu, Sodankylä, Finland

[3]SSC, ESRANGE, Kiruna, Sweden

[4]Finnish Meteorological Institute, Helsinki, Finland

[5]The Kjell Henriksen Observatory, UNIS, Longyearbyen, Norway

[6]University of Tromsø, Tromsø, Norway

[7]School of Electrical Engineering, Royal Institute of Technology, Stockholm, Sweden

[8]Norwegian University of Science and Technology, Trondheim, Norway

[9]Department of Physics, University of Oulu, Oulu, Finland

[10]University of Bergen, Bergen, Norway

[11]Department of Physics, University of Oslo, Oslo, Norway

[12]Aberystwyth University, Aberystwyth, UK

[13]Polar Geophysical Institute, Murmansk, Russia

[14]Polar Geophysical Institute, Apatity, Russia

Correspondence to: B. U. E. Brändström (urban.brandstrom@irf.se)

Abstract. Following the 38th Annual European Meeting on Atmospheric Studies by Optical Methods in Siuntio in Finland, an intercalibration workshop for optical low light calibration sources was held in Sodankylä, Finland. The main purpose of this workshop was to provide a comparable scale for absolute measurements of aurora and airglow. All sources brought to the intercalibration workshop were compared to the Fritz Peak reference source using the Lindau Calibration Photometer built by Wilhelm Barke and Hans Lauche in 1984. The results were compared to several earlier intercalibration workshops. It was found that most sources were fairly stable over time, with errors in the range of 5–25 %. To further validate the results, two sources were also intercalibrated at UNIS, Longyearbyen, Svalbard. Preliminary analysis indicates agreement with the intercalibration in Sodankylä within about 15–25 %.

1 Introduction

Following the first absolute measurement of night airglow by Rayleigh (1930), accurate absolute measurements of airglow and aurora have become increasingly important (see, for example, Trondsen, 1998; Syrjäsuo, 2001; Brändström, 2003; Gustavsson et al., 2006; Dahlgren et al., 2011 and references therein). Such absolute measurements are traditionally expressed in rayleighs, as proposed by Hunten et al. (1956). Further discussions about the definition of the rayleigh unit appear in Chamberlain (1995, App. II) and Baker (1974). In SI units the rayleigh is defined as follows (Baker and Romick, 1976):

$$1 \text{ rayleigh} \equiv 1 \text{ R} \triangleq 10^{10} \frac{\text{photons}}{\text{s m}^2 \text{ column}} \qquad (1)$$

The word column is often inserted in the units above and denotes the concept of an emission rate from a column of unspecified length along the line of sight (Hunten et al., 1956).

The apparent spectral radiant sterance (spectral radiance), $L_\gamma(\lambda)$, can be obtained from the spectral column emission rate, $I(\lambda)$, (in R/Å) according to Baker and Romick (1976):

$$L_\gamma(\lambda) = \frac{10^{10} I(\lambda)}{4\pi} \frac{\text{photons}}{\text{s m}^2 \text{ sr Å}}. \quad (2)$$

Integrating the spectral quantities $L_\gamma(\lambda)$ and $I(\lambda)$ over wavelength yields the, maybe more familiar, quantities radiance and column emission rate. In this work the rayleigh and the ångström ($1\text{ Å} = 10^{-10}$ m) will be used to preserve continuity with earlier intercalibration results, which expressed spectral column emission rate in R/Å.

After removing the instrument signature (bias, dark current, flat field, bad pixels, etc.), optical instruments are usually absolute calibrated by exposing the instrument to a calibration light source with a known spectral radiant sterance corresponding to a certain column emission rate (see, for example, Trondsen, 1998; Mäkinen, 2001; Brändström, 2003, and references therein). Instead of using calibration light sources, some instruments are calibrated by using known spectra of stars (for example Dahlgren et al., 2011).

This work reports the results of comparisons of calibration light sources during 2011. This is part of a long-term international effort to place aurora and airglow measurements taken at various locations around the world on a common calibration (and hence intensity) standard (Torr and Espy, 1981). In addition, a brief description of the intercalibration method in effect since 1985 is provided.

Following initial efforts in the 1960s by Michael Gadsden (Torr, 1983) and by Torr et al. (1976, 1977), regular intercalibration workshops have been organised (see Table 1 and references therein). After the intercalibration workshop in Katlenburg-Lindau in 1983, Lauche and Barke (1986) constructed the Lindau Calibration Photometer for comparison of low brightness sources (Fig. 1). This was done in order to support the work by M. Torr in the European sector. Yet, calibration sources from other countries have participated in some workshops over the years. As seen in Table 1, some intercalibration workshops have also taken place in non-European countries.

When Hans Lauche retired, Widell and Henricson (2003) took over the responsibility for the Lindau Calibration Photometer, and following Ola Widell's retirement in 2011, this responsibility was handed over to the corresponding author of this paper. Table 1 is an attempt to list all known official intercalibration workshops to date.

2 Calibration sources

In this calibration effort nine calibration sources were compared to the Fritz Peak (FP) reference source (this source is labelled "Fritz Peak international standard source"). This radioactive ^{14}C–activated phosphor source is only used at intercalibration workshops. Apart from the FP reference source,

Table 1. Known official intercalibration workshops. The 1967–1972 intercalibrations are mentioned by Torr (1983). Regarding later calibration workshops lacking a literature reference, the results and raw data are archived by the corresponding author of this paper. Copies are available upon request. The column # refers to the number of participating calibration sources.

Year	#	Location	Reference/responsible
1967		Fritz Peak	Gadsden and Marovich
1968		Paris	Weill
1969		Tokyo	Huruhata
1970		Kitt Peak	Broadfoot
1970		Harvard	Noxon
1970		Johns Hopkins	Schaeffer and Fastie
1972		Lindau	Leinert and Klüppelberg
1979	9	Seattle	Torr (1981)
1981	30	Aberdeen	Torr and Espy (1981)
1983	21	Lindau	Lauche
1985	16	Lysebu	Lauche and Barke (1986)
1987	14	Saskatoon	Lauche
1989	1	Lindau	Lauche
1991	6	Wien	Lauche (IAGA)
1995	4	Boulder	Lauche
1999	18	Lindau	Lauche and Widell (2000b)
2000	9	Stockholm	Lauche and Widell (2000a)
2001	10	Oulu	Widell and Henricson (2003)
2003	8	Longyearbyen	Widell and Mämmi (2003)
2006	7	Kiruna	Widell and Henricson (2008)
2007	6	Andøya	Henricson (2008)
2011a	10	Kiruna	This work
2011b	10	Sodankylä	This work
2011c	3	Longyearbyen	This work (prel. results)

the IRF UJO 920B, L1614, Y275 and the MPI-2 sources are also radioactive ^{14}C activated phosphor sources. The spectral output is continuous and depends on the phosphor. The IRF UJO sources are "light standards", probably manufactured by U.S. Radium Corp. in the 1960s and labelled with phosphor type and luminance values, "920B $< 20\,\mu$L", "L1614 $7\,\mu$L $\pm\,10\,\%$" and "Y275 $15\,\mu$L $\pm\,10\,\%$", respectively. The lambert L is a non-SI unit of luminance; 1 L corresponds to $10^4/\pi$ cd m^{-2}. It is furthermore a photometric unit, involving the spectral sensitivity of the human eye. These luminance values have probably never been used for calibration purposes, at least not in recent years.

Several of these sources have participated in intercalibrations dating back to the late 1960s (see Torr, 1983, Fig. 1). Although stable and easy to handle, these sources are nowadays rather difficult to transport due to flight safety regulations.

The ESRANGE tungsten lamp and the IRF Lauche lamp are tungsten lamps that operated at a predefined lamp current. Both were designed by Hans Lauche. The ESRANGE tungsten lamp was powered by an external power supply, while the IRF Lauche lamp has its own constant current supply.

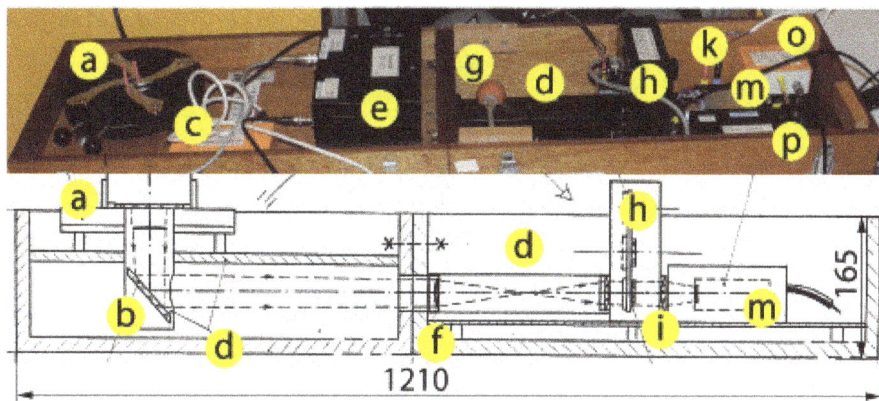

Fig. 1. The Lindau Calibration Photometer built by W. Barke and Hans Lauche at Max Planck Institute for Aeronomy, Katlenburg-Lindau (1984). (a) centering device for source under measurement (b) deflecting mirror, (c) power supply for pulse amplifier, (d) collimators, (e) HV supply. (f) objective lens, (g) dehumidifier, (h) filter wheel, (i) field stop, (k) Peltier cooler connector, (m) PMT Hamamatsu R-632, (o) connection box and (p) pulse amplifier.

These sources are not considered as stable as the radioactive sources, but on the other hand, they are much easier to transport.

The stability of the radioactive sources and the IRF Lauche lamp is discussed in Sect. 6.

Two sources are based on light-emitting diodes (LEDs): the ESRANGE MSP1 and the PGI Chernouss-38AM. The ESRANGE MSP1 has internal current regulators and is powered by a 28 V supply, while the PGI Chernouss-38AM is battery powered. Both participating LED sources consist of several LEDs and none of them has participated in earlier intercalibration workshops.

The FMI sphere (Mäkinen, 2001) consists of an integrating sphere, three identical 30 W internal tungsten lamps, a 75 W external tungsten lamp with a mechanical attenuator and several neutral density (ND) filters. The ND filters are required to decrease the output of the sphere to acceptable levels for low light instrumentation. The output of the sphere is calibrated by the manufacturer in foot-lamberts (an American customary unit for luminance; 1 ft-L corresponds to 3.426 cd m^{-2}). Note that this is a photometric unit involving the spectral sensitivity of the human eye, and that this calibrated luminance value is valid at the exit aperture of the integrating sphere, i.e. before the ND filters. Thus, for the intercalibrating effort described here, the luminance value should only be regarded as a source setting. However, knowing the spectral response of the ND filters, it is possible to compare the calibrated output of the sphere to the results presented in this report. It is hoped that this will be done in the future.

It should be noted that the ESRANGE sources were intercalibrated on 16 September 2011 at the Swedish Institute of Space Physics in Kiruna (referred to as 2011a), while all other sources except the FMI sphere were intercalibrated on 19 October 2011, at Sodankylä Geophysical Observatory in Sodankylä, Finland. The FMI sphere was intercalibrated on

the same date at the calibration laboratory at Finnish Meteorological Institute's Arctic Research Centre (FMI-ARC), also in Sodankylä. Both Sodankylä intercalibrations above are referred to as 2011b. The IRF sources as well as the MPI-2 source were intercalibrated at both locations.

During the course "Optical methods in auroral physics research" held in November 2011 at the University Centre in Svalbard (UNIS), the IRF Lauche lamp and the PGI Chernouss-38AM sources were intercalibrated with an SN-1633 NIST-traceable tungsten lamp in the calibration laboratory at UNIS (Sigernes et al., 2007). This intercalibration is referred to as 2011c.

During earlier intercalibration workshops the source naming conventions have been somewhat different for some sources. To remedy this in the future, a unique source identification number (SID) was introduced in 2011 to simplify future comparisons. Radioactive calibration sources have been assigned SID in the range 1–99; other sources are numbered from 101 (see Table 2).

This report only concerns sources intercalibrated in 2011. A full list of all sources that participated in this long-term calibration series is under preparation. Some of the participating calibration sources are shown in Fig. 2.

3   The Lindau Calibration Photometer

The Lindau Calibration Photometer is described by Lauche and Barke (1986). Furthermore, all technical documentation and design drawings, raw data and results from the calibration photometer as well as previous intercalibration workshops are archived by the corresponding author of this paper and are available upon request. As soon as time permits, this information will be scanned and made available on the Internet.

Table 2. Results of the intercalibration workshop. All values are spectral column emission rates in R/Å. The absolute calibration values at 3914 Å should be considered less reliable (see Sect. 6). Filter transmittance plots are available upon request. SID is source identification number.

Filter position		1	2	3	4	5	6	7		
Filter CW		3914	4280	4866	5573	5882	6299	6562	[Å]	
Filter bandwidth (FWHM)		41	27	25	16	13	12	15	[Å]	
Source name	SID								Settings	Note
FP reference source	1	0.34	5.7	3.2	2.6	5.1	9.2	15	Torr and Espy (1981)	
MPI-2	2			2	173	263	187	93	^{14}C	
IRF UJO 920B	3	4	101	62	22	13	8	4	^{14}C Phosphor 920B	
IRF UJO L1614	4	5	1	38	34	9			^{14}C Phosphor L1614	
IRF UJO Y275	5			4	261	362	201	107	^{14}C Phosphor Y275	
IRF Lauche lamp	101		1	8	54	96	207	352	1.62 V, 198.50 mA	1
IRF Lycksele lamp	102		1	9	72	150	360	489	6.21 V, 22.7 mA	2
ESRANGE tungsten lamp	103	3	10	61	359	544	728	635	10.9 V, 217.5 mA	2, 3
ESRANGE tungsten lamp	103			1	6	12	20	32	5.10 V 141.00 mA	2, 3
ESRANGE MSP1	104	226	335	150	280	308	523	299	LED 28 V supply	1, 3
PGI Chernouss-38AM	105	12	164	382	710	639	1520	1782	LED, setting 3 (max)	4
FMI sphere	106		5	26	72	78	180	353	L:C, A:150, ND:7, 1473.3 ft-L	5
FMI sphere	106		9	49	139	150	348	696	L:C, A:255, ND:7, 3092.0 ft-L	5
FMI sphere	106	1	13	67	170	189	422	809	L:BC, A:100, ND:7, 3388.0 ft-L	5
FMI sphere	106	1	20	100	294	304	682	1311	L:BC, A:255, ND:7, 5947.0 ft-L	5

Notes: 1. Constant current supply, 2. adjustable power supply, 3. 2011a intercalibration (Kiruna 16 September), 4. battery powered, 5. settings refer to lamp(s) in use (L), attenuator setting (A), neutral density filter (ND) and luminance in foot-lamberts (before the neutral density filters).

Fig. 2. Some of the low light sources intercalibrated at this workshop: (a) IRF Lauche lamp (SID 101) with power supply, (b) PGI Chernouss-38AM (SID 105), (c) IRF UJO Y275 (SID 5), (d) IRF UJO L1614 (SID 4), (e) IRF UJO 920B (SID 3), (f) FP reference source (SID 1), (g) MPI-2 source (SID 2) and (h) IRF Lycksele lamp (SID 102).

Figure 1 shows the general layout of the instrument. The source is attached to the centering device (a) and light passes a mirror (b), collimating tubes (d), an objective lens (f), filter wheel (h), with telecentric optics and field stop (i) and finally reaches the Peltier-cooled photomultiplier tube (PMT, Hamamatsu R632 GA37). Datasheets with plots of spectral response and quantum efficiency for this PMT are available on the Internet (www.datasheetcatalog. org/datasheet/hamamatsu/R632.pdf). The length of the instrument is 1210 mm, and the height is 165 mm. The two parts are folded together during transportation.

4   Intercalibration procedure

The Lindau Calibration Photometer was installed in a darkroom and the Peltier cooler was switched on several hours before measurements, so that the photomultiplier tube (PMT) would be sufficiently cooled and thermally stable. One person operated the calibration photometer and sources in the darkroom, while another person recorded the filter position and PMT counts using a filter position display and a precision frequency counter (HP 5328A and HP 53181A for 2011a and 2011b, respectively) located outside the darkroom. The frequency counter was set for a long gate time (3–5 s). In addition, an intercom was available between the darkroom and the outside. Filter position 0 is blocked and corresponds to dark current; the remaining positions correspond to seven filters from 3914 to 6562 Å (listed in Table 2). The filter bandwidths in the table correspond to the full width at half maximum (FWHM). Position 8 corresponds to a filter with centre wavelength 6707 Å. This filter is included in the intercalibration procedure, but the results are traditionally discarded since the FP reference source lacks calibration data for this wavelength. Transmittance curves for each filter exist in the calibration photometer documentation and are available upon request. Each source was then compared to the FP reference source. This was done according to the following procedure:

1. The FP reference source was attached to the centering device of the calibration photometer (Fig. 1a).

2. Three measurements were recorded from the frequency counter for each of the nine filter wheel positions (including dark current). As the filter wheel was rotated manually, the filter changes were announced and verified over the intercom and by using the filter position display.

3. The FP reference source was then replaced with the calibration source and step 2 above was repeated for that source. Metadata was recorded (filter temperature, start and stop times, etc.).

4. Steps 1–3 above were repeated for each of the nine calibration sources.

The spectral column emission rate (I_{Sp}) at filter position p (1 . . . 8) was then calculated from the following equation (by using a spreadsheet):

$$I_{Sp} = \frac{I_{Rp}(\overline{S}_p - \overline{S}_0)}{\overline{R}_p - \overline{R}_0} \frac{R}{Å} \tag{3}$$

where $\overline{S}_p$ and $\overline{R}_p$ are the average measured count rates for the calibration source and the FP reference source, respectively; $\overline{S}_0$ and $\overline{R}_0$ are averaged dark current measurements (filter position 0). I_{Rp} is the FP reference spectral column emission rate for filter p (refer to Table 2). To preserve continuity this procedure has been changed as little as possible since 1985.

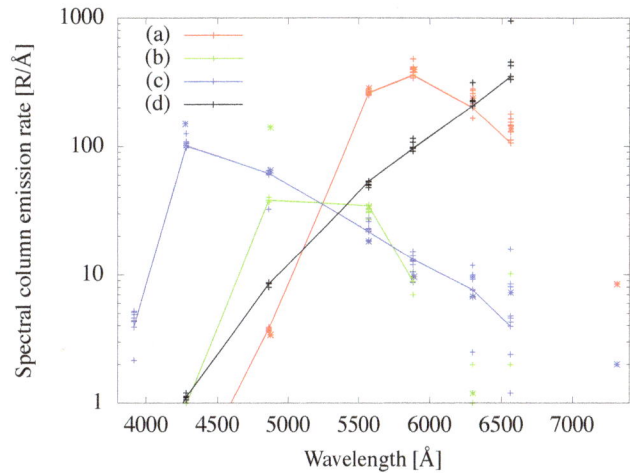

Fig. 3. Intercalibration results for three sources since 1981: (a) IRF UJO Y275 (SID 5), (b) IRF UJO L1614 (SID 4), (c) IRF UJO 920B (SID 3), (d) IRF Lauche lamp (SID 101, since 2000). The 2011b intercalibration results are connected with lines, giving a rough idea of the spectra of these sources. The 1981 intercalibration used different filters indicated by a "*".

5   Results

The results from this intercalibration effort are given in Table 2. Note that spectral column emission rates less than 1 R/Å have been removed in Table 2 due to poor signal-to-noise ratio. All raw data and preliminary results before post-processing are available at http://alis.irf.se/ewoc/2011.

Figure 3 plots all intercalibration results from 1981 until the present time for three radioactive and one tungsten lamp source. Table 3 lists the ratios of this intercalibration to earlier intercalibration workshops, as well as to the mean value of all listed workshops. Sources not appearing in Table 3 have only been intercalibrated once, or earlier intercalibration data have not been located yet. Figure 4 plots selected ratios from Table 3 as a time series. The ratios and wavelengths are selected based on the normal usage of the source for calibration of optical instrumentation.

The intercalibration was done under two assumptions: (1) the spectral radiant sterance of the FP reference source is stable and sufficiently well known, and (2) the calibration photometer is linear and stable during the calibration.

6   Discussion

The FP reference source is traceable to intercalibrations in the late 1960s (Torr, 1983) and the present absolute calibration values, obtained with a national standard source (Q47 tungsten filament lamp, calibrated by the National Bureau of Standards in 1977) from an intercalibration done by Torr and Espy (1981). Since 1981 the FP reference source has been used as reference source for intercalibration workshops in the

Table 3. Ratios of the 2011b intercalibration (Sodankylä) to earlier intercalibrations and to the mean value of all listed measurements. Sources not appearing in this table lack information of earlier calibration workshops. SID is source identification number.

Source name	SID	Year	Filter [Å]						
			3914	4280	4866	5573	5882	6299	6562
IRF UJO 920B	3	1981		0.67	0.95	1.19	1.36	1.12	0.55
		1985	0.77	0.80	1.01	1.17	1.26	1.14	0.49
		1999	0.89	0.99	1.03	0.99	1.10	1.01	1.66
		2000	0.85	0.93	0.97	0.95	1.04	0.77	0.47
		2001	0.75	0.96	0.95	0.99	1.02	0.83	3.32
		2006	0.80	0.95	1.00	0.95	0.94	0.81	0.25
		2007	0.91	1.03	1.01	0.96	0.87	0.65	0.86
		2011a	1.81	1.03	0.99	0.84	1.03	0.78	0.83
		Mean	1.02	0.91	0.99	0.99	1.05	0.87	0.63
IRF UJO L1614	4	1981		1.24	0.27	1.00	0.91		2.20
		1985	1.14	1.36	1.18	1.24	1.02		0.23
		1999	0.13	0.99	1.04	1.04	0.95		0.49
		2000	1.14	0.99	0.96	0.98	0.92		1.74
		2001	0.80	0.90	1.01	1.28	1.26		0.10
		2006	0.57	0.93	1.00	1.10	0.83		0.25
		2007	1.33	0.99	1.01	1.07	0.91		
		2011a	0.80	1.03	1.01	1.13	0.93		
		Mean	0.59	1.03	0.78	1.09	0.96		0.52
IRF UJO Y275	5	1981		6.00	1.12	0.92	0.91	0.84	0.77
		1985	0.33	0.70	1.01	1.04	0.96	0.93	0.94
		1999	1.00	1.05	1.01	0.99	0.94	0.71	0.65
		2000	1.00	0.95	0.98	0.95	0.89	0.71	0.59
		2001	5.00	1.17	1.06	1.01	0.75	0.73	0.69
		2006		0.91	1.03	1.03	0.91	0.89	0.72
		2007	1.00	1.11	1.04	1.02	0.87	0.77	0.82
		2011a	0.33	1.00	1.03	0.98	1.06	1.21	0.73
		Mean	0.88	1.06	1.03	0.99	0.91	0.84	0.75
IRF Lauche lamp	101	2000	0.95	1.02	1.00	1.07	0.98	0.93	0.78
		2001	1.06	0.93	1.06	1.12	0.84	0.91	0.82
		2007	1.20	1.06	0.98	1.06	0.90	0.92	1.06
		2011a	1.06	0.99	0.98	1.02	1.05	0.66	0.37
		Mean	1.05	1.00	1.00	1.05	0.95	0.86	0.70

aurora/airglow optical community. Note that the 1981 calibration did not include 3914 Å and 6707 Å. The origin of the absolute calibration value at 3914 Å (0.34 R/Å) is currently unknown. This is under investigation, and until further notice it should be treated as less reliable (extrapolated). As 1981 is a rather long time ago, doubts can clearly be cast on the stability of the FP reference source. It is thus of great importance to compare the FP reference source to a source traceable to a National Bureau of Standards as soon as possible. Although strongly desired, this has not been possible yet. Some steps have therefore been taken to indirectly assess the stability of the FP reference source.

Preliminary results from the independent 2011c intercalibration (Longyearbyen) of two sources are given in Table 4.

In addition, the spectra of these two sources were measured with a spectrograph. For the IRF Lauche source (SID 101), deviations appear to be less than ±15 % for wavelengths from 5573 Å. For shorter wavelengths this source has a very low output, as should be expected from a tungsten lamp. The ratios for the PGI Chernouss-38AM (SID 105) source are a bit more puzzling and, in particular, the large difference for 4866 Å is still under investigation. The spectrum of the PGI Chernouss-38AM LED source was found to be continuous but with two sharp peaks. One possible preliminary explanation for the discrepancy is that while the former calibrations were done by a filtered photometer, the 2011c calibration was done with a spectrograph. The spectrograph had a bandpass of approximately 100 Å, while the photometer filters have

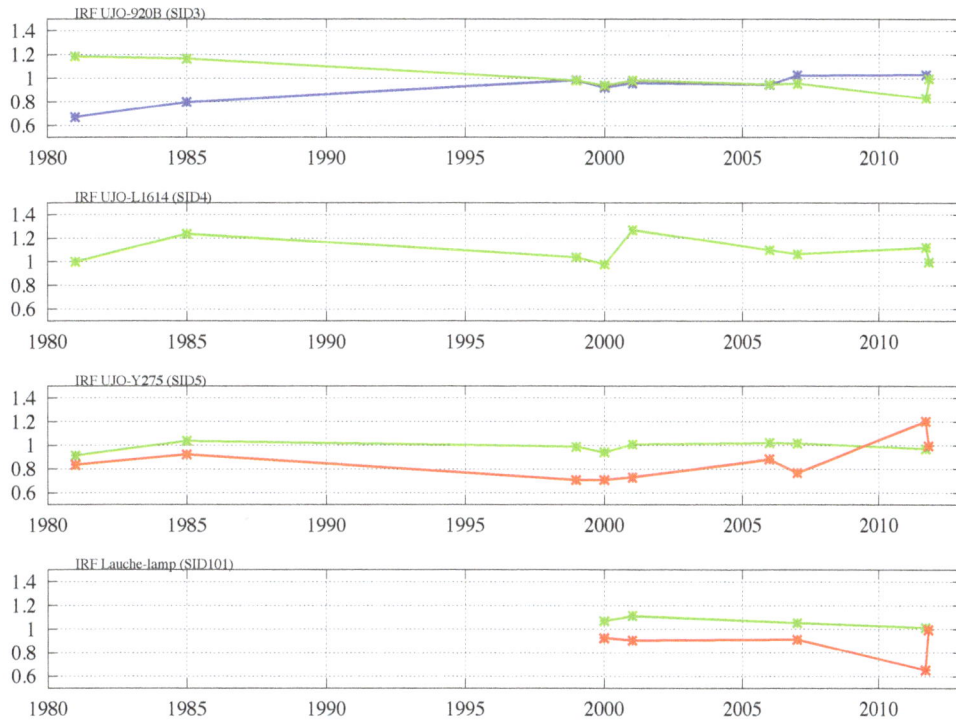

Fig. 4. Ratios of 2011b intercalibration in Sodankylä to earlier intercalibrations for IRF UJO 920B (SID3, top panel), IRF UJO L1614 (SID 4), IRF UJO Y275 (SID5) and IRF Lauche lamp (SID101) (bottom panel): (blue) 4280 Å, (green) 5573 Å and (red) 6299 Å. Note that none of these sources were intercalibrated 1987–1995.

a bandpass around 20 Å. As this source has no earlier intercalibration history, additional measurements are required. A preliminary conclusion from the 2011c intercalibration is that the intercalibration error for the FP reference source is probably less than ±25 % for wavelengths from 5573 Å. This preliminary, but promising, conclusion is to be confirmed by the final results from the 2011c intercalibration session.

The FMI MIRACLE EMCCD imager normally operated at Kilpisjärvi was recently calibrated by the manufacturer, Keo Scientific in Canada (T. S. Trondsen, personal communication, 2011). For further validation, this imager was then calibrated by the FMI integrating sphere and two of the IRF radioactive sources (920B and Y275). Data from this effort are not analysed yet and will appear in a later publication. Then, it will be possible to compare the 2011b intercalibration both to the calibration by Keo Scientific in Canada as well as to the FMI integrating sphere.

Furthermore, if the FP reference source should become unstable over time, it is highly likely that other ^{14}C-activated phosphor sources also would become unstable. This would be noticed as increasing deviations between the intercalibration workshops.

It has been found that Torr and Espy (1981) and Lauche and Barke (1986) did not use exactly the same filter sets. This is under investigation and might explain the difference in ratios for 1981 and 1985 (Fig. 4).

The 2011a intercalibration was mainly a practice run by a new calibration team before the official 2011b intercalibration in Sodankylä. This might explain the larger deviations seen for the 2011a intercalibration (Table 3 and Fig. 4). The 2011a intercalibration should therefore be excluded from the long-term series, if results from later workshops confirm it to be an outlier.

Aging effects of various components (sources, filters, PMT, etc.) will also contribute to the errors. Looking at Fig. 3 it is seen that the intercalibration errors tend to increase towards the red part of the spectra. This is under investigation and is probably related either to aging effects (PMT and/or filters), stray light, or to design compromises of the calibration photometer.

On the other hand, the IRF UJO Y275 (SID 5) source appears very stable over time at 5573 Å (Fig. 4). In fact, recovered fragments of old documentation (1960s) concerning "light calibration by C14 activated light standards from U.S. Radium Corp." appear to indicate 262.65 R/Å for the IRF UJO Y275 (SID 5) at 5600 Å (by conversion of the luminance values stamped onto the source; see Sect. 2). This is to be compared to the 2011b intercalibration that gave 261 R/Å at 5573 Å. To confirm this, the spectra of these sources must be measured. It is hoped that this will be possible in the autumn of 2012.

The mean ratios in Table 3 indicate a typical deviation, ranging from a few percent to around ±10 % for wavelengths

Table 4. Preliminary ratios of this intercalibration to measurements in November 2011 at the calibration laboratory at UNIS, Longyearbyen, Svalbard (2011c). The large differences for wavelengths below 5573 Å are expected since the IRF Lauche lamp is a tungsten lamp. For the PGI Chernouss-38AM source, the large difference at 4866 Å is more puzzling and remains to be explained.

Source name	SID	Filter [Å]						
		3914	4280	4866	5573	5882	6299	6562
IRF Lauche lamp	101	0.02	0.10	0.32	0.85	0.89	0.96	1.11
PGI Chernouss-38AM	105	0.53	0.86	0.30	0.69	0.75	0.95	1.09

4280, 4866, 5573 and 5882 Å. For 6299 Å this value is around ±15 %.

While none of what is said above provides hard evidence concerning the validity of the 30-year-old absolute calibration of the FP reference source, it is probably safe to assume that absolute calibration errors are probably less than 15–25 %, with a few exceptions and not including filters at 3914 and 6707 Å. This is also in agreement with Torr and Espy (1981), who report an accuracy of ±10 % over a 12-year-period. This should be compared to differences up to a factor of six during the early phases of this long-term intercalibration effort (Torr et al., 1977). Finally, even in the case that the absolute calibration values are completely wrong, the relative intercalibration is not affected by this, and thus it would be possible to correct these errors in the future.

7 Conclusions

This work presents the official results from the intercalibration workshop following the 38th Annual European Meeting on Atmospheric Studies by Optical Methods (in Table 2). Ratios of this intercalibration to earlier work are presented in Table 3 and Fig. 4. Preliminary results of the independent 2011c intercalibration (Longyearbyen) of two sources are given in Table 4.

A brief description of the intercalibration method, in effect since 1985, is provided. Furthermore, a large set of documentation and publications regarding this long-term intercalibration effort has been collected. As much as possible of this information will be made available on the Internet (http://alis.irf.se/ewoc/).

It is concluded that well-justified doubts exist about the validity of the absolute calibration of the FP reference source after 30 years. On the other hand, preliminary results from the 2011c intercalibration (Table 4) suggest errors of around ±15 % for wavelengths from 5573 Å and possibly also at 4280 Å. This is to be confirmed by the final results of the 2011c intercalibration as well as to be compared to the calibrations of the FMI MIRACLE EMCCD, performed in Canada by Keo Scientific and to the certified luminance values of the FMI sphere. Until this is done the absolute calibration error is estimated at 15–25 % and the relative intercalibration error at 5–25 %.

Future work

Following the intercalibration efforts in 2011, several radioactive calibration light sources have been found in Norway (Y. L. Andalsvik, personal communication, 2012). Many of these sources appear in earlier intercalibration workshops, in particular at the Lysebu 1985 workshop (Lauche and Barke, 1986). In addition, at least two calibration sources have been found at University of Oulu, Finland. Therefore, it would be desirable to include these sources in the intercalibration workshop planned for the autumn of 2012 in Sodankylä.

For the next workshop it will hopefully also be possible to measure the spectra of all participating sources. This is of general importance for improving the quality of this long-term intercalibration effort, but, in particular, it might help resolve problems related to LED-based sources, such as the PGI Chernouss-38AM source (SID 105).

This intercalibration effort should also be compared to absolute calibration methods involving the known spectra of stars.

The intercalibration procedure from 1985 is a rather tedious and manual nature. To automate the filter wheel operation and data acquisition would probably both improve the accuracy and speed up the intercalibration procedure.

Last but not least, it is of the utmost importance to perform an intercalibration of the FP reference source to a source traceable to a National Bureau of Standards source as soon as possible.

Acknowledgements. This work is presented in memory of Ingrid Sandahl who passed away in 2011. This work was funded by a University of Oulu grant for short-term international research visits. The comparison at UNIS was financed by a grant from the Nordic Council of Ministers. The authors also wish to thank two anonymous referees for their unusually helpful and thorough review of this work.

Edited by: A. Benedetto

References

Baker, D. J.: Rayleigh, the Unit for Light Radiance, Appl. Optics, 13, 2160–2163, 1974.

Baker, D. J. and Romick, G. J.: The Rayleigh: interpretation of the unit in terms of column emission rate or apparent radiance expressed in SI units, Appl. Optics, 15, 1966–1968, 1976.

Brändström, U.: The Auroral Large Imaging System – Design, operation and scientific results, Ph.D. thesis, Swedish Institute of Space Physics, Kiruna, Sweden, (IRF Scientific Report 279), ISBN: 91-7305-405-4, 2003.

Chamberlain, J. W.: Physics of the aurora and airglow, Classics in geophysics, AGU (American Geophysical Union), (A reprint of the original work from 1961), 1995.

Dahlgren, H., Gustavsson, B., Lanchester, B. S., Ivchenko, N., Brändström, U., Whiter, D. K., Sergienko, T., Sandahl, I., and Marklund, G.: Energy and flux variations across thin auroral arcs, Ann. Geophys., 29, 1699–1712, doi:10.5194/angeo-29-1699-2011, 2011.

Gustavsson, B., Leyser, T. B., Kosch, M., Rietveld, M. T., Åke Steen, Brändström, B. U. E., and Aso, T.: Electron Gyroharmonic Effects in Ionization and Electron Acceleration during High-Frequency Pumping in the Ionosphere, Phys. Rev. Lett., 97, 195002, doi:10.1103/PhysRevLett.97.195002, 2006.

Henricson, H.: Results from the intercalibration of low light level sources at Andøya 2007, in: Proceedings of the 33rd Annual European Meeting on Atmospheric Studies by Optical Methods, edited by: Sandahl, I. and Arvelius, J., no. 292 in IRF Scientific report, p. 131, Swedish Institute of Space Physics, Kiruna, http://www.irf.se/publications/proc33AM, 2008.

Hunten, D. M., Roach, F. E., and Chamberlain, J. W.: A photometric unit for the aurora and airglow, J. Atmos. Terr. Phys., 8, 345–346, 1956.

Lauche, H. and Barke, W.: A calibration photometer for low brightness sources, in: Proceedings of the 13th annual Meeting on Upper Atmosphere Studies by Optical Methods, edited by: Måseide, K., University of Oslo, Department of Physics, 86–28, 364–370, 1986.

Lauche, H. and Widell, O.: Intercalibration of low light level sources, in: Proc. of 27th Annual European Meeting on Atmospheric Studies by Optical Methods, Stockholm, Sweden, Meteorological institution, Stockholm university, Sweden, 2000a.

Lauche, H. and Widell, O.: Intercalibration of low light level sources, Phys. Chem. Earth, B25, 483–483, 2000b.

Mäkinen, S.: All-sky camera calibration, Master's thesis, Helsinki University of Technology (now Aalto University), 2001.

Rayleigh, L.: Absolute Intensity of the Aurora Line in the Night Sky, and the number of Atomic Transitions Required to Maintain it, P. R. Soc. London, A129, 458–467, 1930.

Sigernes, F., Holmes, J. M., Dyrland, M., Lorentzen, D. A., Chernous, S. A., Svenøe, T., Moen, J., and Deehr, C. S.: Absolute calibration of optical devices with a small field of view, J. Opt. Technol., 74, 669–674, 2007.

Syrjäsuo, M. T.: Auroral monitoring network: From all-sky camera system to automated image analysis (D.Sc.(Tech.)-thesis), Finnish Meteorological Institute, Helsinki, Finland, Contribution series 32, ISBN: 951-697-551-8, 2001.

Torr, M. R.: Intercalibration of instrumentation used in the observation of atmospheric emissions: A progress report 1976–1979, Tech. Rep. 100, Utah State University, Center for atmospheric and space sciences, Logan Utah, 1981.

Torr, M. R.: Report on a project to intercalibrate instrumentation used in the observation of visible atmospheric emissions, Tech. rep., Utah State University, Center for atmospheric and space sciences, Logan Utah, 1983.

Torr, M. R. and Espy, P.: Intercalibration of instrumentation used in the observation of atmospheric emissions: Second progress report, Tech. Rep. 101, Utah State University, Center for atmospheric and space sciences, Logan Utah, 1981.

Torr, M. R., Hays, P. B., Kennedy, B. C., and Torr, D. G.: Photometer calibration error using extended standard sources, Appl. Optics, 15, 600–602, doi:10.1364/AO.15.000600, 1976.

Torr, M. R., Hays, P. B., Kennedy, B. C., and Walker, J. C. G.: Intercalibration of airglow observatories with the Atmosphere Explorer satellite, Planet Space Sci., 25, 173–184, 1977.

Trondsen, T. S.: High spatial and temporal resolution auroral imaging, Ph.D. thesis, University of Tromsø, 1998.

Widell, O. and Henricson, H.: Intercalibration of low light level sources, in: Proc. of 28th Annual European Meeting on Atmospheric Studies by Optical Methods, 19–24.8.2001, Oulu, Finland, edited by: Kaila, K. U., Jussila, J. R. T., and Holma, H., Sodankylä Geophysical Observatory, 92, 125–125, 2003.

Widell, O. and Henricson, H.: Results from the intercalibration of low light level sources at IRF 2006, in: Proceedings of the 33rd Annual European Meeting on Atmospheric Studies by Optical Methods, edited by: Sandahl, I. and Arvelius, J., no. 292 in IRF Scientific report, Swedish Institute of Space Physics, Kiruna, p. 130, http://www.irf.se/publications/proc33AM, 2008.

Widell, O. and Mämmi, S.: Results from the intercalibration of low light level sources at Svalbard 2003, in: Proceedings of the 30th Annual European Meeting on Atmospheric Studies by Optical Methods, edited by: Sigernes, F. and Lorentzen, D., The University Centre on Svalbard, Longyearbyen, p. 121, 2003.

2

Calibration of non-ideal thermal conductivity sensors

N. I. Kömle, W. Macher, G. Kargl, and M. S. Bentley

Space Research Institute, Austrian Academy of Sciences, Graz, Austria

Correspondence to: N. I. Kömle (norbert.koemle@oeaw.ac.at)

Abstract. A popular method for measuring the thermal conductivity of solid materials is the transient hot needle method. It allows the thermal conductivity of a solid or granular material to be evaluated simply by combining a temperature measurement with a well-defined electrical current flowing through a resistance wire enclosed in a long and thin needle. Standard laboratory sensors that are typically used in laboratory work consist of very thin steel needles with a large length-to-diameter ratio. This type of needle is convenient since it is mathematically easy to derive the thermal conductivity of a soft granular material from a simple temperature measurement. However, such a geometry often results in a mechanically weak sensor, which can bend or fail when inserted into a material that is harder than expected. For deploying such a sensor on a planetary surface, with often unknown soil properties, it is necessary to construct more rugged sensors. These requirements can lead to a design which differs substantially from the ideal geometry, and additional care must be taken in the calibration and data analysis.

In this paper we present the performance of a prototype thermal conductivity sensor designed for planetary missions. The thermal conductivity of a suite of solid and granular materials was measured both by a standard needle sensor and by several customized sensors with non-ideal geometry. We thus obtained a calibration curve for the non-ideal sensors. The theory describing the temperature response of a sensor with such unfavorable length-to-diameter ratio is complicated and highly nonlinear. However, our measurements reveal that over a wide range of thermal conductivities there is an almost linear relationship between the result obtained by the standard sensor and the result derived from the customized, non-ideal sensors. This allows for the measurement of thermal conductivity values for harder soils, which are not easily accessible when using standard needle sensors.

1 Introduction

Thermal conductivity is one of the key parameters required for modeling the thermal evolution of a planetary body and the interaction between the solid surface and subsurface layers and the atmospheric and radiative environment. While there exist methods to determine thermal parameters of a surface layer by remote measurements, e.g. by analyzing the irradiation emitted from the surface, these methods usually demand "ground truth" measurements that have to be performed inside the material, i.e. by an in situ method to allow for proper evaluation. The simplest way to do this is to insert a long and thin needle into the material to be measured and to heat this needle with a constant electrical power for a specified time. The thermal conductivity of the surrounding material can then be determined directly from the temperature increase of the needle as a function of time. According to the classical hot needle theory (see e.g. Healy et al., 1976), the temperature response of a needle inside a material which is heated by a constant power consists of two parts: an initial nonlinear phase which depends on conductivity and heat capacity and later on a phase where the temperature versus logarithm of time graph rises linearly and the inclination of the graph depends on heat conductivity only. Thus if the measurement time is long enough (from minutes to hours, depending on the material to be measured) the thermal conductivity can be evaluated without knowing the heat capacity of the material. The heat conductivity k of the material can simply be determined by the formula

$$k = \frac{Q}{4\pi} \left(\frac{\mathrm{d}T}{\mathrm{d}\ln t} \right)^{-1}, \qquad (1)$$

where Q is the heating power supplied to the sensor in $[\mathrm{Wm}^{-1}]$ and $\left(\frac{\mathrm{d}T}{\mathrm{d}\ln t} \right)$ is the measured temperature rise of the sensor as a function of the natural logarithm of time.

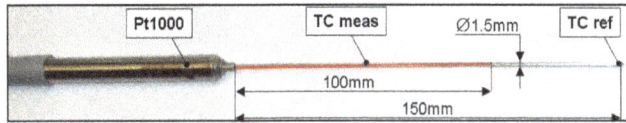

Fig. 1. The commercial TP02 (long needle) thermal conductivity probe produced by the Dutch company Hukseflux. The heated part is indicated in red.

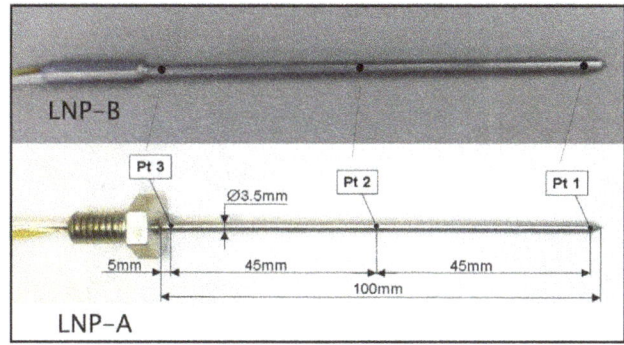

Fig. 2. The custom-made LNP-sensors fabricated by Hukseflux for use on planetary surfaces like on the Moon or Mars. They are heated over the whole length.

However, such a simple evaluation is only possible for a very long and thin sensor needle with a length-to-diameter ratio of 60 or more, since the theory behind Eq. (1) assumes an infinitely thin and infinitely long line heat source. The reference sensor used here meets this requirement.[1] However, sensors of this type are not rugged enough to be directly used in harder materials without pre-drilled holes (which may cause other errors for the evaluation of thermal conductivity) or on planetary surface missions, where the properties of the soil to be tested are generally unknown. A more rugged sensor is necessarily thicker and shorter and can no longer be considered as "long and thin". Therefore, Eq. (1) is no longer directly applicable.

In order to derive thermal conductivity values from measurements with this type of sensors, there are in principle two possibilities. Either a much more complicated formalism is used, applying the theory of "short and thick" sensors, or the same simple theory is used with an additional calibration function valid over the desired range of conductivities. The first method has been described in much detail in a recent paper by Hütter and Kömle (2012), in Hütter (2011) and most recently in Macher et al. (2013). The second method is the topic of the current paper. A more detailed description of the theoretical background can be found in Wechsler (1992) and in Kömle et al. (2011).

Very few thermal conductivity sensors have so far been successfully deployed in planetary missions. In the framework of the Apollo missions in the early 1970s, a few thermal conductivity measurements were performed on the lunar surface. However, evaluation was largely done along the first line, using the theory of multi-layered hollow cylindrical sensors (Langseth et al., 1972, 1973). Later on, in the frame of the Huyghens–Cassini mission, the thermal conductivity of Titan's atmosphere was determined by a probe working on the basis of the hot needle method (Hathi et al., 2007). However, the only space instrument that has measured thermal conductivity in the solid material of an extraterrestrial body other than the Moon was the TECP-instrument aboard the NASA *Phoenix* spacecraft, which landed on the Martian polar plains in 2008 (Zent et al., 2009, 2010). The method used to evaluate the thermal conductivity measurements obtained from the TECP-instrument is described in Cobos et

al. (2006). The needles used for the TECP measurements had an even greater deviation from the "ideal" geometry than the sensors described in this paper. It may be useful to compare their results with our findings in order to further validate the use of this calibration method for future heat conduction measurements on planetary surfaces. In the following sections we give a description of the sensors used for our calibration measurements, characterize the samples used and discuss the results obtained.

2 Description of sensors

2.1 Reference sensor

As reference sensor we have used an off-the-shelf thermal conductivity probe manufactured by the Dutch company Hukseflux (Type TP02). This sensor is shown in Fig. 1. According to the handbook, it is suitable for standard measurements in the range 0.1–$6\,\mathrm{W\,m^{-1}\,K^{-1}}$ with an accuracy of $\pm 3\,\%$ in the final thermal conductivity value. The needle has a diameter of 1.5 mm and a total length of 15 cm. The uppermost 10 cm is actively heated during a measurement. The needle temperature in the heated part and in the unheated tip is measured by two thermocouples. With a length-to-diameter ratio of 66, this sensor fulfils the requirements for an "easy" evaluation of the thermal conductivity without the need for additional calibration.

2.2 Prototype ruggedized sensors

Two slightly different prototypes of custom-made sensors (LNP-A and LNP-B) were tested, and are shown in Fig. 2. They differ only in one detail: LNP-A has a mounting stud with a screw thread at the top which could be used to mount it onto a deployment device (for example a robotic arm on a planetary lander spacecraft). Because of the small dimensions of the needle such a part could influence the measurements due to its relatively large mass and heat capacity.

---

[1] Hukseflux Thermal Sensors – TP02 NON-STEADY STATE PROBE FOR THERMAL CONDUCTIVITY MEASUREMENTS: User Manual www.hukseflux.com

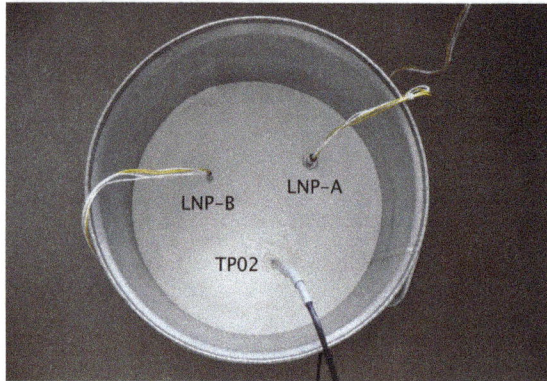

Glass beads sample

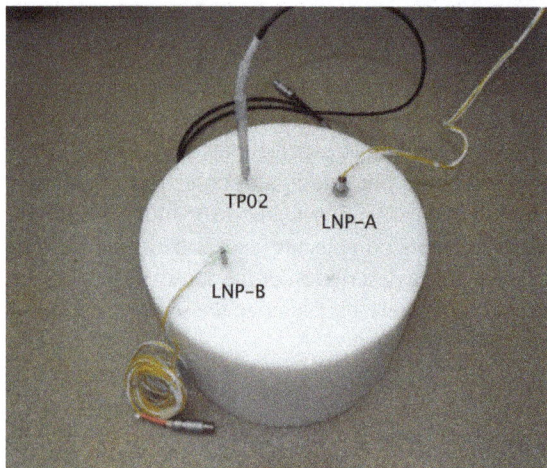

PE sample

Fig. 3. Experimental setup used for the calibration measurements. Top: sensors inserted into glass beads sample; bottom sensors inserted into PE sample.

Therefore, a second prototype (LNP-B) was built, which consisted only of the needle and the necessary connection wires without such a mounting stud. Both sensors were also built by Hukseflux and had a needle length of 100 mm and a diameter of 3.5 mm. This implies a length-to-diameter ratio of 28. In these sensors the heating wire inside the needle extends over the whole needle length of 100 mm. Temperature is recorded at three positions, as indicated in Fig. 2: in the center of the needle, close to the tip and close to the upper end. The readings of the central sensor are used for the evaluation of the thermal conductivity value. The temperature sensors are platinum resistance thermometers (PT1000) and temperatures are measured with a 4-wire technique.[2] Due to their larger diameter and their shorter length these sensors

_____

[2]For an explanation of the four wire measurement technique with platinum resistance thermometers refer, for example, to the National Instruments webpage http://www.ni.com/white-paper/7115/en.

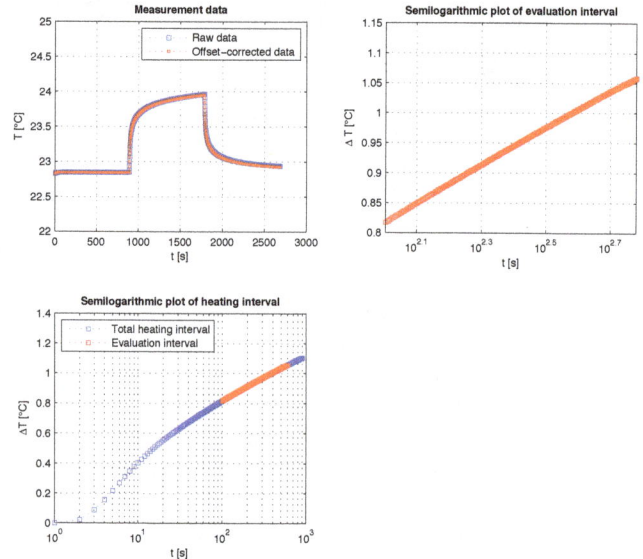

Fig. 4. Illustration of the thermal conductivity evaluation procedure using a Kerafol-KP96 measurement with sensor LNP-B as an example (heating time: 900 s; heating power: 0.215 W). Top left: raw data and offset-corrected data of the total measurement (including the decline phase of the temperature which is not used for the evaluation). Bottom left: semi-logarithmic plot of the heating interval. Top right: semi-logarithmic plot of the interval used for the thermal conductivity evaluation.

are by far more robust than the TP02 and easily withstand penetration into harder soils without damage.

3   Characterization of calibration samples

The calibration materials used for our measurements have been selected in order to cover the range of thermal conductivities from 10^{-1} to about $2\,\text{Wm}^{-1}\,\text{K}^{-1}$. The lower end corresponds to granular materials under normal pressure. As a representative of such a kind of material we have chosen silica glass beads with a grain size in the range 0.25–0.5 mm. For the range 0.3–$0.5\,\text{Wm}^{-1}\,\text{K}^{-1}$ the solid plastic material polyethylene (PE) was used. The range 0.5–$0.6\,\text{Wm}^{-1}\,\text{K}^{-1}$ is typical for the thermal conductivity of water. However, since water is a fluid, it may undergo convection when heated by the sensor, which strongly increases the heat transfer between sensor and sample and therefore would lead to large errors in the determined thermal conductivity. To circumvent this problem, a small amount of agar (50 grams per liter of water) is added and dissolved in the water. This mixture is then heated up and boiled for several minutes. Upon cooling the solution back to room temperature a transparent, highly viscous gel is obtained, in which any form of convection is suppressed. However, its thermal properties are the same as those of water. When the agar is frozen and kept in a thermally stable environment, one obtains another useful

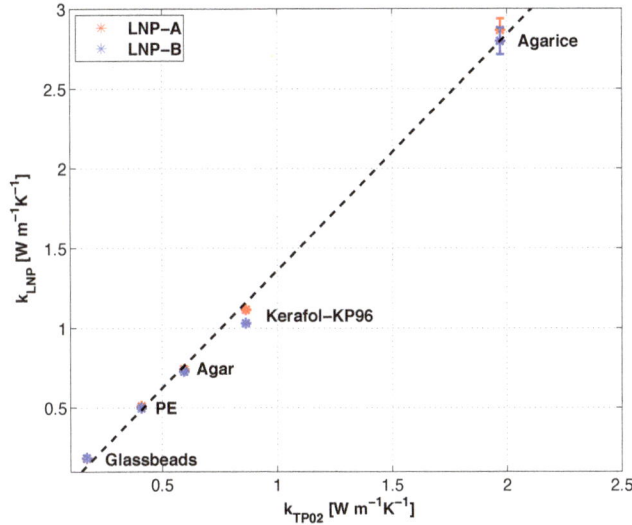

Fig. 5. Calibration of the custom-made LNP-sensors versus the commercial Hukseflux TP02 thermal conductivity sensor, which is considered as the reference sensor. The symbols represent the average measurement values obtained from the two LNP probes, while the dotted line is a linear fit between the measurement values using the data from both LNP-sensors.

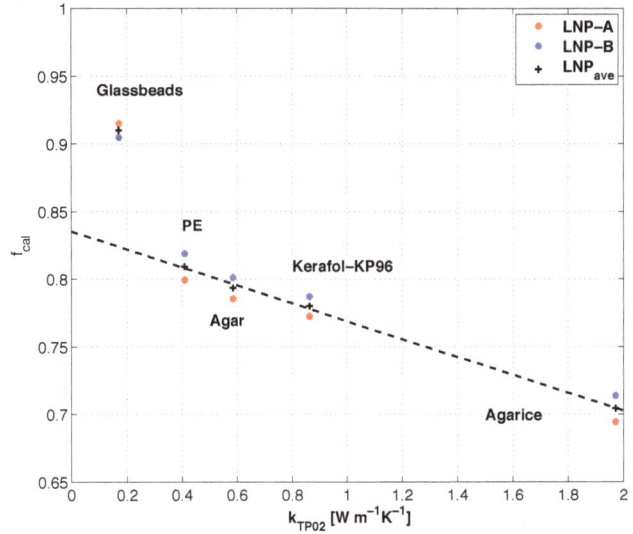

Fig. 6. Plot of the calibration factors for the different materials as listed in Table 2. The dashed line is a linear fit through the measurement values excluding the glass beads. LNP$_{ave}$ is the average of the measured LNP-A and LNP-B values for each calibration material.

calibration material, with $k \approx 2 \, \mathrm{Wm^{-1} \, K^{-1}}$, i.e. at the upper limit of our range of interest. To obtain the frozen agar sample, the agar samples measured before at room temperature were placed into a deep-freezer with the measurement needles inserted and kept there for several days, until a constant homogeneous temperature around $-20 \, ^\circ\mathrm{C}$ was reached. This led to a slight expansion of the sample analogous to the freezing of pure water, but the thermal contact of the needles to the surrounding ice remained well established, no major cracks were formed during the freezing process which would have compromised the thermal contact. This was possible because the containers used were somewhat flexible so that, in addition to the upward expansion, a volume expansion both towards the sides and towards the bottom was possible. Since the temperature changes due to sensor heating never exceeded a few degrees, there was no danger that local phase changes of the material along the sensor/sample could have occurred.

There is a lack of easily accessible materials in the range between 0.7 and 2 $\mathrm{Wm^{-1} \, K^{-1}}$. To fill this gap, we used a bulk of thermally conductive grease as sample material (Kerafol-KP96). This grease is viscous enough that convection is suppressed, but at the same time soft enough that sensors can be easily inserted. In our calibration measurements we have used a thermal grease with a bulk thermal conductivity around 1 $\mathrm{Wm^{-1} \, K^{-1}}$ to bridge the gap between PE and water (agar) ice.

4 Calibration of the ruggedized sensors

For the calibration of the custom-made sensors the following procedures were performed:

- Reasonably large-sized samples were prepared, which were big enough in diameter and height that all three sensors could be inserted without disturbing each other during a measurement and making sure that no influence from the sample boundaries could disturb the measurements. For estimating minimum sample sizes, refer to the formulae given in Hütter and Kömle (2012).

- The samples (with sensors inserted) were kept for at least several hours in a thermally stable environment to make sure that they were isothermal at the beginning of a measurement series. All measurements were performed at room temperature, i.e. at an ambient temperature in the range 20–25 $^\circ\mathrm{C}$, apart from those in agarice, where the samples were stored in a deep-freezer at $-22 \, ^\circ\mathrm{C}$.

- Thermal conductivity measurements were made by heating each sensor separately, allowing for long enough time periods between two subsequent measurements (at least several hours). The chosen heating periods of the sensors were between 300 and 900 s, depending on the sample used.

- The thermal conductivity was evaluated using the following standard procedure, as illustrated in Fig. 4:

    1. Removal of any temperature trend from the data not associated with the active heating of the sensor.

Table 1. Thermal conductivity measurement results for the different calibration materials and sensors.

Sensor/λ	Glassbeads $\mathrm{Wm^{-1}K^{-1}}$	PE $\mathrm{Wm^{-1}K^{-1}}$	Agar $\mathrm{Wm^{-1}K^{-1}}$	Kerafol KP96 $\mathrm{Wm^{-1}K^{-1}}$	Agarice $\mathrm{Wm^{-1}K^{-1}}$
TP02	0.1684	0.4126	0.5874	0.8477	2.1064
	0.1707	0.4479	0.6251	0.8709	1.9281
	0.1688	0.3755	0.5938	0.8743	1.9631
	......	0.4560	0.6145	0.8554	1.8860
	......	0.3680	0.6033	......	......
	......	0.3779	0.6165	......	......
	......	0.4190	......	......	......
	......	0.4127	......	......	......
	0.1693	*0.4087*	*0.6068*	*0.8621*	*1.9709*
LNP-A	0.1841	0.5125	0.7396	1.1125	2.7702
	0.1859	0.5119	0.7529	1.1201	2.8757
	0.1850	0.5092	0.7462	1.1161	2.7835
	......	......	0.7373	1.1139	2.9208
	0.1850	*0.5112*	*0.7440*	*1.1157*	*2.8375*
LNP-B	0.1865	0.4979	0.7328	1.0957	2.7807
	0.1868	0.4998	0.7228	1.1000	2.7039
	0.1879	0.4994	0.7336	1.0918	2.6502
	......	......	0.7255	1.0931	2.8554
	......	......	......	......	2.8146
	0.1871	*0.4990*	*0.7274*	*1.0951*	*2.7609*

Table 2. Calibration factors derived for the different materials and sensors.

Sensor/Material	Glassbeads	PE	Agar	Kerafol-KP96	Agarice
LNP-A	0.9151	0.7995	0.7856	0.7727	0.6946
LNP-B	0.9048	0.8190	0.8013	0.7872	0.7139

2. Identification of the suitable interval of the measured temperature profile (linear part on the T versus $\ln t$ graph).

3. Calculation of the thermal conductivity according to Eq. (1).

The general setup of the measurements is shown in Fig. 3 for two of the used samples, the glass beads and the PE-block. The agar sample was prepared in the same 30 cm diameter steel container as the glass beads sample. For performing the measurements in the agar-ice, this sample was placed in a deep freezer with the sensors inserted and frozen at a temperature of $-22\,°C$. In this way it was ensured that the sensor needles were firmly frozen into the ice and thus had a good contact to the sample with negligible thermal resistance. For the Kerafol-KP96 sample the sample container was smaller (diameter of 16 cm) but still large enough to ensure that there was no influence of the container walls on the measurement results. The active heating times of the sensors used for the thermal conductivity measurements were typically 900 s. The heating powers were in the range

100–500 mW, depending on the estimated conductivity of the sample material.

5 Results

The results of our calibration measurements with the ruggedized thermal conductivity sensors are summarized in Table 1 and in Fig. 5. All measurements were repeated two times or more. The average value from the individual results was used to calculate the calibration factor (italic numbers in the table). The scattering of the measurement results can be seen from Table 1 and from the error bars in Fig. 5. Note that the 2-sigma error bars are based on only a few data samples (3–6) per measurement point. Thus we do not claim that they are the result of a significant statistical analysis, which would have demanded an unrealistically large number of measurements for each material. Rather we show them here to illustrate the good repeatability of the performed measurements and the fact that both of the two ruggedized sensors (despite of having a slightly different geometry) give consistent results. Actually, the error bars are only visible on the graph for the agar-ice sample, where the quality is poorer than for the other measurements, which were done at room temperature in a thermally stable room (The reason for this is not clear, but it may have to do with the temperature regulation cycles of the the cooling device, which overlay the heating of the sensors during a measurement or with other unknown disturbances).

Using these measured average values from both types of sensors (the reference sensor TP02 and the ruggedized sensors LNP-A and LNP-B) a correction factor was calculated per sample and hence conductivity range. This allows the true thermal conductivity to be obtained from a measurement with one of the ruggedized sensors according to the formula:

$$k_{TP02} = f_{cal} \cdot k_{LNP}. \tag{2}$$

The calibration factors derived for the different materials and sensors (LNP-A and LNP-B) as calculated from the measurements are listed in Table 2. As can be seen from Fig. 5, in general the measured thermal conductivity values over the range of interest can be well fitted by a linear relationship (constant calibration factor, dotted line). The average value of the individual calibration factors as given in Table 2 is $f_{cal} = 0.799 \approx 0.8$. This would lead to a maximum error of 15 % over the thermal conductivity range covered by the measurements. However, this is not a statistical error only, as can be seen by a closer look on Table 2. Plotting the individual f_{cal}-values versus the measured k_{TP02}-values (Fig. 6) indicates a linear decreasing trend in the range 0.4–2 $Wm^{-1} K^{-1}$, but the derived values for the glass beads tend to be closer to the reference value than the linear trend in the rest of the curve would suggest. Further tests exploring the trend at lower thermal conductivities are therefore necessary and planned in a next step.

6 Conclusions

The measurements reveal that the prototype sensors give consistently higher values of the thermal conductivity when evaluated in the same way as the measurements with the standard sensor. However, we found that the data produced by the ruggedized sensors could be well fitted by assuming a linear relation between the values obtained by the standard sensors and the ruggedized sensors. This result confirms that measurements with the rugged prototype sensors, which have strongly non-ideal geometry, can be made for any unknown material (in the appropriate thermal conductivity range) by applying a constant calibration factor.

Acknowledgements. This paper is a late result of the project L317–N14 supported by the Austrian *Fonds zur Förderung der wissenschaftlichen Forschung*. The custom-made sensors used for the measurements described here were designed and manufactured under the funding of this project.

Edited by: P. Falkner

References

Cobos, D. R., Campbell, G. S., and Campbell, C. S.: Modified line heat source for measurement of thermal properties on Mars, in: Proceedings of the 28th International Thermal Conductivity Conference, edited by: Dinwiddie, R. D., White, M. A., and McElroy, D. L., 331–338, Destech. Publ., Lancaster, Pa, 2006.

Hathi, B., Daniell, P. M., Banaszkiewicz, M., Hagermann, A., Leese, M. R., and Zarnecki, J. C.: Thermal conductivity instrument for measuring planetary atmospheric properties and data analysis technique, J. Therm. Anal. Calorim. 87, 585–590, 2007.

Hütter, E. S.: Development and testing of thermal sensors for planetary applications, PhD Thesis, Karl-Franzens- Universität Graz, 2011.

Hütter, E. S. and Kömle, N. I.: Performance of thermal conductivity probes for planetary applications, Geosci. Instrum. Method. Data Syst., 1, 53–75, doi:10.5194/gi-1-53-2012, 2012.

Kömle, N. I., Hütter, E. S., Macher, W., Kaufmann, E., Kargl, G., Knollenberg, J., Grott, M., Spohn, T., Wawrzaszek, R., Banaszkiewicz, M., and Hagermann, A.: In situ methods for measuring thermal properties and heat flux on planetary bodies, Planet. Space Sci., 59, 639–660, 2011.

Langseth, M. G. J., Clark, S. P. J., Chute, J. L. J., Kheim, S. J. J., and Wechsler, A. E.: Heat flow experiment, in: Apollo 15: Preliminary science report, (NASA SP-289), 1972.

Langseth, M. G. J., Kheim, S. J. J., and Chute, J. L. J.: Heat flow experiment, in: Apollo 17: Preliminary science report, (NASA SP-330), 1973.

Macher, W., Kömle, N. I., Bentley, M. S., and Kargl, G.: The heated infinite cylinder with sheath and two thermal surface resistance layers, Int. J. Heat Mass Transf., 57, 528–534, 2013.

Wechsler, A. E.: The probe method for measurement of thermal conductivity, in: Compendium of Thermophysical Property Measurement Methods, edited by: Maglić, H. D., Cezairliyan, A., and Peletsky, V. E., 161–185, Plenum Press, New York, 1992.

Zent, A. P., Hecht, M. H., Cobos, D. R., Campbell, G. S., Campbell, C. S., Cardell, G., Foote, M. C., Wood, S. E., and Mehta, M.: Thermal and Electrical Conductivity Probe (TECP) for Phoenix, J. Geophys. Res., 114, E00A27, doi:10.1029/2007JE003052, 2009.

Zent, A. P., Hecht, M. H., Cobos, D. R., Wood, S. E., Hudson, T. L., Milkovich, S. M., DeFlores, L. P., and Mellon, M. T.: Initial results from the thermal and electrical conductivity probe (TECP) on Phoenix, J. Geophys. Res., 115, E00E14, doi:10.1029/2009JE003420, 2010.

3

In-flight calibration of the Cluster/CODIF sensor

L. M. Kistler[1], **C. G. Mouikis**[1], **and K. J. Genestreti**[1,*]

[1]University of New Hampshire, Durham, NH, USA
[*]now at: Southwest Research Institute, San Antonio, TX, USA

Correspondence to: L. M. Kistler (lynn.kistler@unh.edu)

Abstract. The Cluster/CODIF sensor is a time-of-flight instrument that measures the ion composition over the energy range $40\,\mathrm{eV\,e^{-1}}$ to $40\,\mathrm{keV\,e^{-1}}$. It operated for 4 yr on S/C 1, 9 yr on S/C 3, and is still operational on S/C 4, after more than 12 yr. During this time the total ion detection efficiency has decreased by a factor of 50. In this paper, we describe the methods used to track the efficiency changes throughout the mission for the three different spacecraft and for the different ion species. The methods include calculations of the efficiencies using rate data collected in the instrument, comparisons with other instruments on the Cluster satellites, and checks based on geophysically reasonable assumptions.

1 Introduction

The Cluster/CODIF instrument measures ion composition over the energy range $\sim 40\,\mathrm{eV}$ to $40\,\mathrm{keV\,e^{-1}}$. The CODIF instrument operated on S/C 1 from 1 February 2001 to 24 October 2004, on S/C 3 from 1 February 2001 to 11 November 2009, and on S/C 4 from 1 February 2001 through the present time, and is still operating. The instrument never operated on S/C 2. The Cluster/CODIF instrument is a combination of an electrostatic analyzer followed by postacceleration of 15 kV, and then a time of flight section, as described in Rème et al. (2001). Details of the instrument are also discussed in Möbius et al. (1998). The entrance into the electrostatic analyzer is divided into two 180° sections. Grids in the entrance define the geometric factor of the two sides. One side, with highly transparent grids, is called the "high side", or HS. The other side, with grids that drop the flux by a factor of 100, is called the "low side", or LS. Only one side is used at any time.

After exiting the analyzer, ions go through a thin carbon foil at the entrance to the time-of-flight section. Electrons knocked out of the foil are steered to a microchannel plate (MCP). The MCP pulse from the electrons is used to create two signals: a "start signal", which is used to initiate the time-of-flight calculation, and a "position" signal that indicates the direction from which the ion entered the instrument. The position is determined by detecting the signal in one of eight 22.5° sections. After passing through the foil, the ion traverses the flight path and hits an MCP to create the "stop" pulse. A "valid event" within the instrument requires a "start" and a "stop" pulse, to get the time-of-flight, and a "position" pulse. The ion species (mass per charge) is determined by the combination of the energy per charge of the ion, from the electrostatic analyzer, and the time-of-flight of the ion over the known flight path length.

The number of ions detected for a given input flux depends on a number of factors. First, it is a function of the geometric factor, which includes the geometry of the electrostatic analyzer and the transparency of any grids in the optics path. Then it is determined by the efficiency of the three required signals. The start pulse efficiency depends on the number of electrons emitted from the carbon foil, which varies statistically with the particle energy and species (Ritzau and Baragiola, 1998; Allegrini et al., 2003), the fraction of the electrons that are steered to the detector, and the active area of the MCP itself ($\sim 50\,\%$). The start and position signals are obtained from the electron pulse from the MCP by a grid that separates the pulse into two parts. The start signal comes from the grid, while the position signal comes from the electrons that go through the grid to an anode (Fig. 11 in Rème et al., 2001). Thus the difference in efficiency of the detection of the two signals depends on the transparency of the grid, and on the thresholds in the two sets of electronics that detect the signals. The size of the pulse from the MCP has some statistical variation. If operating at optimum gain, each ion would give a pulse large enough to be measured. However if the gain is not optimum, some fraction of the signals will be below the detection threshold. The stop pulse efficiency depends on the fraction of ions that reach the detector, and again, the active area of the MCP and the electronics detection efficiency. The fraction of ions that reach the detector is a function of the amount that the ions scatter in the foil, which depends on the particle species and energy (e.g., Gonin et al., 1992). Finally, to be counted as a particular ion species, the ion must have a time-of-flight that falls between defined limits. Because of energy loss in the foil (e.g., Allegrini et al., 2006), the peaks in time of flight are not symmetric, but have a tail towards longer times of flight (Fig. 16 in Rème et al., 2001). Some fraction of the ions will fall outside the defined limits (Fig. 17 in Rème et al., 2001). This also needs to be taken into account.

Most of these dependencies are not expected to change with time. The instruments were well calibrated before launch over the full range of energies using the four major ion species expected in the instrument: H^+, O^+, He^+, and He^{++}. The results of the ground calibrations are summarized in Kistler (2000a,b,c). However, the gain of the microchannel plates does decrease significantly over time (Sandel et al., 1977; Drake et al., 1998; Kishimoto et al., 2006). In principle, the gain can be increased by increasing the MCP voltage. However there are two aspects of the instrument that limit the effectiveness of raising the voltage. The first aspect is the MCP geometry. The instrument uses large MCPs with the same MCP covering both the start and stop signals. In addition, the instrument has 4 quadrants of MCPs, covering the full 360° entrance, both the LS and the HS, and there is only one MCP power supply for the four quadrants. Thus, while the efficiency may decrease on one part of an MCP, or on one set of MCPs, increasing the voltage will increase the gain over the whole MCP area. The second aspect is that the MCPs are located at high voltage (HV), and the electron pulse from the MCP goes across the high-voltage gap (~ 12 kV) to the anode plane (Fig. 11 in Rème et al., 2001). If the gain gets too high, there can be a large current across the high-voltage gap that can trigger the HV power supply to shut down. Since the gain of the MCPs does not change uniformly, part of an MCP may require a higher voltage, but that makes the gain too high in other areas. Thus there is a limit to how high the MCP voltage can be raised. While the MCP voltages have been increased to improve the gain over the course of the mission, the instrument has also had to operate with a reduced gain, and tracking this non-uniform gain change, and determining how it affects the efficiencies of the different species, is a significant calibration effort.

An overview of some of the in-flight calibration techniques used for Cluster/CODIF, and the resulting calibrations through 2003 are presented in McFadden et al. (2007). In this paper we provide an updated description of the calibration techniques, including additional cross-calibration techniques that have been applied later in the mission, and show how the calibrations have changed over the 12 yr time period.

2 H^+ calibration

Because H^+ dominates the flux of ions almost all the time, H^+ can be calibrated using the widest range of techniques. The "engineering rate" counters, which count the start, stop, and position pulses, are dominated by H^+. This allows the efficiency of H^+ to be determined directly. Geophysically reasonable assumptions can also be used to check the determined efficiency. The ion pressure is usually dominated by H^+, so in some cases the H^+ calibration can be checked by monitoring for pressure balance as the spacecraft transitions through different regions in the magnetosphere. In addition, the ion and electron densities in a plasma should be the same, so in time periods where the CODIF energy range carries the dominant density, the CODIF-measured density can be compared with that measured by the electrons with the PEACE instrument (Johnston et al., 1997), with the electron density determined through wave measurements by the WHISPER instrument (Décréau et al., 2001; Trotignon et al., 2010), as well as with the ion density calculated by the all-ion instrument, Hot-Ion Analyzer (HIA; Rème et al., 2001).

2.1 High side signal efficiencies using engineering rates

A valid event in the CODIF instrument requires a start and a stop pulse in coincidence, and one and only one position signal. The efficiencies for getting a start pulse and a stop pulse can be determined using the start rate (SF), the stop rate (SR), and the start/stop coincidence rate (SFR). The fraction of ions that give a coincidence for each measured start count gives a measure of the stop efficiency. Similarly, the fraction of ions that give a coincidence for each measured stop count is a measure of the start efficiency. The fraction of the ions that give a single position, SEV, for each measured coincidence, SFR, is the "single position" efficiency. To summarize,

$$\text{Stop\_Efficiency} = \text{SFR/SF} \qquad (1)$$

$$\text{Start\_Efficiency} = \text{SFR/SR} \qquad (2)$$

$$\text{Single\_Position\_Efficiency} = \text{SEV/SFR}. \qquad (3)$$

The product of these three efficiencies gives the total efficiency. Again, we note that these rates count all species, but are normally dominated by H^+.

There is one final caveat that applies to the stop rate. While SF and SFR directly measure the signal used to start the time-of-flight and the coincidence rate, the SR is a separate signal derived from the stop rate, but is not directly the signal used by the time-of-flight circuit. Thus, the threshold on the SR signal may be different from the threshold of the "stop" signal used in the time of flight. While an effort was made to adjust the threshold prior to launch, the actual thresholds ended up different on the three operating sensors. S/C 1 had the SR that most reliably tracked the real stop, while the instruments on the other two spacecraft had thresholds set somewhat too high, so that in general, the SR is lower than the actual "stop" rate. As a result, calculations of the start efficiency using this method can be greater than 1. When CODIF on S/C 1 operated, we used S/C 1 for our baseline calibration, and then verified the other two spacecraft calibrations through cross-calibration with S/C 1. After 2004, when CODIF on S/C 1 no longer operated, we began to rely on the pressure balance technique in the magnetotail and cusp as a method to cross-check the calibration. This will be discussed in Sect. 2.2

Figure 1 shows the "stop", "start", and "single position" efficiencies for 1 keV ions from 2001 into 2012 for the HS of CODIF on S/C 4. The bottom panel shows the MCP set values over the mission. Clearly, the sharp increases in efficiencies correspond with time periods when the MCP voltage was increased. The changes in the stop efficiency (top panel) are relatively modest. It started at ~ 0.65 at the beginning of mission, and dropped to 0.4 at its lowest, in 2008 and 2009. In July 2009 a final attempt to increase the voltage was performed, which brought the stop efficiency up to ~ 0.5, where it has remained, quite stable, since then. We believe that the relative stability of the stop pulse efficiency is due to the spread of ions that create the stop signal on the MCP. Because they are spread over a large area of the MCP, due to scattering in the foil, no single area of the MCP has had a large decrease in efficiency.

The start efficiency, shown in the second panel, has changed more dramatically. As noted above, the numbers greater than one are due to the thresholds of the SR. But the efficiency overall has dropped to 25 % of its initial value. The final MCP increase brought it back up slightly, but it is still significantly lower than what it had been at the start of the mission. The greater decrease compared to the "stop" signal is because the electrons from the foil are steered and focused onto the MCP, towards the center of each position. Thus, only a small area of the MCP receives the majority of the flux resulting in a larger decrease in efficiency.

The single position efficiency (third panel) also began decreasing after launch. The reason for this was initially not clear. Because the position signals in general have a lower threshold than the start signals, any signal that gives a start should also give a position, even when the pulse height is small. Thus any start/stop coincidence (SFR) should have had a corresponding position signal, and the expected SEV / SFR ratio should be close to 1. We found that the reason that the single position efficiency decreases is an error in the event logic. The registers that record that a position signal has been observed are not cleared at the beginning of an event. As a result, if an ion generates a "position" signal but not a start signal, that position is saved in the register, but not counted as an event. When a new event occurs that does have a start signal and a position signal, the logic checks whether one and only one position signal has been detected and finds

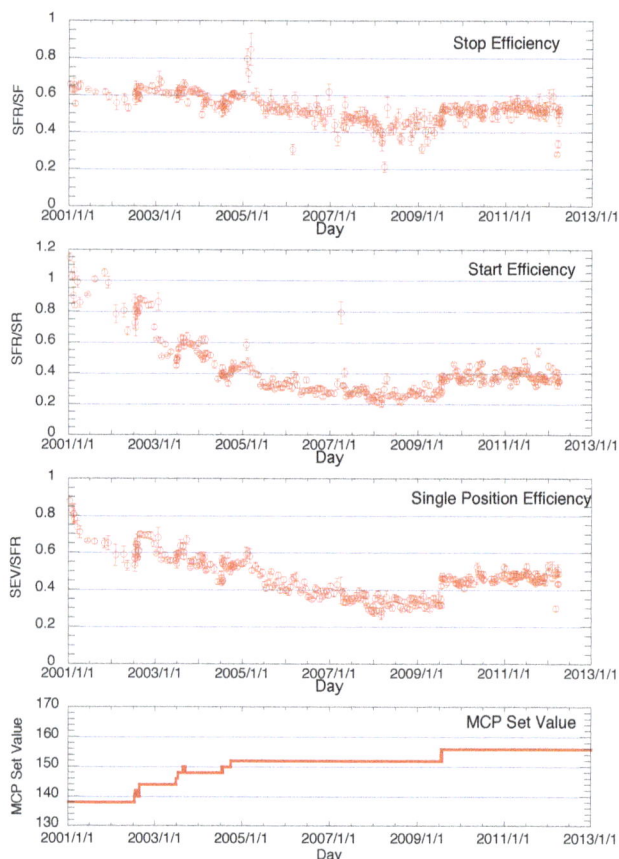

Fig. 1. Stop, start, and single position efficiencies for SC4 from 2001 to 2012 for 1 keV ions. The bottom panel shows the setting for the voltage across the MCP.

two positions. Thus the event is rejected. At the beginning of the mission, almost all events that had a position signal also had a start signal. But as the MCP gain decreased, it increased the probability that a position signal would be measured with no corresponding start signal, which generates the invalid events. Thus, when the start efficiency goes down, the single position efficiency also goes down, essentially squaring the effect of the decrease in start efficiency in the total efficiency calculation.

2.2 High side total efficiencies

As discussed above, because of issues with the SR on S/C 4, the engineering rates give an indication of how the efficiencies have changed, but do not give the absolute value. Starting towards the end of 2004, when CODIF on S/C 1 stopped operating, we use pressure balance in the magnetotail and cusp to validate the changes in efficiency. Figure 2 shows an example illustrating the technique. In the magnetotail, the pressure changes from being dominated by the magnetic field close to the lobes to being dominated by the plasma pressure close to the neutral sheet, with the overall pressure

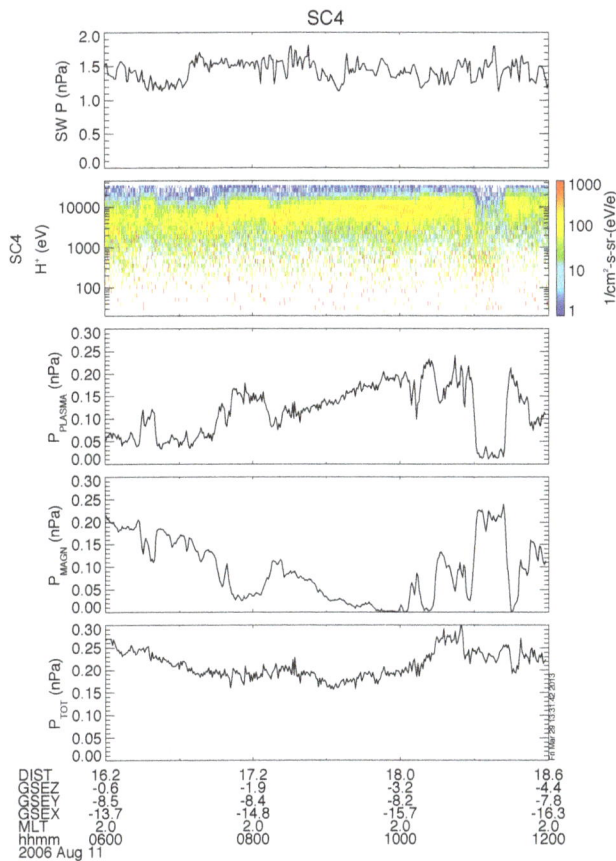

Fig. 2. An example time period when pressure balance in the tail can be used to check the H$^+$ calibration. The panels, from top to bottom, give the solar wind pressure, the H$^+$ energy spectrogram, the plasma pressure, the magnetic field pressure, and the total pressure.

remaining approximately constant (Fairfield et al., 1981). During a plasma sheet encounter, the tail flaps or waves, so that the spacecraft moves from areas dominated by magnetic pressure to areas dominated by particle pressure. In Fig. 2, the top panel shows the solar wind pressure at this time. This information is used in order to avoid time periods that have a large change in the solar wind pressure. The second panel shows the H$^+$ energy spectrum during the plasma sheet encounter. The next three panels show the plasma pressure, the magnetic field pressure, and the total pressure. In this example, the spacecraft starts close to the lobe, with the pressure dominated by the magnetic field. It then moves closer to the neutral sheet, reaching the neutral sheet at $\sim$ 10 UT. During this time, the magnetic pressure in general decreases while the plasma pressure increases. However there are a number of brief excursions to regions of higher plasma pressure, notably at 06:30 and 07:45 UT. We have adjusted the CODIF efficiency by hand for each event so that the total pressure, shown in the last panel, maintains the smoothest possible profile during these sharp transitions.

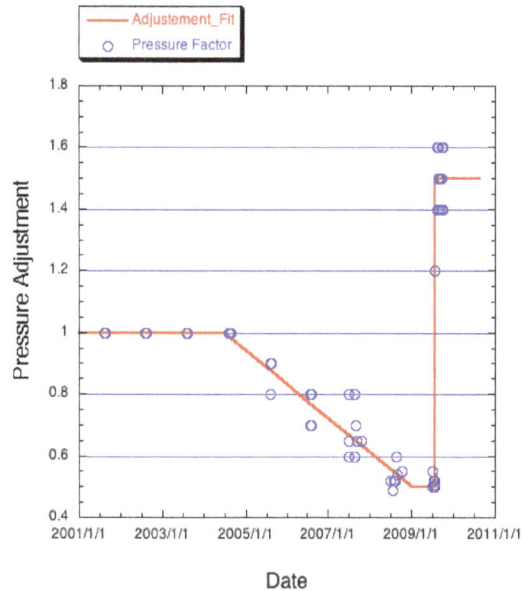

Fig. 3. Adjustment factors to the efficiencies, determined using pressure balance. The large increase corresponds to when the MCP voltage was increased in 2009.

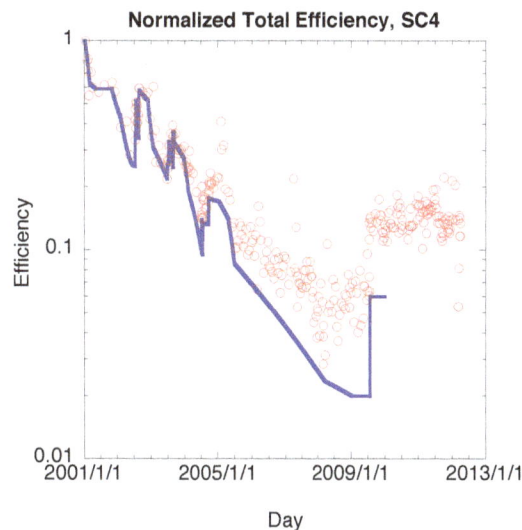

Fig. 4. Normalized total efficiencies derived using rate data (circles) and after the adjustments based on the pressure balance technique (blue line). The efficiencies are normalized to 1.0 at the start of mission.

Figure 3 shows the adjustment parameters that were determined for time periods after 2004, and the adjustment curve that was derived. This curve was used to modify the total efficiency determined from the rate data. The results are shown in Fig. 4. The red circles show the normalized total efficiency determined by taking the product of the three rate-based efficiencies shown in Fig. 1, and normalizing the result to 1.0 at the start of the mission. The blue line shows

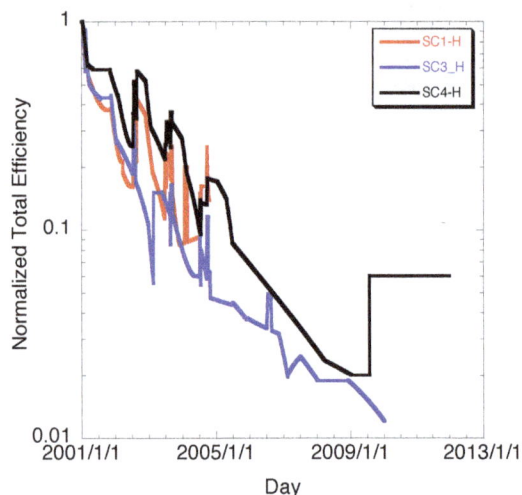

Fig. 5. H^+ normalized total efficiencies for the high side of the three CODIF instruments.

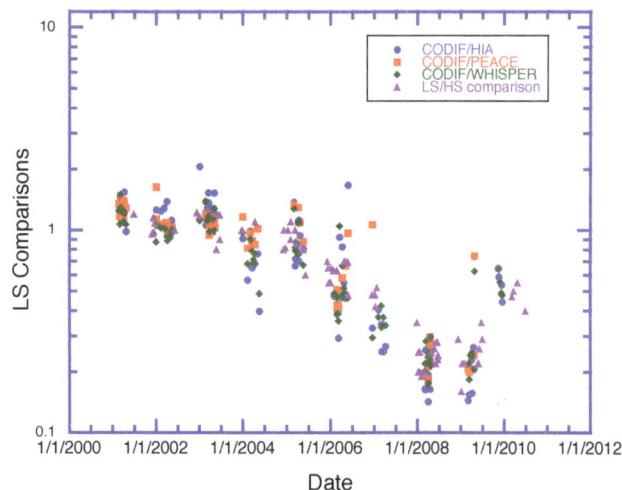

Fig. 6. CODIF SC/4 LS comparisons with the density measurements from HIA, PEACE (electrons) and WHISPER (electrons), and with pressure measurements from the CODIF HS.

the final normalized total efficiency, determined by adjusting the curve determined from the rate data with the pressure adjustment curve from Fig. 3. Again, the reason that an adjustment is required is due to the inaccurate measure of the stop rate, SR.

Figure 5 shows the H^+ normalized total efficiencies for the HS on all three instruments. The overall changes were similar on the three spacecraft.

2.3   Low side total efficiencies

As discussed in the introduction, the LS has a geometric factor that is a factor of 100 smaller than the HS. Thus the amount of flux on its MCPs is dramatically less. As a result, its efficiencies remained essentially constant for the first four years of the mission. However, after that, they also started to decline. The low side of the instrument is used predominantly on S/C 4 in the magnetosheath and in the solar wind. In these regions, the high flux saturates the electronics on the high side, and so the low side is used. In the magnetosheath, the WHISPER instrument is able to derive the total plasma density by identifying the plasma frequency in the data. The HIA all-ion instrument has been well cross-calibrated against WHISPER in this region (see Blagau et al., 2013). The dominant particle energy in the magnetosheath is $\sim 1\,\mathrm{keV}$, so the distribution is well contained within the CODIF energy range, as well as in the energy ranges covered by HIA and PEACE. Thus it is fitting to also cross-calibrate the Cluster LS against these instruments.

The LS calibration has been done in two ways. First, the pressure measured by the CODIF instrument is compared before and after the time when CODIF switches from the HS to the LS. This normally occurs in the outer magnetosphere when the flux is low, and is dominated by high-energy ($\sim 10\,\mathrm{keV}$) ions. Second, the CODIF densities are compared

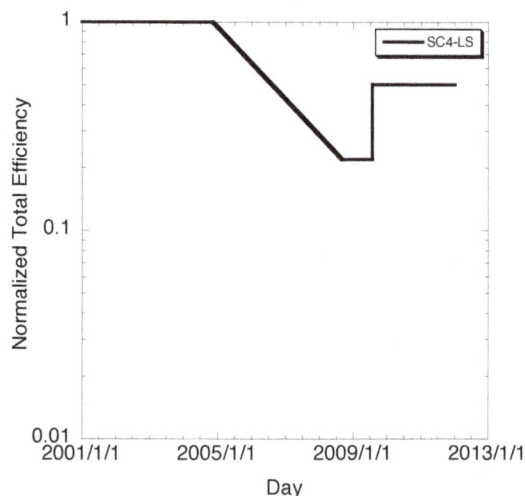

Fig. 7. H^+ normalized total efficiencies for the low side of SC 4.

with electron densities from WHISPER and PEACE, and ion densities from HIA in the magnetosheath. All data were obtained from the Cluster Active Archive (CAA). In this case, the dominant energy is $\sim 1\,\mathrm{keV}$. Figure 6 shows the ratio of the CODIF LS measurement with the density measurements by HIA, PEACE and WHISPER, and the pressure measurements using the HS of CODIF. The comparisons with the HS on CODIF were done after the HS efficiency corrections had been finalized. All methods consistently show that the LS side efficiency began decreasing in 2005. When the MCP voltage was increased in 2009, the efficiency then increased, but only up to 50 % of its initial value. From these measurements, the normalized total efficiency curve for the LS, shown in Fig. 7, was determined.

a)

	1/4 spin				1/2 spin				3/4 spin				1 spin		
32 Sweeps 0 1 2 3	4 5 6 7	8 9 10 11	12 13 14 15	16 17 18 19	20 21 22 23	24 25 26 27	28 29 30 31								
Anode 1	0			1			2			3					
Anode 2	4		5		6		7		8		9		10		11
Anode 3	12	13	14	15	16	17	18	19	20	21	22	23	24	25	26	27
Anode 4	28	29	30	31	32	33	34	35	36	37	38	39	40	41	42	43
Anode 5	44	45	46	47	48	49	50	51	52	53	54	55	56	57	58	59
Anode 6	60	61	62	63	64	65	66	67	68	69	70	71	72	73	74	75
Anode 7	76		77		78		79		780		81		82		83
Anode 8	84				85				86				87		

b)

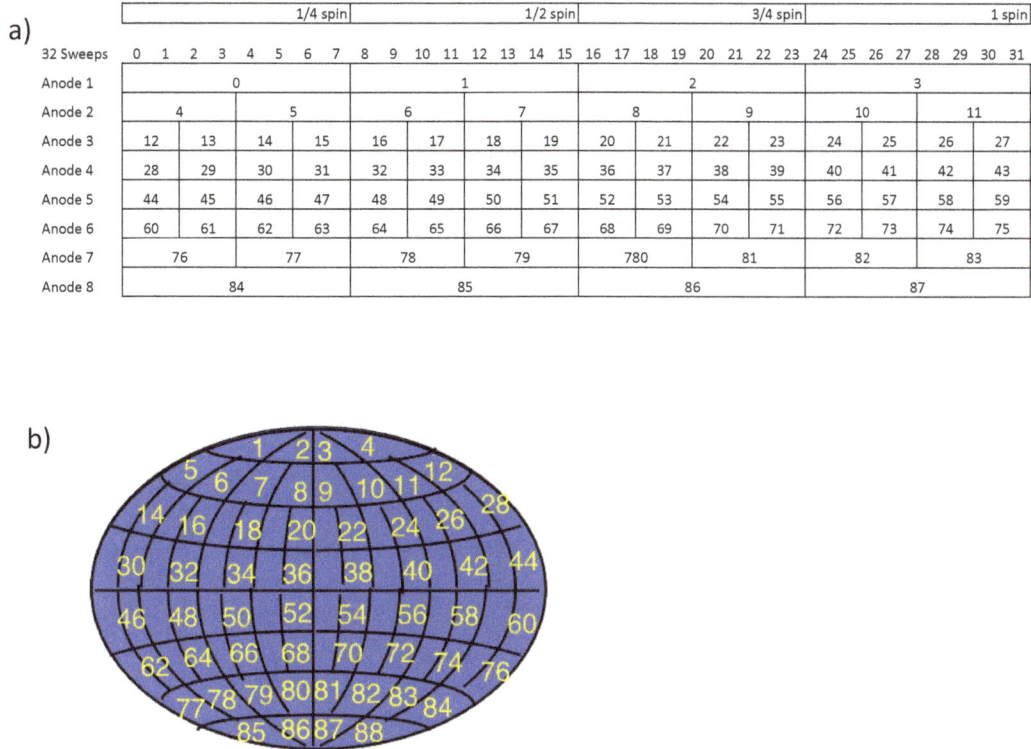

Fig. 8. (a) Angle map showing how the eight position anodes in the instrument are combined with the spacecraft spin to create an 88-angle product. (b) A Mollweide projection showing how the 88 angles cover the full sphere.

2.4 Relative anode efficiencies

As discussed above, the change in the efficiency is often not uniform across the MCPs. The efficiencies as determined from the rates, or from pressure balance or density comparisons only give information about the whole side (HS or LS) of the instrument. However, the relative efficiencies of different positions (or anodes) are also important. If an anisotropy is introduced due to non-uniform efficiencies, a calculation of the velocity in the spin-axis direction will result in an anomalous velocity. In order to calculate the velocity vector accurately, the relative calibrations within the instrument must also be corrected.

The method for calculating the relative efficiencies is to normalize the efficiencies at different positions, or anodes, in the instrument by assuming gyrotropy during selected time periods. The selected time periods are either from the outer magnetosphere, where the distribution is normally peaked at 90 degrees, or in the plasma sheet, where the distribution is normally isotropic. This normalization is done using the 3D distribution function products that are included in the science data from the instrument. The data accumulated in the 3-D distribution function products are binned by both species and angle. The eight instrument anodes on each side (HS or LS) give eight 22.5° polar angles, as described above. The azimuthal angle, in the spin plane, is determined by syncing the data accumulation to the spacecraft spin. The full spin is divided into 16 sectors of 22.5°. Figure 8a shows how the eight polar angles and 16 azimuthal angles are combined to give 88 total angles. Figure 8b shows how these 88 angles cover the sphere in a Mollweide projection. Depending on the magnetic field orientation, one anode (corresponding to one row in Fig. 8a) can cover a range of pitch angles during a spacecraft spin. The flux is calculated for each of the 88 angles, and is plotted at the pitch angle determined for that position. Each anode contributes 4 to 16 individual points on the pitch angle plot. Then each anode is individually adjusted so that the flux is the same at a given pitch angle.

Figures 9–11 illustrate an example from a time when the spacecraft are in the plasma sheet. Figure 9 shows the pitch angle distribution for H^+ at 11 different energies. In each panel, there is one point for each of the 88 angles in the 3-D distribution. The different colors/symbols indicate which anode was used. A fit is performed to the points around 90° (90° ± 45°) in order to determine the average pitch angle distribution, as shown with a black line in Fig. 9. Then a normalization factor is determined for each anode at each energy that will make the flux measured by that anode, over the range of pitch angles that it covers, best match the average fit. In this case, the flux of anode 8 (red) is systematically high, while the flux for anode 5 (greenish yellow) is low. The resulting normalized pitch angle plots are shown in Fig. 10.

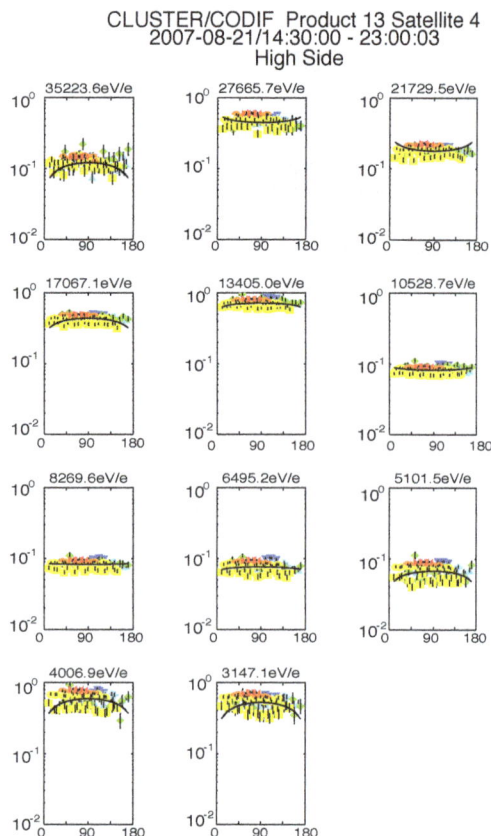

Fig. 9. The flux as a function of pitch angle for individual anodes on the CODIF high side. Each panel shows a different energy channel, ranging from 35 keV (top left panel) to 3 keV (bottom right panel).

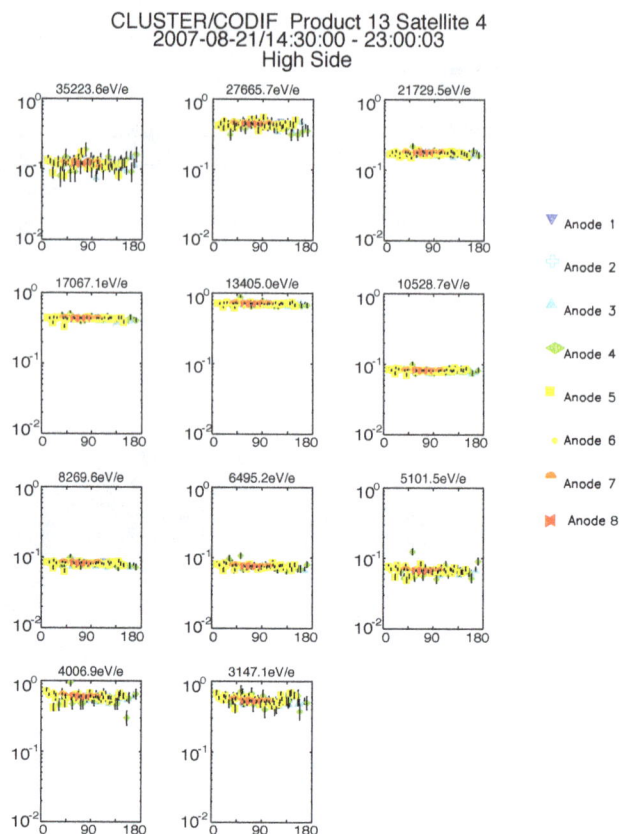

Fig. 10. The flux as a function of pitch angle for individual anodes on the CODIF high side after each individual anode has been normalized.

Now, for each energy and pitch angle, all anodes give the same flux. Then, the adjustment parameters determined for each anode and energy are used to redefine the calibration curves. The final result is shown in Fig. 11. The top panel shows the original curves used to make the first pitch angle plot, and the bottom panel shows the revised curves. To make the revised curves, the points for each anode at each energy where there are sufficient counts to make a pitch angle fit are shifted based on the normalization factor. Then additional points are added at 60 and 14 keV to force the fit to follow the same trend as the data at the outer limits of the instrument energy range. Then a fit is done to this new set of points to generate the curve. This procedure is performed about once per month, and more often if the distributions show a significant change.

After the new files with the relative anode efficiencies have been determined, the efficiencies are checked by calculating the average velocities in the plasma sheet. Figure 12 shows a comparison of orbit averages of the density and the three components of the velocity for CODIF on SC4, and HIA on SC1 and SC3, in GSE coordinates. The x and y coordinates are approximately in the spin plane, and so the relative anode efficiency does not significantly affect these velocities. The

z coordinate is almost along the spin plane, so any discrepancies in the relative anode efficiencies are observed here. It is expected that the average velocity in z should be 0. The relative anode efficiencies are adjusted to keep the error to less than $15 \, \mathrm{km \, s^{-1}}$.

3 O$^+$ calibration

O$^+$ is the second most abundant species in the magnetosphere, and is the key tracer for ionospheric input. Thus tracking the O$^+$ efficiencies is critical for measuring the importance of the ionosphere as a source of plasma in the magnetosphere over the mission.

3.1 Relative anode efficiencies

The relative anode efficiencies for O$^+$ can be determined in exactly the same way as they are determined for H$^+$. Because the method uses data that are classified by species in the instrument, it does not require time periods where O$^+$ is dominant. We still need to identify time periods when there is a significant flux of O$^+$, so that we have sufficient statistics to compare the different anodes, but H$^+$ can still be the

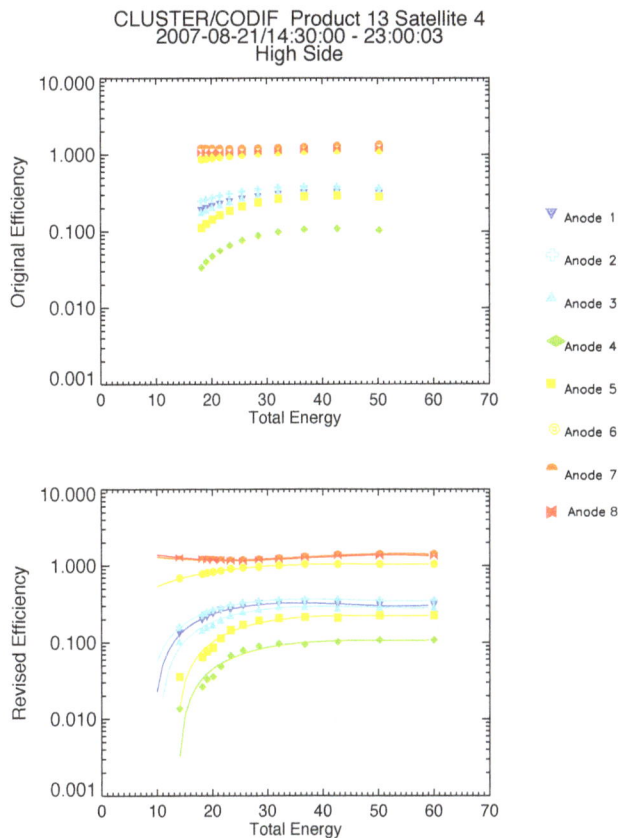

Fig. 11. The top panel shows the relative anode efficiency as a function of energy for each of the eight anodes on the CODIF high side. The bottom panel shows revised anode efficiencies, using the normalization factors determined assuming a gyrotropic distribution.

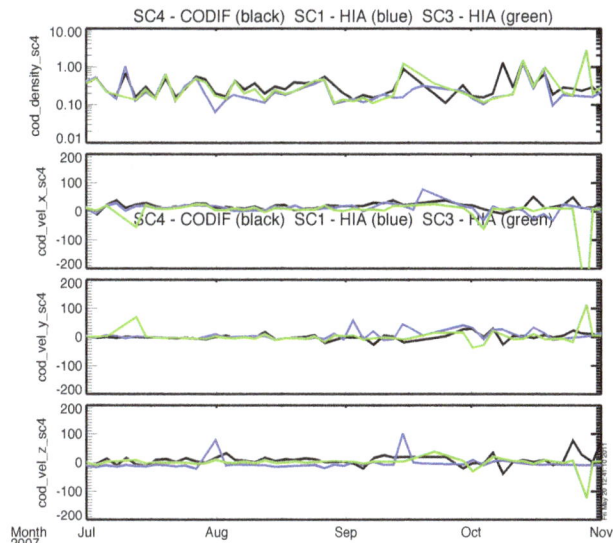

Fig. 12. The average densities and velocities in the plasma sheet from CODIF, on S/C 4, and from HIA on S/C 1 and S/C 3.

dominant ion. An example of an O^+ relative calibration is shown in McFadden et al. (2007).

3.2 Total efficiencies

Determining O^+ total efficiencies in-flight is difficult because all the techniques applied for H^+ are only useful for the dominant species. In order to use the engineering rate techniques, we need to find time periods when O^+ is actually the dominant species, at least at a particular energy per charge, so that the start, stop, and coincidence rates that we measure are due to O^+. The only time periods that we have found to do this reliably are times when the spacecraft is over the polar cap and in the lobes. In these regions, we sometimes observe narrow energy "beams" of O^+. The origin of these beams is outflow from the cusp. Because the ions move up along the field lines, as the field line is convected into the tail, the original particle distribution, which can be broad in energy, becomes separated by velocity, because the higher velocity ions are moving further along the field line than the lower velocity ions. This is referred to as the "velocity filter effect". Since O^+ and H^+ at the same velocity are separated

in energy per charge by a factor of 16, the O^+ beam is observed at a distinctly different energy step in the instrument than the H^+ beam. Thus we can use the energy steps where the O^+ is observed to determine the O^+ efficiency. Because the O^+ beams are generally field-aligned, they are only observed in one or two anodes at a time. Thus we can obtain the total efficiencies for a particular anode. We then use the data from the relative anode O^+ calibration to determine an overall change to the total efficiency.

An example of these O^+ beams is shown in Fig. 13. During this time period on 21 December 2002, a strong O^+ beam is clearly observed as the spacecraft approaches the cusp. Figure 13 shows the O^+ energy spectra for the 8 individual anodes on the HS. In this case, the beam is initially observed in anode 1. It then moves to anode 2, and then into 3. The beams are observed for more than an hour in each of the anodes 1 and 2. Thus we can accumulate sufficient statistics to use the rate data to calculate the O^+ efficiencies. For 2001–2004, we used this method to determine the O^+ calibration for SC/1. The final S/C 1 O^+ normalized total efficiencies derived using this method are shown in Fig. 14. The H^+ normalized total efficiency (green line) is shown for comparison. The calibrations for S/C 3 and S/C 4 were then done by comparison with S/C 1. Stable time periods were chosen when the three spacecraft would be expected to observe the same flux, and the efficiencies were adjusted to bring the O^+ fluxes on S/C 3 and S/C 4 to the same level as S/C 1. About one time period per month is used for this adjustment.

After 2004, the same method was used, using beams observed by S/C 4. However, this method was only useful through 2006. After 2006, the flux of O^+ in these regions decreased significantly, because there is less ionospheric outflow during solar minimum (Yau and André,

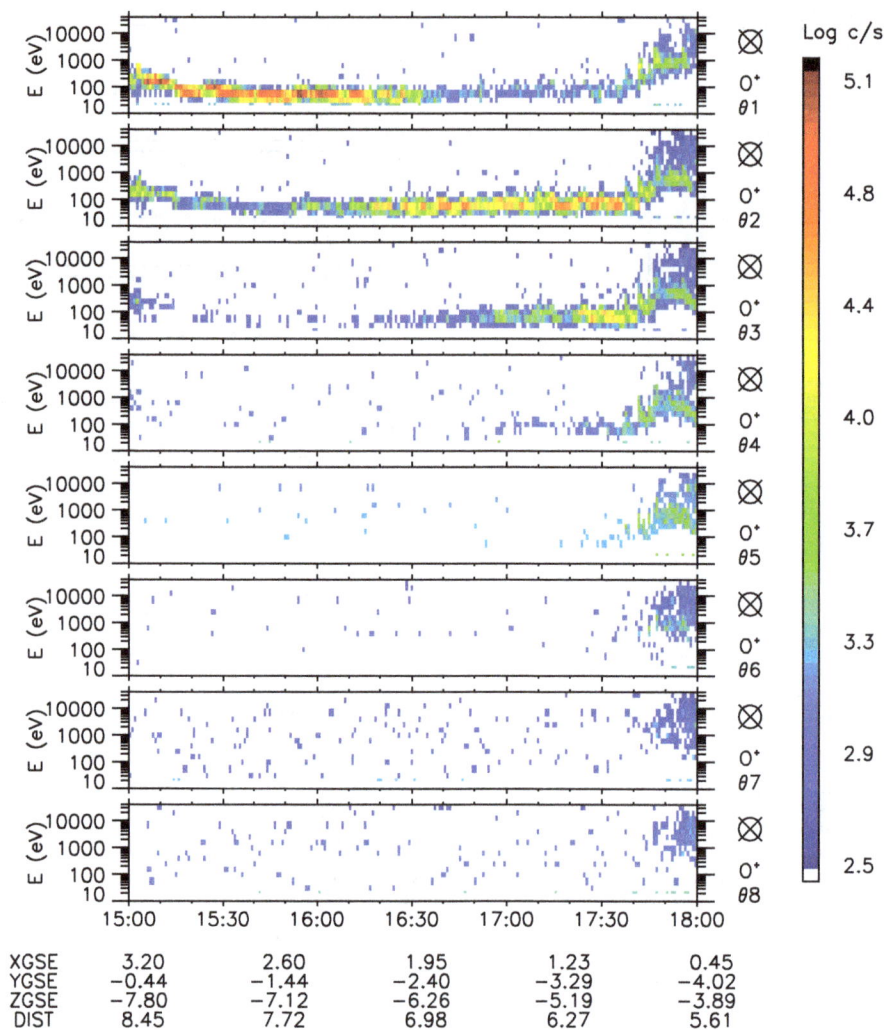

Fig. 13. Example of O^+ beams observed on 21 December 2002, on SC 1, as the spacecraft approaches the cusp. Each panel shows the O^+ energy spectrum for one position anode in the instrument. In this case, the beam is first predominantly in anode 1, and then moves to anode 2, and then 3.

1997). Therefore, we were no longer able to do this measurement. Now that the solar cycle is approaching solar maximum, we again observe O^+ beams. Using new time periods, we will again be able to measure the total efficiency, and adjust the O^+ efficiency values. At this point, the O^+ efficiencies are considered valid through 2006, and are estimated after that. A new release of calibration files will update the O^+ total efficiency values.

4   Other species

Because the He^{++} suffers from significant contamination from H^+, as discussed in the companion paper by Mouikis et al. (2013), no separate calibration for He^{++} has been determined. He^{++} fluxes are calculated using the initial calibration curves determined pre-launch, and with the total efficiencies following the H^+ changes.

He^+ is also difficult to calibrate because there are not many time periods when its flux is significant. From the preflight calibrations, the He^+ efficiency was found to be very similar to the O^+ efficiency, except at the low energies, where O^+ has a lower efficiency due to the greater scattering in the foil. Thus we have used the energy dependence for He^+ that we established pre-flight, but then assumed that the total efficiency changes with time following the O^+ efficiency.

5   Conclusions

Using a variety of methods, we have determined the calibrations of H^+ through the duration of the mission. We were

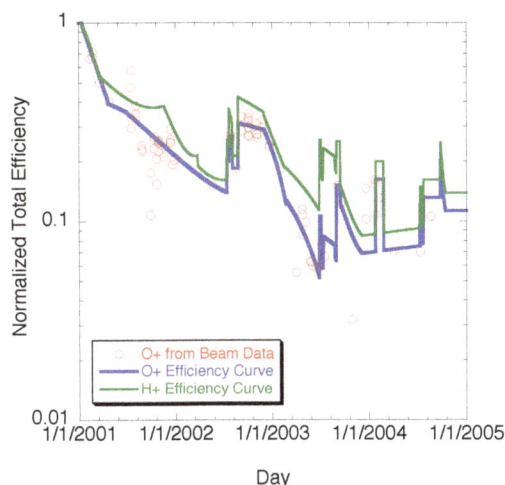

Fig. 14. O^+ normalized total efficiency derived using the O^+ beam data for S/C 1. The final curve used for the efficiency is shown with the blue line, and the H^+ efficiency is shown in green, for comparison.

able to calibrate the total efficiencies of O^+ successfully through 2006 using the O^+ "beams" observed in the lobe. Due to the diminished ionospheric outflow during the solar minimum, we have only been able to determine the relative anode calibrations after 2006, while the total efficiency calibration is only an estimate. However, now that the mission is again approaching solar maximum, the beams are observed again, and we will use them to confirm the O^+ calibration. After the last MCP voltage increase in 2009, the efficiencies have remained relatively stable for more than three years. We anticipate that CODIF on S/C 4 will continue returning data with reasonable efficiencies through the extended mission.

Acknowledgements. We are grateful to the many engineers and scientists from UNH, MPE, CESR, MPS, IFSI, IRF, UCB and UW who made the development of the CIS instrument possible. The calibration work at UNH was supported by NASA under grant NNX11AB65G. We also thank the other Cluster teams, in particular the WHISPER team (Principle Investigator, Jean Louis Rauch; former Principle Investigators, Pierrette M. E. Decreau, Jean Gabriel Trotigen; and Gabor Fascko), the PEACE team (Principle Investigator, A. Fazakerly), and the MAG team (Principle Investigator, E. Lucek) for making their data available for the cross-calibration work.

Edited by: M. Genzer

References

Allegrini, F., Wimmer-Schweingruber, R. F., Wurz, P., and Bochsler, P.: Determination of low-energy ion-induced electron yields from thin carbon foils, Nucl. Instrum. Meth. Phys. Res. B, 211, 487–494, doi:10.1016/S0168-583X(03)01705-1, 2003.

Allegrini, F., McComas, D. J., Young, D. T., Berthelier, J. J., Covinhes, J., Illiano, J. M., Riou, J. F., Funsten, H. O., and Harper, R. W.: Energy loss of 1–50 keV H, He, C, N, O, Ne, and Ar ions transmitted through thin carbon foils, Rev. Sci. Instrum., 77, 4501, doi:10.1063/1.2185490, 2006.

Blagau, A., Dandouras, I., Barthe, A., Brunato, S., Facskó, G., and Constantinescu, V.: In-flight calibration of Hot Ion Analyser onboard Cluster, Geosci. Instrum. Method. Data Syst. Discuss., 3, 407–435, doi:10.5194/gid-3-407-2013, 2013.

Décréau, P. M. E., Fergeau, P., Krasnoselskikh, V., Le Guirriec, E., Lévêque, M., Martin, Ph., Randriamboarison, O., Rauch, J. L., Sené, F. X., Séran, H. C., Trotignon, J. G., Canu, P., Cornilleau, N., de Féraudy, H., Alleyne, H., Yearby, K., Mögensen, P. B., Gustafsson, G., André, M., Gurnett, D. C., Darrouzet, F., Lemaire, J., Harvey, C. C., Travnicek, P., and Whisper experimenters (Table 1): Early results from the Whisper instrument on Cluster: an overview, Ann. Geophys., 19, 1241–1258, doi:10.5194/angeo-19-1241-2001, 2001.

Drake, V. A., Eparvier, F. G., McClintock, W. E., Woods, T. N., Ucker, G. J., and Hill, C.: Microchannel plate performance and life-test results for the TIMED Solar EUV experiment, Proc. SPIE, 3445, 603–614, 1998.

Fairfield, D. H., Lepping, R. P., Hones, E. W. J., Bame, S. J., and Asbridge, J. R.: Simultaneous measurements of magnetotail dynamics by IMP spacecraft, J. Geophys. Res., 86, 1396–1414, doi:10.1029/JA086iA03p01396, 1981.

Gonin, M., Buergi, A., Oetliker, M., and Bochsler, P.: Interaction of solar wind ions with thin carbon foils: Calibration of time-of-flight spectrometers, in: ESA, Proceedings of the First SOHO Workshop: Coronal Streamers, Coronal Loops, and Coronal and Solar Wind Composition (SEE N93-31343 12-92), 348, 381–384, 1992.

Johnstone, A. D., Alsop, C., Burge, S., Carter, P. J., Coates, A. J., Coker, A. J., Fazakerley, A. N., Grande, M., Gowen, R. A., Gurgiolo, C., Hancock, B. K., Narheim, B., Preece, A., Sheather, P. H., Winningham, J. D., and Woodliffe, R. D.: Peace: a Plasma Electron and Current Experiment, Space Sci. Rev., 79, 351–398, doi:10.1023/A:1004938001388, 1997.

Kishimoto, N., Nagamine, M., Inami, K., Enari, Y., and Ohshima, T.: Lifetime of MCP–PMT, Nucl. Instrum. Meth. Phys. Res. A, 564, 204–211, doi:10.1016/j.nima.2006.04.089, 2006.

Kistler, L. M.: Cluster CODIF Calibration Report, Part I, Ion Efficiency Analysis, available at: http://caa.estec.esa.int/documents/CLUSTER_CODIFEff.pdf (last access: 25 April 2013), 2000a.

Kistler, L. M.: Cluster CODIF Calibration Report, Part II, Geometric Factor, Energy and Angle Response, available at: http://caa.estec.esa.int/documents/CODIF_ESAAlpha.pdf (last access: 25 April 2013), 2000b.

Kistler, L. M.: Cluster CODIF Calibration Report, Part III, TOF Peak Analysis, available at: http://caa.estec.esa.int/documents/CODIF_CalTOFv2.pdf (last access: 25 April 2013), 2000c.

McFadden, J. P., Evans, D. S., Kasprzak, W. T., Brace, L. H., Chornay, D. J., Coates, A. J., Dichter, B. K. H. W. R., Holeman, E., Kadinsky-Cade, K., Kasper, J. C., Kataria, D., Kistler, L., Larson, D., Lazarus, A. J. M. F., Mukai, T., Ogilvie, K. W., Paschmann, G., Rich, F., Saito, Y., Sudder, J. D., Steinberg, J. T., Wuest, M., and Wurz, P.: In-Flight Instrument Calibration and Performance Verification, Calibr. Part. Instrum. Space Phys., 7, 277–385, 2007.

Möbius, E., Kistler, L. M., Popecki, M. A., Crocker, K. C., Granoff, M., Jiang, Y., Sartori, E., Ye, V., Rème, H., Sauvaud, J. A., Cros, A., Aoustin, C., Camus, T., Medale, J. L., Rouzaud, J., Carlson, C. W., McFadden, J. P., Curtis, D., Heetder, H., Croyle, J., Ingraham, C., Klecker, B., Hovestadt, D., Ertl, M., Eberl, F., Kästle, H., Künneth, E., Laeverenz, P., Seidenschwang, E., Shelley, E. G., Klumpar, D. M., Hertzberg, E., Parks, G. K., McCarthy, M., Korth, A., Rosenbauer, H., Gräve, B., Eliasson, L., Olsen, S., Balsiger, H., Schwab, U., and Steinacher, M.: The 3-D Plasma Distribution Function Analyzers with Time-of-Flight Mass Discrimination for Cluster, FAST, and Equator-S, in: Measurement Techniques in Space Plasmas:Particles, edited by: Pfaff, R. F., Borovsky, J. E., and Young, D. T., American Geophysical Union, Washington, D. C., doi:10.1029/GM102p0243, 2013.

Mouikis, C. G., Kistler, L. M., Wang, G., and Liu, Y.: Background subtraction for the Cluster/CODIF plasma ion mass spectrometer, Geosci. Instrum. Method. Data Syst. Discuss., 3, 567–589, doi:10.5194/gid-3-567-2013, 2013.

Rème, H., Aoustin, C., Bosqued, J. M., Dandouras, I., Lavraud, B., Sauvaud, J. A., Barthe, A., Bouyssou, J., Camus, Th., Coeur-Joly, O., Cros, A., Cuvilo, J., Ducay, F., Garbarowitz, Y., Medale, J. L., Penou, E., Perrier, H., Romefort, D., Rouzaud, J., Vallat, C., Alcaydé, D., Jacquey, C., Mazelle, C., d'Uston, C., Möbius, E., Kistler, L. M., Crocker, K., Granoff, M., Mouikis, C., Popecki, M., Vosbury, M., Klecker, B., Hovestadt, D., Kucharek, H., Kuenneth, E., Paschmann, G., Scholer, M., Sckopke, N., Seidenschwang, E., Carlson, C. W., Curtis, D. W., Ingraham, C., Lin, R. P., McFadden, J. P., Parks, G. K., Phan, T., Formisano, V., Amata, E., Bavassano-Cattaneo, M. B., Baldetti, P., Bruno, R., Chionchio, G., Di Lellis, A., Marcucci, M. F., Pallocchia, G., Korth, A., Daly, P. W., Graeve, B., Rosenbauer, H., Vasyliunas, V., McCarthy, M., Wilber, M., Eliasson, L., Lundin, R., Olsen, S., Shelley, E. G., Fuselier, S., Ghielmetti, A. G., Lennartsson, W., Escoubet, C. P., Balsiger, H., Friedel, R., Cao, J.-B., Kovrazhkin, R. A., Papamastorakis, I., Pellat, R., Scudder, J., and Sonnerup, B.: First multispacecraft ion measurements in and near the Earth's magnetosphere with the identical Cluster ion spectrometry (CIS) experiment, Ann. Geophys., 19, 1303–1354, doi:10.5194/angeo-19-1303-2001, 2001.

Ritzau, S. M. and Baragiola, R. A.: Electron emission from carbon foils induced by keV ions, Phys. Rev. B, 58, 2529–2538, doi:10.1103/PhysRevB.58.2529, 1998.

Sandel, B. R., Broadfoot, A. L., and Shemansky, D. E.: Microchannel plate life tests, Appl. Optics, 16, 1435–1437, doi:10.1364/AO.16.001435, 1977.

Trotignon, J. G., Décréau, P. M. E., Rauch, J. L., Vallières, X., Rochel, A., Kougblénou, S., Lointier, F., Facskó, G., Canu, P., Darrouzet, F., and Masson, A.: The WHISPER Relaxation Sounder and the Cluster Active Archive, in: The Cluster Active Archive, C. P. Astrophysics and Space Science Proceedings, edited by: Laakso, H., Taylor, M. G. T. T., and Escoubet, Springer, Berlin, 185–208, doi:10.1007/978-90-481-3499-1_12, 2010.

Yau, A. W. and André, M.: Sources of Ion Outflow in the High Latitude Ionosphere, Space Sci. Rev., 80, 1–25, doi:10.1023/A:1004947203046, 1997.

4

Contribution to solving the orientation problem for an automatic magnetic observatory

A. Khokhlov[1,2,3], **J. L. Le Mouël**[3], **and M. Mandea**[4]

[1]International Institute of Earthquake Prediction Theory and Mathematical Geophysics 79, b2, Warshavskoe shosse, 113556 Moscow, Russia
[2]Geophysical Center of RAS, 3 Molodezhnaya St., 119296 Moscow, Russia
[3]Institut de Physique du Globe de Paris, UMR7154, CNRS – 1 Rue Jussieu, 75005 Paris, France
[4]Centre National d'Etudes Spatiales, 2, Place Maurice Quentin, 75001 Paris, France

Correspondence to: M. Mandea (mioara.mandea@cnes.fr)

Abstract. The problem of the absolute calibration of a vectorial (tri-axial) magnetometer is addressed with the objective that the apparatus, once calibrated, gives afterwards, for a few years, the absolute values of the three components of the geomagnetic field (say the Northern geographical component, Eastern component and vertical component) with an accuracy on the order of 1 nT. The calibration procedure comes down to measure the orientation in space of the three physical axes of the sensor or, in other words, the entries of the transfer matrix from the local geographical axes to these physical axes. Absolute calibration follows indeed an internal calibration which provides accurate values of the three scale factors corresponding to the three axes – and in addition their relative angles. The absolute calibration can be achieved through classical absolute measurements made with an independent equipment. It is shown – after an error analysis which is not trivial – that, while it is not possible to get the axes absolute orientations with a high accuracy, the assigned objective (absolute values of the Northern geographical component, Eastern component and vertical component, with an accuracy of the order of 1 nT) is nevertheless reachable; this is because in the time interval of interest the field to measure is not far from the field prevailing during the calibration process.

1 Introduction

The geomagnetic field is continuously measured in a network of magnetic observatories, which, however, has significant gaps in the remote areas and over the oceans. This uneven distribution is linked to the fact that currently it is not possible to operate fully automated observatories which do not require manual operation of any instrument. Already, some fifty years ago, Alldregde planned an automatic standard magnetic observatory (ASMO) – (Alldregde, 1960; Alldregde and Saldukas, 1964), i.e. a device providing at each time the absolute values of – say – the Northern geographical component, Eastern component and vertical component of the geomagnetic field, without extra independent absolute measurements, at least for a long enough timespan. This idea has remained in the geomagnetism community and, over the last years, attempts have been made to automate a DI-theodolite (van Loo and Rasson, 2006; Rasson and Gonsette, 2011), a proton vector magnetometer (Auster et al., 2006), or to build a device which can perform discrete absolute measurements automatically (Auster et al., 2007). The idea to use absolute measurements for solving orientation problems has been used by Schott and Leroy (2001) when developing a DIDD magnetometer.

In a former paper (Gravrand et al., 2001), the question was addressed of the internal calibration of a vectorial (or tri-axial) magnetometer, such as the ^{4}He pumped magnetometer built by the Laboratoire d'Electronique et de Technique de l'Information (LETI) of the French Commissariat à l'Energie Atomique (CEA), or the Oersted space mission fluxgate magnetometer, built by the Institute for automatization of the Danish Technical University (DTU). The problem of the internal calibration is to determine the six (^{4}He magnetometer) or nine (fluxgate magnetometer) intrinsic parameters: the three scale values (one for each modulation coil physical axis in the ^{4}He case, or each fluxgate sensor axis in the DTU case), the three angles between the three physical axis in both cases, and the three offset values in the case of the fluxgate variometer. The calibration problem can be solved by rotating in space the triaxial magnetometer while making simultaneously absolute intensity measurements with a proton or optical pumping magnetometer (Olsen et al., 2003), and it is essentially linear (Gravrand et al., 2001). The calibration algorithm was developed for ground operation but can be – and has been – extended to in flight calibration. But this internal calibration is not enough if we want to install somewhere a genuine automatic magnetic observatory. In the following u_1, u_2, u_3 are the unit vectors corresponding to these north east down directions.

It was stated in Gravrand et al. (2001) that such an absolute calibration should not be too difficult in a place where standard absolute measurements can be performed. This is the question (of quite practical interest) that we address in the present paper. We show that the above statement is both valid and invalid, depending on the objective. We also show that only the determination of the scale factors provided by the internal calibration process are of interest for the absolute calibration (the accurate determination of all angles between physical axes is not necessary however provided by the internal calibration; see also Appendix).

2   The practical problem

Suppose we want to install an automatic observatory in some new place, say a remote island in the Pacific ocean. What is required is, to refresh Alldregde's statement, to obtain one minute absolute values of the field components X, Y, Z in, say, the north east vertical down frame, fitting INTERMAGNET standards (see http:/www.intermagnet.org), without needing an observer to visit the place in the few years following the installation.

One first builds a pillar (the permanent pillar) in a location propitious to install the ^{4}He magnetometer, and an auxiliary pillar a few meters apart. The calibration process can start. The observer determines the differences ΔX, ΔY, ΔZ between the absolute values of X, Y, Z at the two pillars. This is classical observatory work, not negligible, but which can be completed in a few days using a DI-flux theodolite and a proton magnetometer; modern devices for determining azimuths are welcome. The magnetometer-variometer, as we call it, can now be installed on the permanent pillar (in fact after a non magnetic house has been built around it; we do not develop here these practical aspects). By construction,

the unit vectors e_1, e_2, e_3 of the physical axes, or coil axes, of the apparatus are nearly orthogonal, and its installation on the pillar is generally made in such a way that e_1 is close to u_1, e_2 close to u_2 and e_3 close to u_3; although this is by no way a necessary condition. The observer makes at the auxiliary pillar a series of absolute measurements of the magnetic field at time moments $t_1, t_2, ...t_k, ...$ and corrections ΔX, ΔY, ΔZ are applied to get the corresponding absolute values on the permanent pillar. At the same time moments t_k, the magnetometer-variometer to be calibrated provides the values $\{V_k^1, V_k^2, V_k^3\}, k = 1, 2, ...K$ of the components of the magnetic vector $V(t_k)$ along its physical axes $\{e_i\}$ whose orientations with respect to $\{u_1, u_2, u_3\}$ are not exactly known.

The observer, with his equipment, now leaves the place. The magnetometer-variometer in place continues to provide the values $\{V^1(t), V^2(t), V^3(t)\}$ of the (contravariant) components of V along its physical axes. The problem to solve is the following: how, relying on the set of absolute measurements made previously at times $t_1, t_2, ...t_k$, to compute the geographical components $(X^1(t), X^2(t), X^3(t))$ of $V(t)$ at any following time t (in fact depending on the sampling rate), and estimate the error on those computed values?

This error, as we will see it, is a direct function of the errors on the absolute measurements made at the times $t_1, t_2, ...$ We call it the calibration error. Let us say that we note $B(t_k) = B_k$ the absolute measurements at time t_k and $V(t)$ the measurements provided by the magnetometer-variometer.

3   The principle of the calibration

To compute the e_i vectors in the u_i frame, we go through the B_k. Obviously one B_k is not enough; but as a linear operator in R^3 is uniquely defined by its action on three linearly independent vectors, we take three of them, that we note B_1, B_2, B_3, to present the algorithm of the calibration. In practice, several triplets among the K measurements available, if $K > 3$, are used.

Let d_k^j be the geographical components of of $B_k(k = 1, 2, 3)$ (as measured by the observer), and f_k^j the values of the (contravariant) components of B_k along the physical axes e_1, e_2, e_3 provided at the same times by the magnetometer-variometer. We have

$$B_k = \hat{d}_k^1 u_1 + \hat{d}_k^2 u_2 + \hat{d}_k^3 u_3$$
$$B_k = \hat{f}_k^1 e_1 + \hat{f}_k^2 e_2 + \hat{f}_k^3 e_3 \qquad (1)$$

where the *hat* symbol is to stress the error-free nature of the corresponding quantities. The solution of the calibration is trivial, the e_i being straightforwardly obtained in function of the u_i through the B_k:

$$\begin{pmatrix} e_1 \\ e_2 \\ e_3 \end{pmatrix} = \hat{F}^{-1} \hat{D} \begin{pmatrix} u_1 \\ u_2 \\ u_3 \end{pmatrix} = \hat{C} \begin{pmatrix} u_1 \\ u_2 \\ u_3 \end{pmatrix} \qquad (2)$$

where $\hat{\mathbf{F}}$ and $\hat{\mathbf{D}}$ are the matrices of coefficients (components) $\hat{f}_k^j$ and $\hat{d}_k^j$.

The magnetometer-variometer provides the values ($V^1(t)$, $V^2(t)$, $V^3(t)$) of the physical components of $\boldsymbol{V}$:

$$
\boldsymbol{V}(t) = \left(V^1, V^2, V^3 \right) \begin{pmatrix} \boldsymbol{e}_1 \\ \boldsymbol{e}_2 \\ \boldsymbol{e}_3 \end{pmatrix} = \left(V^1, V^2, V^3 \right) \hat{\mathbf{C}} \begin{pmatrix} \boldsymbol{u}_1 \\ \boldsymbol{u}_2 \\ \boldsymbol{u}_3 \end{pmatrix}
$$

$$
= \left(X^1, X^2, X^3 \right) \begin{pmatrix} \boldsymbol{u}_1 \\ \boldsymbol{u}_2 \\ \boldsymbol{u}_3 \end{pmatrix}. \tag{3}
$$

The problem is solved in the error-free case; we have obtained the geographical components (X_1, X_2, X_3) of $\boldsymbol{V}$, at each measurement time. But the problem is in dealing with errors.

4 The errors

4.1 General statements

The absolute measurements of $\boldsymbol{B}_1$, $\boldsymbol{B}_2$, $\boldsymbol{B}_3$ are affected by errors which can be viewed as errors on the geographical components d_k^j of $\boldsymbol{B}_k$ vectors (Eq. 1). They could be discussed at some length; but it is a very well-known topic. For the purpose of the present study, we suppose the magnitude of the errors on $\boldsymbol{B}_k$ to be ε in relative value and randomly distributed in direction. In other words:

$$
d_k^j \text{ (measured)} = \hat{d}_k^j \text{ (true)} + \varepsilon^j b\, r_k^j.
$$

Here $\varepsilon \ll 1$ and r_k^j are $O(1)$. In matrix form:

$$
\mathbf{D} = \hat{\mathbf{D}} + \varepsilon b\, \mathbf{R}.
$$

$\mathbf{R}$ is the matrix of the r_k^j, and b a characteristic value of the $\boldsymbol{B}$ intensity at the station and the epoch of interest (see infra).

The values of the components along the physical axes $\{\boldsymbol{e}_i\}$, f_k^j are also not error-free. Nevertheless, in the case of the ^{4}He magnetometer that we have especially in mind, these f_k^j are measured with a very high accuracy, better than $0.1\,\text{nT}$ (Léger et al., 1992; Gravrand et al., 2001) after the internal calibration has been performed. To simplify the writing, we consider the values f_k^j as error-free, i.e. $f_k^j = \hat{f}_k^j$. Indeed, the f_k^j, the components of $\boldsymbol{B}_k$ along the physical axes of the variometer-magnetometer, are determined with a high accuracy: $0.1\,\text{nT}$, i.e. a relative error of a few 10^{-6}. The poor conditioning of $\mathbf{F}$ matrix comes from the $\boldsymbol{B}_1$, $\boldsymbol{B}_2$, $\boldsymbol{B}_3$ of the calibration being only $\delta b\omega$ apart, with the values considered in this paper ($|\omega_i| = 0.1$, $\delta b \approx 50\,\text{nT}$), i.e. $\delta \approx 10^{-3}$. Therefore, the measurement error of f does not significantly distort the inverse matrix $\mathbf{F}^{-1}$. It is in fact possible to develop the theory taking into account errors on f (supposed larger than above). It makes the writing larger and heavier. We have preferred to present the simplified version, essentially relevant, in this paper.

It now comes immediately:

$$
\mathbf{D} \begin{pmatrix} \boldsymbol{u}_1 \\ \boldsymbol{u}_2 \\ \boldsymbol{u}_3 \end{pmatrix} = \hat{\mathbf{D}} \begin{pmatrix} \boldsymbol{u}_1 \\ \boldsymbol{u}_2 \\ \boldsymbol{u}_3 \end{pmatrix} + \varepsilon b\, \mathbf{R} \begin{pmatrix} \boldsymbol{u}_1 \\ \boldsymbol{u}_2 \\ \boldsymbol{u}_3 \end{pmatrix} \tag{4}
$$

$$
\mathbf{D} \begin{pmatrix} \boldsymbol{u}_1 \\ \boldsymbol{u}_2 \\ \boldsymbol{u}_3 \end{pmatrix} = \begin{pmatrix} \boldsymbol{B}_1 \\ \boldsymbol{B}_2 \\ \boldsymbol{B}_3 \end{pmatrix} + \varepsilon b \begin{pmatrix} \omega_1 \\ \omega_2 \\ \omega_3 \end{pmatrix} = \begin{pmatrix} \boldsymbol{B}_1' \\ \boldsymbol{B}_2' \\ \boldsymbol{B}_3' \end{pmatrix} \tag{5}
$$

where $\varepsilon b\omega_i$, $i = 1, 2, 3$ denote the error along the i-th direction, and $|\omega_i| = O(1)$.

Multiplying Eq. (4) by $\mathbf{F}^{-1}$ (recall that $\mathbf{F}^{-1} = \hat{\mathbf{F}}^{-1}$) and using Eq. (2):

$$
\begin{pmatrix} \boldsymbol{e}_1 \\ \boldsymbol{e}_2 \\ \boldsymbol{e}_3 \end{pmatrix} = \hat{\mathbf{C}} \begin{pmatrix} \boldsymbol{u}_1 \\ \boldsymbol{u}_2 \\ \boldsymbol{u}_3 \end{pmatrix} - \varepsilon b\, \hat{\mathbf{F}}^{-1} \begin{pmatrix} \omega_1 \\ \omega_2 \\ \omega_3 \end{pmatrix}. \tag{6}
$$

In other words, when computing the physical unit vectors $\boldsymbol{e}_i$ using the "measured" transformation matrix $\mathbf{C} = \mathbf{F}^{-1}\mathbf{D}$ (instead of $\hat{\mathbf{C}} = \mathbf{F}^{-1}\hat{\mathbf{D}}$), an error is made which depends on $\mathbf{F}^{-1}$. The difficulty to be expected is rather obvious. We go from the orthogonal frame $\boldsymbol{u}_i$ to the nearly tri-orthogonal frame $\boldsymbol{e}_i$ through the $\boldsymbol{B}_k$ frame. But the three vectors $\boldsymbol{B}_1$, $\boldsymbol{B}_2$, $\boldsymbol{B}_3$ have directions close to one another (remember that they are measurements made at the station during a timespan of say a week; see Sect. 4 for numerical values). The matrix $\mathbf{F}$ whose lines are close to one another is a priori poorly conditioned; its inverse $\mathbf{F}^{-1}$ may have large eigenvalues, and a strong amplification of error εb might affect the directions of $\boldsymbol{e}_i$, and the error on $\boldsymbol{V}(t)$ might be much larger than the error εb on $\boldsymbol{B}_k$ (see Appendix). But, in fact, the practical conditions of the calibration process ($\boldsymbol{B}_k$) and of the following measurements of the current magnetic field $\boldsymbol{V}(t)$ by the magnetometer-variometer discard such error amplification as shown later. We now build a simple algorithm allowing a statistical modeling and providing realistic error estimation, sufficient for the present study.

4.2 A simple algorithm

Let us now consider the vector $\boldsymbol{V}$ at time t. From Eqs. (1), (3) and (5):

$$
\boldsymbol{V}(t) = \left(V^1(t), V^2(t), V^3(t) \right) \mathbf{F}^{-1} \begin{pmatrix} \boldsymbol{B}_1 \\ \boldsymbol{B}_2 \\ \boldsymbol{B}_3 \end{pmatrix}
$$

$$
\boldsymbol{V}'(t) = \left(V^1(t), V^2(t), V^3(t) \right) \mathbf{F}^{-1} \begin{pmatrix} \boldsymbol{B}_1' \\ \boldsymbol{B}_2' \\ \boldsymbol{B}_3' \end{pmatrix} \tag{7}
$$

The $\boldsymbol{B}_k$ are the true values, the $\boldsymbol{B}_k'$ the erroneous absolute measurements of the field at times t_k. $\boldsymbol{V}(t)$ is the true value of $\boldsymbol{V}(t)$ at time t and $\boldsymbol{V}'(t)$ the erroneous measurement of $\boldsymbol{V}(t)$

Table 1. The magnetic observatories used in this study.

N		IAGA code	Latitude	Longitude
1	Resolute Bay	RES	74.69	265.11
2	Chambon la Forêt	CLF	48.02	2.27
3	Bangui	BNG	4.33	18.57
4	Hermanus	HER	−34.43	19.23

provided by the magnetometer-variometer due to the error on the determination of the physical axes directions e_i. The calibration error on $V(t)$ appears directly as a linear form of the measurement errors on the B_k, without explicit reference to the frames u_i and e_i:

$$V' - V = \left(V^1, V^2, V^3\right) \mathbf{F}^{-1} \begin{pmatrix} B'_1 - B_1 \\ B'_2 - B_2 \\ B'_3 - B_3 \end{pmatrix}$$

$$= \alpha_1 \left(B'_1 - B_1\right) + \alpha_2 \left(B'_2 - B_2\right) + \alpha_3 \left(B'_3 - B_3\right). \quad (8)$$

Of course we do not know the true values of B_1, B_2, B_3, but to estimate the error $|\Delta V| = |V' - V|$, it is enough to replace in Eq. (8) the quantities $(B'_i - B_i)$ by their estimates. For that, for a given triplet B_k and a given vector V we resort to a statistical estimate. We consider that the measurement errors $B'_i - B_i$ are randomly and uniformly distributed in a ball of center B'_i and radius εb. The small balls are drawn in dark gray in Fig. 1. Note that the B_i and B'_i can be interchanged, in this error estimation.

5 Numerical results

In this section we estimate the calibration error using the simple algorithm presented above, for different configurations of $V(t)$ and B_k, $k = 1, 2, 3$. These vectors are essentially taken or simulated from observatory records. In other words, we estimate the calibration error which would affect the vector data provided by the variometer-magnetometer, in different locations at the Earth's surface.

5.1 Observatory data

We use one-minute values of the three components of the field recorded during the year 1999, as available on the IN-TERMAGNET CDROM 1999, from four observatories: a high-latitude observatory, Resolute-Bay (RES), an equatorial observatory, Bangui (BNG), and two middle-latitude observatories, one in the Northern and one in the Southern hemispheres, Chambon la Forêt (CLF) and Hermanus (HER). Their coordinates are given in Table 1.

From the theory developed above, it is obvious that the more diverse in direction the B_k are, the better the configuration is for calibration. Then, at a given observatory, the larger the magnetic activity, the larger the probability for the

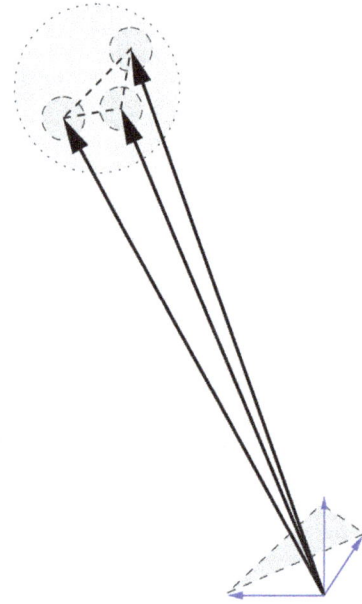

Fig. 1. Calibration triplet B_1, B_2, B_3 and the geographical north east down frame $\{u_i\}$. The unit vectors e_1, e_2, e_3 (defining the physical axes) are nearly orthogonal, and each e_i is close to corresponding u_i. The large gray ball represents the variation of vector V (the one to be measured after calibration); small balls radii represent the measurement error εb; value δb is the upper bound for all $|B_i - B_k|$, $i, k = 1, 2, 3$; b is a typical value of the intensity of the geomagnetic field at the site and epoch of measurements.

triplet B_k to be open. To evidence this effect, we select, in 1999, for each observatory O_i, two subsets of 60 days each, $\mathcal{Q}_i$ containing the five quietest days of each month of 1999, and $\mathcal{D}_i$ containing the five most disturbed days. The day selection is made using the Kp indices (Mayaud, 1980).

5.2 Effect of the (B_1, B_2, B_3) configuration

To study this effect, we take a full day of one-minute values of X, Y, Z from, for example, CLF, specifically the day 6 September 1999, a quiet day belonging to $\mathcal{Q}_{\text{CLF}}$. Figure 2 presents two illustrations of the path of the vector $B(X, Y, Z)$ during this day (see figure caption).

We form at each minute t the triplet $B_1 = B(t)$, $B_2 = B(t + t_0)$, $B_3 = B(t + 2t_0)$. And, along the lines indicated supra, we associate to each of these triplets a set of vectors (B'_1, B'_2, B'_3), B'_i being in the ball of center B_i and radius εb (Fig. 1).

Note that we have $(1441 - 2t_0)$ triplets B_k (t_0 in minutes). We then compute the calibration error – through formula Eq. (8) – affecting a set of vectors $V = W + v$, W being the mean value of the (recorded) field for day 6 September 1999, and v a vector uniformly distributed in a ball of center W and radius δb (the big light gray ball of Fig. 1); note that the set of vectors V is partly simulated. We compute, for a given calibration triplet $B_k(t)$, (Eq. 9) the value $|V' - V|$ for all

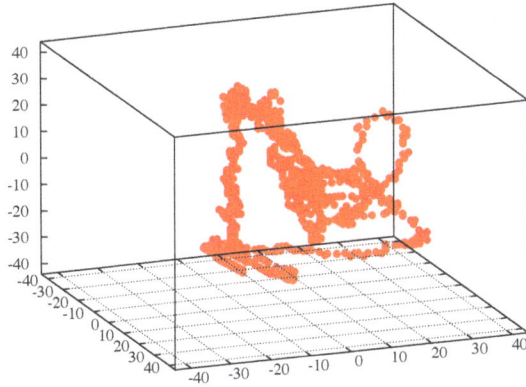

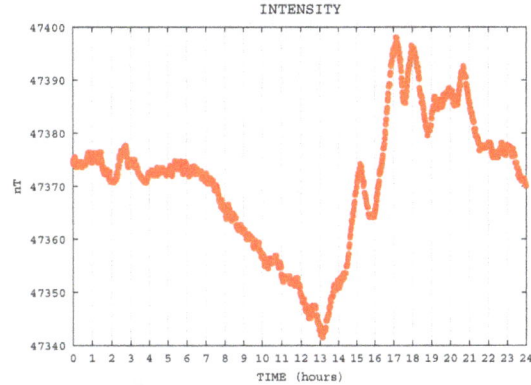

Fig. 2. Magnetic vector evolution (in nT) for the record CLF (6 September 1999): in 3-D frame centered to its mean (left panel), intensity only (right panel).

vectors $V = W + v$ and all triplets B'_1, B'_2, B'_3 built around the considered triplet $B_k(t)$, in the manner made explicit just above. We compute the average $|\Delta V|_{\mathrm{av}}$ and pick up the maximum value $|\Delta V|_{\max}$ of this collection of $|\Delta V| = |V' - V|$. We make this computation for all B_k, $t = 1, 2, \ldots 1441 - 2t_0$. We order the collection $|\Delta V|_{\mathrm{av}}$ and $|\Delta V|_{\max}$ using a parameter η which grossly characterizes the quality of the configuration B_k, i.e. the aperture of this triplet. We choose

$$
\eta = \left| \left\langle \frac{B_1}{|B_1|}, \frac{B_2 - B_1}{|B_2 - B_1|}, \frac{B_3 - B_1}{|B_3 - B_1|} \right\rangle \right|
$$

$$
= \frac{\left| \left\langle B_1, B_2, B_3 \right\rangle \right|}{|B_1| \cdot |B_2 - B_1| \cdot |B_3 - B_1|}. \tag{9}
$$

<> is for the vector triple product. Figure 3 represents the distribution of $|\Delta B|_{\mathrm{av}}(t)$ and $|\Delta B|_{\max}(t)$ versus η. The parameter η is not discriminant enough to rank unequivocally the $|\Delta B|(t)$ distribution; many points of the plot have the same abscissa. Nevertheless, it clearly appears that the calibration errors $|\Delta B|_{\mathrm{av}}(t)$ and $|\Delta B|_{\max}(t)$ decrease when η increases.

5.3 Histograms of the calibration error

We now present some reciprocal numerical experiments, closer to the real situation to be met, using again minute data of day 6 September 1999. This time we choose a single absolute measurement triplet B_k, $k = 1, 2, 3$, picked up in the observatory records, specifically at $t = 0300, 0600, 1500$ on 6 September 1999, and retain as current vectors $V(t)$ all the one-minute values recorded at CLF over the 1999 year (instead of the simulated vectors in the ball of center W). Again a set of triplets (B'_1, B'_2, B'_3) is associated with (B_1, B_2, B_3), uniformly distributed in a ball of radius εb centered respectively at (B_1, B_2, B_3). For each vector $V(t)$ (1440×365 of them) we compute the average and maximum values of $|\Delta V|$ over the set of (B'_1, B'_2, B'_3). The histograms of the set of $|\Delta V|_{\mathrm{av}}(t)$ and $|\Delta V|_{\max}(t)$ are shown in Figs. 4 and 5 for $\varepsilon b = 0.75, 1, 2$ nT. It appears that $|\Delta V|_{\mathrm{av}}(t)$ (the most

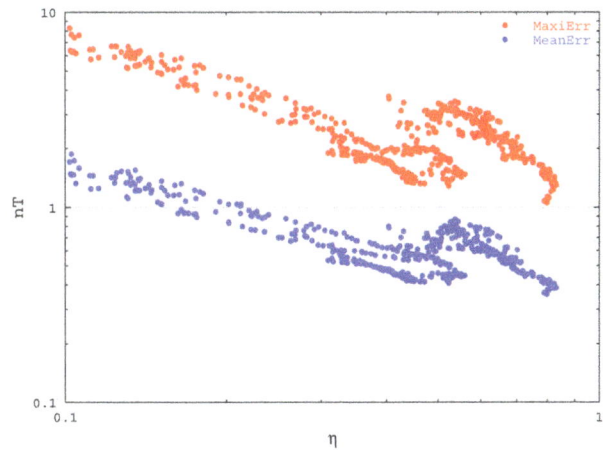

Fig. 3. Data CLF (6 September 1999). Calibration errors (maximum and average) in nT for $|v| = 50$ nT, $\varepsilon b = 0.75$ nT, delay $t_0 = 6$ h (see main text).

realistic estimate), for $\varepsilon b = 1$ nT, is most of the time smaller than 2 nT (Fig. 5).

In Fig. 6 $|\Delta V|_{\max}(t)$ values are simply ranked versus time t. An examination of this figure in regard of the magnetic situation shows that, as expected, the largest values of $|\Delta V|_{\max}(t)$ are associated with magnetic storms. $V(t)$, during these events, leaves the ball of centre W and radius δb (Fig. 1; $\delta b = 50$ nT). Note in passing that it is not important, in general, to know with a high accuracy the absolute value of $V(t)$ at each minute of a magnetic storm.

5.4 Time tables

We now give, for each of our four observatories, a different, more practical presentation of the calibration error, which gives the timespans during which this error is smaller than a given threshold of α nT. We choose again values of year 1999, consider the triplets $B_1 = B(t)$, $B_2 = B(t+t_0)$, $B_3 = B(t+2t_0)$, and compute the corresponding calibration errors

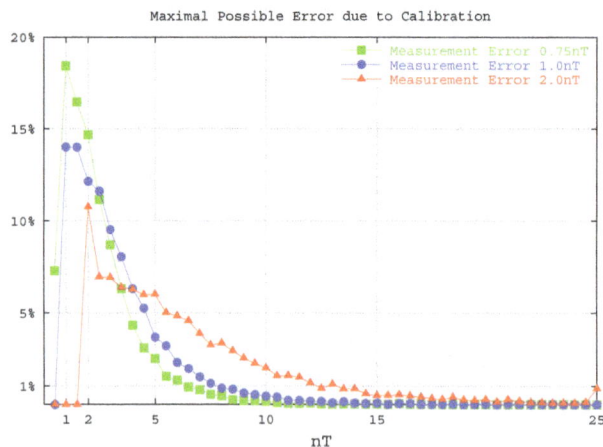

Fig. 4. Normalized histograms of the calibration errors for CLF observatory during 1999: the maximal possible error for $\varepsilon b = 0.75, 1, 2$ nT.

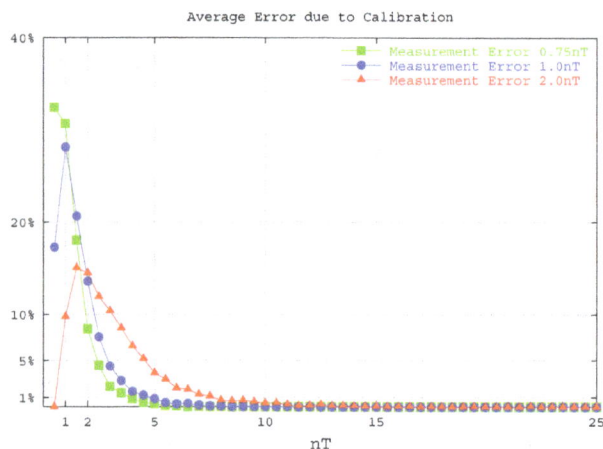

Fig. 5. Normalized histograms of the calibration errors for CLF observatory during 1999: the average possible error for $\varepsilon b = 0.75, 1, 2$ nT.

$|\Delta \boldsymbol{B}|_{\text{av}}(t)$ as explained in Sect. 5.2. Figures 7–11 present the results for the four observatories. All graphs – that we call time tables – are to be read in the following way: the upper sub-panel (in blue) is for the 60 most disturbed days of the year, the lower one (in red) for the 60 quietest days. In each of the sub-panels the days are ranked as follows: the five quietest days of January, according to their calendar date are at the bottom of the lower sub-panel, the five quietest days of December at its top. The same for the upper sub-panel.

In Figs. 7–10 the value of α is 2 nT. In Fig. 11, relative to Bangui, a time table for $\alpha = 4$ nT is also presented. Everywhere $t_0 = 7.5$ h for the left panels, $t_0 = 6.0$ h for the right panels. Of course, $t < 0900$ for $t_0 = 7.5$ h $(24 - 2 \times 7.5)$, and $t < 1200$ for $t_0 = 6.0$ h. In all the computations $\varepsilon b = 1$ nT. We plot a characteristic function which is equal to zero at time t (white) if the triplet $(\boldsymbol{B}_1 = \boldsymbol{B}(t), \boldsymbol{B}_2 = \boldsymbol{B}(t + t_0)$,

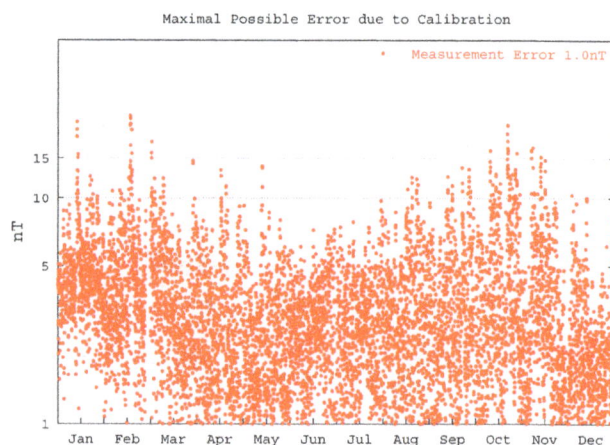

Fig. 6. Sequential observations of the maximal possible calibration error for CLF data, during the year 1999; the value εb is supposed to be 1 nT. The largest values are associated with known magnetic storms of the year 1999 as those of 13 January, 18 February, or 20 October.

$\boldsymbol{B}_3 = \boldsymbol{B}(t + 2t_0)$) leads to a $|\Delta \boldsymbol{B}|$ error > 2 nT; otherwise, a colored tiret, red or blue, is drawn. Continuous red or blue time intervals are such that, for any first measurement with t in this interval leads to a calibration error smaller than 2 nT (or 4 nT) on $\boldsymbol{V}(t)$.

Looking at graphs of Figs. 7–10, we remark, as expected, that it is easier to get time intervals with $\Delta < 2$ nT for the disturbed days than for the quiet days, and for a high latitude observatory (RES) than for an equatorial one (BNG). Figure 11 is an example of the effect of changing α.

6   Discussion and conclusion

Stability in absolute values, and particularly long-term stability – say up to a few years – used to be the most difficult requirement to fulfill in magnetic observatories. Let us adopt the standards of the INTERMAGNET program, which are up to now essentially intended to classical observatories with regular (generally weekly) man-made absolute measurements. The one minute magnetic field values provided by the magnetometer-variometer should be characterized by a resolution of 0.1 nT and a long-term stability of 5 nT yr^{-1}.

From the results of Sect. 4, it appears that, after the calibration performed as in Sects. 2 and 3, the magnetometer-variometer – as already said, we have especially in mind the LETI (CEA) apparatus – can function as an automatic observatory, fitting INTERMAGNET standards for a time-span of one to a few years, depending on the amplitude of the secular variation. A special study is necessary in the case of the highest latitude observatories. The necessity of a visiting the station every other year or so to renew the calibration is not so hard a constraint; in any case, such visits should be necessary for other purposes and checking of instruments and

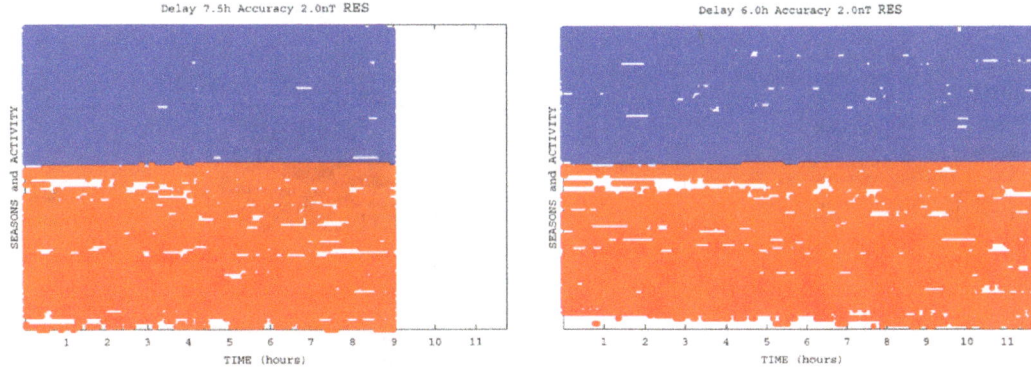

Fig. 7. Time table for RES data, with $t_0 = 7.5$ h (left panel) and $t_0 = 6$ h (right panel) periods for the $\boldsymbol{B}_1(t)$. Note the distinction between subsets $\mathcal{D}_i$ and $\mathcal{Q}_i$: blue lines for disturbed days, red lines for quiet days. Yearly variation reads from down to top (starting with the first five days from January, up to the last five days of December, separately for $\mathcal{D}_i$ and $\mathcal{Q}_i$ families).

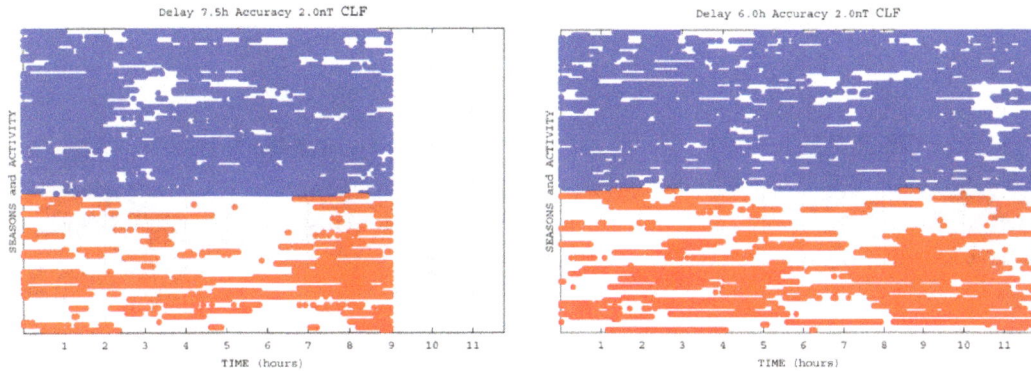

Fig. 8. Time table for CLF data. Same legend as for Fig. 7.

environmental conditions. We conclude with four remarks, of different nature, which could not be developed in this paper.

First, at each new calibration, a step in the series provided by the observatory will happen; but it is only of the magnitude discussed above, i.e. small. Smoothing the small steps may be considered; in the absence of extra information, no better technique than a linear correction between the two calibrations, distant by one year or so, exists.

Second, we have to stress that we only discussed the effect of the inaccuracy of the absolute measurements of $\boldsymbol{B}_k$ on the values V given subsequently by the magnetometer supposed to remain identical to itself, in particular geometrically invariant. The ^{4}He magnetometer is built in such a way as to ensure this stability. We do not discuss either the important question of the stability of the pillar. To our knowledge, there are no available data to make any good estimation of the stability of the pillars. Therefore, it is important to build the best possible pillar and retain an adequate magnetometer.

Third, we stress again that we only made an excursion in the (calibration) error space, using the simple algorithm described in Sect. 4. A full exploration of this space would be a heavier task; in the Appendix we give a glimpse of it.

The fourth remark, which we already touched upon in Sect. 5 is, although relevant to the problem at hand, more general. Long-term stability is generally required for the study of long time scale phenomena (secular variation of the main field, solar cycle related variations, seasonal variations). For this kind of studies, what is relevant is not the absolute accuracy of one-minute values, but of some means (annual, monthly, daily, hourly); and, briefly speaking, averaging reduces the error.

A fuller understanding of the Earth's magnetic field will come from improvements in measuring it and separating its different components with a better spatial and temporal resolution. The upcoming ESA Swarm mission will provide the best-ever survey of the geomagnetic field and its temporal evolution. This constellation will benefit from a new generation of instruments, as each satellite will carry two ^{4}He magnetometers; these Absolute Scalar Magnetometers (ASM) are the nominal instruments for measuring the magnetic field intensity, but it is planned to operate them in vector mode, as demonstrator.

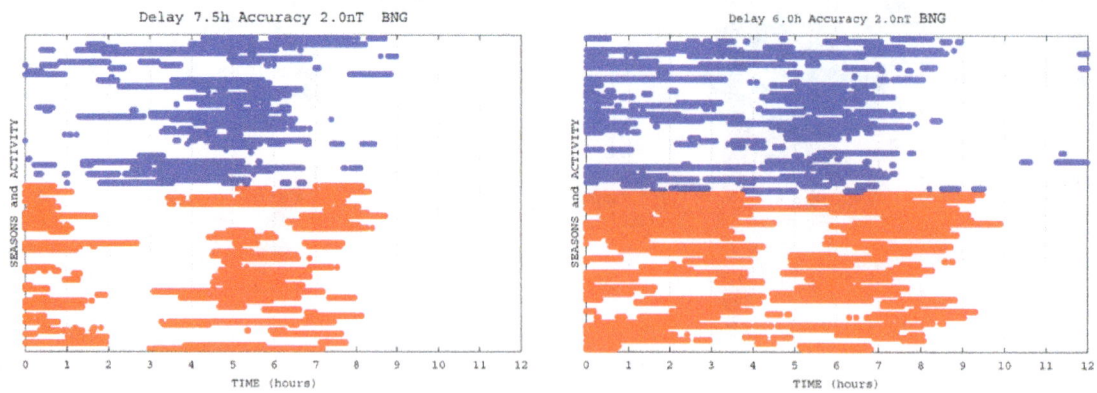

Fig. 9. Time table for BNG data. Same legend as for Fig. 7.

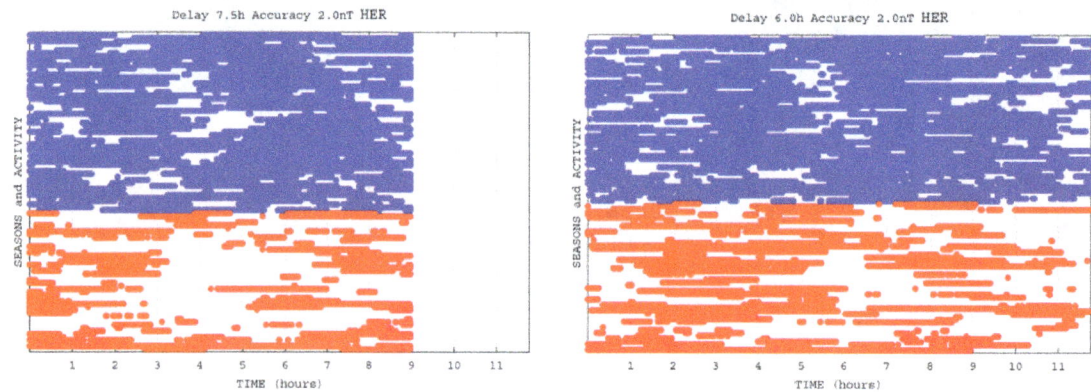

Fig. 10. Time table for HER data. Same legend as for Fig. 7.

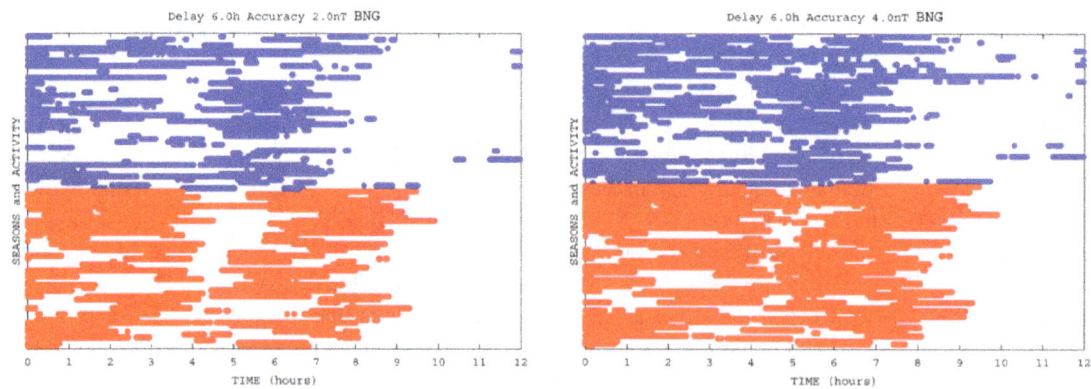

Fig. 11. Time table for BNG data, with $t_0 = 6$ h, $\alpha = 2$ nT (left panel) and $\alpha = 4$ nT (right panel) periods for the $\boldsymbol{B}_1(t)$. Note the distinction between subsets $\mathcal{D}_i$ and $\mathcal{Q}_i$: blue lines for disturbed days, red lines for quiet days. Yearly variation reads from top to down (starting with the first five days from January, up to the last five days of December, separately for $\mathcal{D}_i$ and $\mathcal{Q}_i$ sets).

Appendix A

First we give some upper bound of the error amplification in the case where there is no restriction on the current vector $V(t)$. We start from Eq. (1), $B_k = \mathbf{F}(e_1, e_2, e_3)^T$ (T for transposed) and define here the amplification as the ratio of the directional error on the physical axes e_i to the directional error on the B_k vectors of the calibration triplet. Consider the two triangles whose summits are the extremities of (e_1, e_2, e_3) and (B_1, B_2, B_3), respectively (Fig. 1). From Eq. (1), we can express the e_i versus the B_k through $\mathbf{F}^{-1}$ matrix, and vectors ($e_l - e_m$) in terms of vectors ($B_i - B_j$) (in fact two of them, since the sum of the three differences is zero). The lengths of vectors ($e_l - e_m$) are $\approx \sqrt{2}$, and the lengths of the $B_i - B_j$ of the order of δb, as given in the main text. Therefore, $\mathbf{F}^{-1}$ transforms (B_1, B_2, B_3) triangle sides into (e_1, e_2, e_3) triangle sides (Fig. 1 of the main text) through factors of the order of $(\delta b)^{-1}$. If the direction of some of the measurement errors $\varepsilon b \omega_i$ (Eq. 5) happens to be close to that of one of the sides ($B_i - B_j$), the corresponding error on the $|e_l - e_m|$ will be multiplied by a factor $(\delta b)^{-1}$. Directional errors on the e_i result which are of the order of $(\delta b)^{-1}\varepsilon b = \varepsilon/\delta \approx 50^{-1}$; the amplification, as defined supra, of the directional error ε on the B_k is then $(\delta)^{-1} \approx 10^3$, with the value of the (δb) adopted in the main text. This estimate of the maximum amplification can be obtained through a more rigorous analysis using operator theory. We do not present it.

In the numerical experiments of the main text, we did not observe strong amplifications of the error on the current vector $V(t)$ compared to the error εb on the calibration vectors B_k (see e.g. histograms of Figs. 4 and 5). The reason is as follows: all vectors $V(t)$ are supposed to belong to a rather small neighborhood of the vector B_k which can also be characterized by the quantity δb. In other words, (V^1, V^2, V^3) is close, within εb, of $(f_1^1, f_1^2, f_1^3), (f_2^1, f_2^2, f_2^3), (f_3^1, f_3^2, f_3^3)$, like these three triplets are close to one another. The result is that the coefficients $\alpha_1, \alpha_2, \alpha_3$ of Eq. (8) of our practical algorithm are close enough to 1. No large amplification of error arises, even if the e_i are not accurately determined. These considerations shed light on the statement of the introduction that absolute calibration should not be too difficult: it is true for the objectives of an automatic magnetic observatory, not in general.

Acknowledgements. The results presented in this paper rely on data collected at geomagnetic observatories. We thank the national institutes that support them and INTERMAGNET for promoting high standards of magnetic observatory practice. We also thank Mike Rose who agreed to serve as Associate Editor, and Jean Rasson, Valery Korepanov and Chris Turbitt for suggestions in improving the manuscript.

Edited by: M. Rose

References

Alldregde, L. R.: A proposed automatic standard magnetic observatory, J. Geophys. Res., 65, 3777–3786, 1960.

Alldregde, L. R. and Saldukas, I.: An automatic standard magnetic observatory, J. Geophys. Res., 69, 1963–1970, 1964.

Auster, H. U., Mandea, M., Hemshorn, A., Pulz, E., and Korte, M.: Automation of absolute measurement of the geomagnetic field, Earth Planet. Space, 59, 1007–1014, 2007.

Auster, V., Hillenmaier, O., Kroth, R., and Weidemann, M.: Proton Magnetometer Development, XIIth IAGA Workshop on Geomagnetic Observatory Instruments, Data Acquisition and Processing, Abstract Volume 56, Instytut Geofizyki Polskiej Akademii, Poland, 2006.

Gravrand, O., Khokhlov, A., Le Mouël, J.-L., and Léger, J.-M.: On the Calibration of a Vectorial ^{4}He Pumped Magnetometer, Earth Planet. Space, 53, 949–958, 2001.

Léger, J.-M., Kervnevez, N., and Morbieu, B.: Developement of a laser pumped helium magnetometer, Undersea Defence Technology conference proceedings, 831–835, UDT 92, London, UK, 1992.

Mayaud, P. N.: Derivation, meaning, and use of geomagnetic indices, Geophysical Monograph Series of American Geophysical Union, edited by: Spilhaus Jr., A. F., Geophysical Monograph 22, Am. Geophys. Union, Washington, DC, 1980.

Olsen, N., Toffner-Clausen, L., Sabaka, T., Brauer, P., Merayo, J. M., Jorgensen, J. L., Léger, J., Nielsen, O. V., Primdahl, F., and Risbo, T.: Calibration of the Ørsted vector magnetometer, Earth Planet. Space, 55, 11–18, 2003.

Rasson, J. L. and Gonsette, A.: The Mark II Automatic DIFlux, Data Science Journal, 10, 169–173, ISSN:1683-1470, 2011.

Schott, J. J. and Leroy, P.: Orientation of the DIDD magnetometer, Contributions to Geophysics and Geodesy, 31, 43–50, 2001.

Van Loo, S. A. and Rasson, J. L.: Presentation of the prototype of an automatic DI-Flux, XIIth IAGA Workshop on Geomagnetic Observatory Instruments, Data Acquisition and Processing, Abstract Volume 21, Instytut Geofizyki Polskiej Akademii, Poland, 2006.

5

An initial investigation of the long-term trends in the fluxgate magnetometer (FGM) calibration parameters on the four Cluster spacecraft

L. N. S. Alconcel, P. Fox, P. Brown, T. M. Oddy, E. L. Lucek, and C. M. Carr

Department of Physics, The Blackett Laboratory, Imperial College London, Prince Consort Road, London, SW7 2BW, UK

Correspondence to: L. N. S. Alconcel (l.alconcel@imperial.ac.uk)

Abstract. Over the course of more than 10 years in operation, the calibration parameters of the outboard fluxgate magnetometer (FGM) sensors on the four Cluster spacecraft are shown to be remarkably stable. The parameters are refined on the ground during the rigorous FGM calibration process performed for the Cluster Active Archive (CAA). Fluctuations in some parameters show some correlation with trends in the sensor temperature (orbit position). The parameters, particularly the offsets, of the spacecraft 1 (C1) sensor have undergone more long-term drift than those of the other spacecraft (C2, C3 and C4) sensors. Some potentially anomalous calibration parameters have been identified and will require further investigation in future. However, the observed long-term stability demonstrated in this initial study gives confidence in the accuracy of the Cluster magnetic field data. For the most sensitive ranges of the FGM instrument, the offset drift is typically 0.2 nT per year in each sensor on C1 and negligible on C2, C3 and C4.

1 Introduction

The Cluster mission (Escoubet et al., 1997) consists of four Earth-orbiting spacecraft flying in formation at variable separations (100–10 000 km). The science phase of the mission began in February of 2001 and is presently scheduled to continue until December 2016 (pending final confirmation by the European Space Agency). Mission scientists study small-scale plasma structures in space and time in key regions of the magnetosphere, including the solar wind, the bow shock, the magnetopause, the polar cusps, the magnetotail and the auroral zones (Walsh et al., 2010). Each spacecraft carries the same set of eleven instruments which detect spatial and temporal changes in the magnetosphere by measuring ambient electromagnetic fields and particle populations. FGM is a DC (direct current) magnetometer used to measure the magnetic field vector at the instrument's position (Balogh et al., 1997).

Each FGM instrument consists of two triaxial fluxgate sensors. They are boom-mounted to minimise interference from the spacecraft's background magnetic field, and the outboard sensor at the end of the 5 m boom is designated as the primary sensor for science data. The sensors can be operated in several ranges depending on the spacecraft's location in the magnetosphere, covering magnetic field magnitudes from less than 1 nT to over 65 000 nT (see Table 1). Data are normally obtained at a rate of ~ 22 vectors per second (Hz), designated as "normal mode", although this can be increased to ~ 67 Hz for short periods to investigate a region or event of particular interest ("burst mode").

After the raw data are downlinked, they are processed into a usable format and the time at which the data were measured is reconstructed. They are subsequently calibrated, validated and processed into the final FGM data products which appear on the Cluster Active Archive (CAA) (Laakso et al., 2010). Submission to the CAA occurs once all of these procedures have been performed on 1 month's worth of data, which is divided into orbits, defined as the periods between successive periapses. Orbit period varies from 51 to 57 h depending on the phase of the mission, with the orbits shortening as the mission progresses.

The four Cluster spacecraft are magnetically very clean, giving a high level of confidence in the DC magnetic field

Table 1. FGM instrument ranges.

Range number	B
2	−64 to 63.97 nT
3	−256 to 255.87 nT
4	−1024 to 1023.5 nT
5	−4096 to 4094 nT
6	−16 384 to 16 376 nT
7	−65 536 to 65 504 nT

data obtained by the FGM instruments. The combination of measurement and modelling on the ground with a rigorous magnetic cleanliness programme and final compensation for magnetic contributions means that the spacecraft field at the outboard magnetometer sensors should be less than 0.25 nT (Balogh et al., 1997). It is not possible to verify this in-flight, however.

The accurate calibration of the FGM instrument is critical for scientific investigations requiring high-accuracy vector magnetic field data, for the production of some data sets by other instruments (PEACE, RAPID) and for the calibration of other instruments aboard the Cluster spacecraft (EFW, STAFF, WHISPER).

2 FGM calibration

In order to place the parameter trends into context, it is useful to describe briefly the calibration, validation and archiving procedures. These are described in detail elsewhere (Gloag et al., 2010).

2.1 Theory

The FGM magnetic field data are subject to several significant sources of error that must be corrected to yield the best results for use in scientific studies and for use by other Cluster instruments. In the coordinate transformation of the magnetic field data from the sensor measurement frame to the spacecraft reference frame (and hence to a geophysical frame), errors may arise due to incomplete knowledge of (i) the orientation of the sensors' axes, (ii) the sensor offsets, and (iii) the sensor gains.

The relationship between the measurement frame and the spacecraft reference frame is specified by a set of 12 parameters for each spacecraft, as shown in Eq. (1) below. The set consists of the sensor angles, gains and offsets. The calibration parameters then define a transformation of the following form:

$$\begin{pmatrix} B_{S_1} \\ B_{S_2} \\ B_{S_3} \end{pmatrix} = \begin{pmatrix} G_1 \sin\vartheta_1 \cos\phi_1 & G_1 \sin\vartheta_1 \sin\phi_1 & G_1 \cos\vartheta_1 \\ G_2 \sin\vartheta_2 \cos\phi_2 & G_2 \sin\vartheta_2 \sin\phi_2 & G_2 \cos\vartheta_2 \\ G_3 \sin\vartheta_3 \cos\phi_3 & G_3 \sin\vartheta_3 \sin\phi_3 & G_3 \cos\vartheta_3 \end{pmatrix}$$

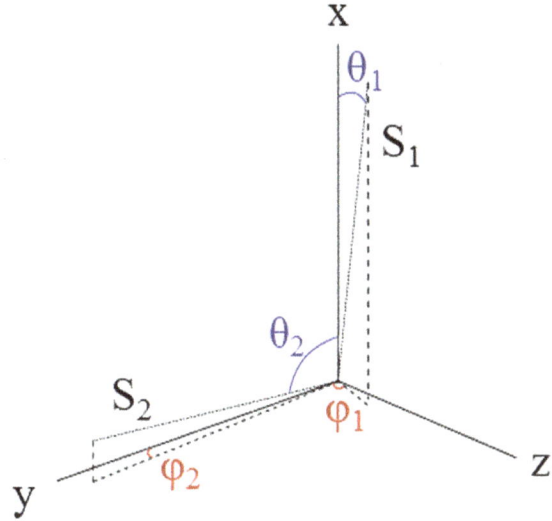

Figure 1. The relation between the orthogonal (x, y, z) and sensor (S_1, S_2, S_3) coordinate systems. The elevation and azimuthal angles θ and φ for each sensor coordinate are defined in the same way. S_3 has been omitted for clarity.

$$\begin{pmatrix} B_x \\ B_y \\ B_z \end{pmatrix} + \begin{pmatrix} O_1 \\ O_2 \\ O_3 \end{pmatrix}. \qquad (1)$$

(B_x, B_y, B_z) is the magnetic field vector in the spinning spacecraft coordinate system, where x is aligned along the spin axis of the spacecraft and y, z are located in the spin plane, forming an orthogonal triad. In this equation, (B_{S1}, B_{S2}, B_{S3}) represents the magnetic field vector as measured in the non-orthogonal sensor coordinate system, where S_1, S_2 and S_3 point approximately along the spacecraft x, y and z axes respectively. The parameters describing the transformation are the offsets (O_i), gains (G_i), elevation angles (θ_i) and azimuthal angles (φ_i). The elevation angle is measured with respect to the spacecraft spin axis x; the azimuthal angle is measured around from the y axis in the spin plane y–z. Figure 1 illustrates the relationship between two reference frames. The gains and angles in the coupling matrix orthogonalise, scale and orient the field measured by the sensors, while the offsets handle zeroing the sensors.

These calibration parameters were accurately measured on the ground at the Technical University of Braunschweig as part of the pre-flight calibration of FGM. *However, these parameters cannot be expected to remain constant over the time scale of the mission; thus, in order to maintain the quality of the measured magnetic field data, an in-flight calibration process is required.* As a mission consisting of multiple spinning spacecraft which spend significant portions of their time in the solar wind, Cluster represents an opportunity to bring several magnetometer calibration methods to bear.

The in-flight calibration technique is based upon two distinct methods: a Fourier analysis method (Kepko et al., 1996), which recovers 8 of the 12 calibration parameters, and a solar wind analysis method, which recovers the spin-axis offsets O_1 (Hedgecock, 1975). A brief description of the theory underlying these methods, together with a discussion of their limitations and constraints on their application, is given below. Note that in addition to being orthogonalised and transformed into spacecraft coordinates, the magnetic field components must also be despun. For the sake of brevity, the despinning procedure will not be outlined here.

2.2 Fourier analysis

The Fourier analysis is based on the procedure detailed in Kepko et al. (1996). When the magnetic field data are despun, errors in particular calibration parameters will produce coherent monochromatic signals at the first and second harmonics of the spin frequency (approximately 0.25 and 0.5 Hz for the Cluster spacecraft). More specifically: errors in the spin-plane elevation angles (θ_2, θ_3) and spin-plane offsets (O_2, O_3) produce signals at the first harmonic in the spin-plane components of the field; errors in the relative spin-plane azimuthal angles ($\Delta\varphi_{32}$) and relative spin-plane gains (ΔG_{32}) produce signals at the second harmonic in the spin-plane components of the field; and errors in the spin-axis elevation angle θ_1 and spin-axis azimuthal angle φ_1 produce signals at the first harmonic in the spin-axis component of the field.

Fourier-transforming the despun data produces a set of equations containing the errors in the above calibration parameters, which can then be inverted to recover the values of those parameters. The errors in the remaining four parameters (G_1, O_1, G_3, φ_3) do not produce coherent signals in the despun data and so they cannot be recovered by this method. After the Fourier analysis, the residual signal power at the first and second harmonics of the spin frequency provides one of the measures by which the accuracy of the calibration can be judged.

2.3 Solar wind analysis

In general, the four Cluster spacecraft sample the solar wind from mid-December to mid-April, a period which is known as the "dayside" season. During this period, the magnetic field in the solar wind is used to adjust the offset (O_1) associated with the axis of the sensor that is aligned with the spin axis of the spacecraft. FGM is nearly always in range 2 during these periods, so this is the only range for which this method can be used to refine the spin-axis offset. This procedure is based on the observation that fluctuations in the solar wind magnetic field are primarily rotational, which means that there should be no correlation between the spin-axis component of the magnetic field and the total field magnitude (Hedgecock, 1975).

The procedure works by searching through the spin-averaged data for rotational discontinuities. At these discontinuities, O_1 is adjusted to minimise the correlation between B_1 and $|B|$. In general, 1 month's worth of data is divided in half and adjustments are applied separately to the first and second halves of the month. The implementation of this procedure was originally developed by FGM co-investigators at UCLA (University of California Los Angeles; personal communication with H. K. Schwarzl, K. Khurana, and M. Kivelson, 2005) who have collaborated with the FGM team on its implementation at Imperial College. A complete description of the theory underlying this method can be found in Hedgecock (1975).

From mid-April to mid-December the four Cluster spacecraft sample Earth's magnetotail, a period which is known as the "nightside" or "tail" season. The technique described above cannot be applied to this data to adjust the spin-axis offset. A simple linear interpolation of the offset between the last solar wind measurement in mid-April and the first solar wind measurement in mid-December is performed instead. This method likely masks the natural variation in the offset during these periods.

2.4 Range changes

When the FGM switches between ranges (Table 1), the magnetic field components are not precisely equal on either side of the change, due to differences in calibration between different ranges. In order to mitigate this, adjustments are performed to the remaining parameters not determined by either of the above procedures; namely O_1 (ranges 3 and above), G_1, G_3 and φ_3. These parameters are adjusted from their measured values on the ground in order to minimise the discontinuities in the field components that occur at the range change. In common with the solar wind analysis, the implementation of this procedure was originally developed by FGM co-investigators at UCLA (personal communication with H. K. Schwarzl, K. Khurana, and M. Kivelson, 2005) who have also collaborated with the FGM team on its implementation at Imperial College.

2.5 Validation and archiving

Once the calibration procedure has been completed, visual inspection of the calibrated data is carried out as a quality-control step. The accuracy of the calibration parameters recovered by the Fourier analysis method manifests itself in the signal power at the spin frequency in the processed data. The accuracy of the spin-axis offset recovered by the solar wind method manifests itself in the spread exhibited between the four spacecraft's spin-axis data in the solar wind. The limitations of the calibration procedures mean that the quality of the final calibration can vary from month to month. Data intervals which do not meet the minimum standard for calibration quality are flagged in caveat files which accompany

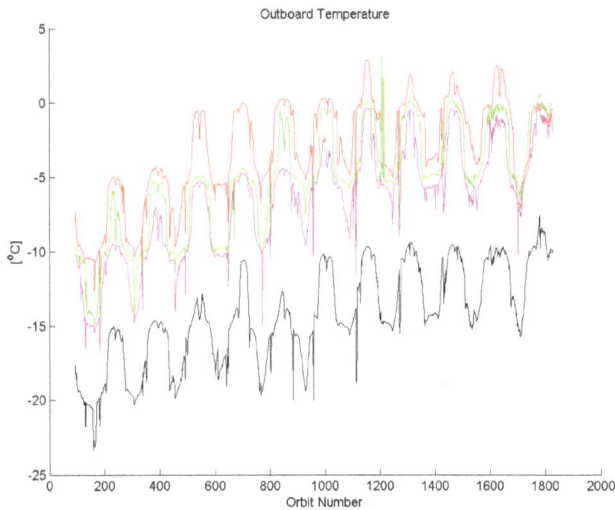

Figure 2. Outboard sensor temperatures in degrees Celsius for each spacecraft for orbits 93–1889 (February 2001–August 2012). C1 – black, C2 – red, C3 – green, C4 – magenta.

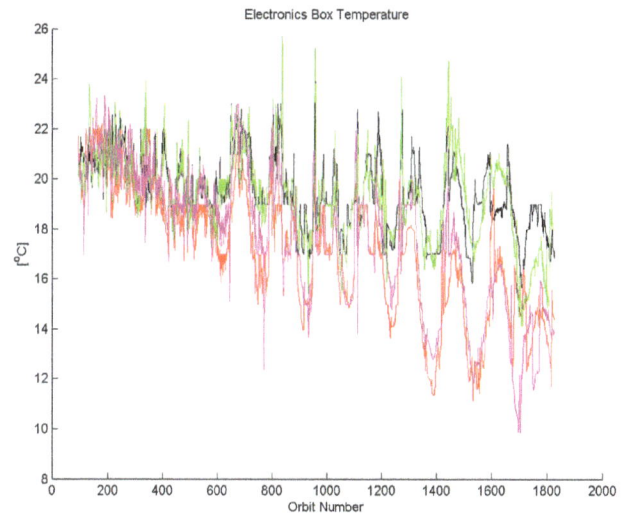

Figure 3. Electronics box temperatures in degrees Celsius for each spacecraft for orbits 93–1889 (February 2001–August 2012). C1 – black, C2 – red, C3 – green, C4 – magenta.

the FGM data products on the CAA. Additionally, a calibration file for each orbit is produced. They are made available to investigators on the CAA; however, since the FGM data products are already calibrated, they simply list the calibration parameters for each range in the orbit. The CAA web site (http://caa.estec.esa.int/caa/) gives researchers access to the data from all of the instruments on board Cluster from the start of the mission. Documentation and software tools are also downloadable.

2.6 Application, limitations and uncertainties of the calibration procedures

Having presented a brief account of the theory underlying the FGM calibration procedure above, it is also worthwhile to discuss some of the practicalities involved in applying these procedures to the FGM data, together with limitations of the calibration procedures and quantification of remaining uncertainty in the calibration parameters.

The parameters recovered by the Fourier analysis method are resolved most frequently, in practice, once per orbit. The remaining parameters are determined less often. Accurate determination of the spin-axis offset by the solar wind method requires a minimum of 20 h of good quality solar wind data. Accordingly, during the dayside season, the spin-axis offset is typically only determined twice per month. The range jump correction is usually performed once per month.

There are two primary criteria used to determine the accuracy of the calibration results. The first is the spread in the spin-axis components of the magnetic field in the solar wind, as measured by each of the four spacecraft. Under quiet solar wind conditions, it is to be expected that the distance between the Cluster spacecraft is small compared to the dis-

tance scales over which the solar wind magnetic field varies significantly. Accordingly, a small spread in B_x is the criterion by which the accuracy of the determined values of O_1 are measured.

The second criterion is the signal power in the spin-axis and spin-plane components of the magnetic field, measured at the first and second harmonics of the spin frequency. This quantity is used to measure the quality of the Fourier analysis component of the calibration procedure. It is frequently the case that the spin-axis parameters recovered by the Fourier analysis method have produced an inferior calibration to that obtainable by using calibration parameters from a previous orbit. Accordingly, in such cases, multiple spin-axis calibrations were substituted, and the set of parameters which produced the minimum spin power in the spin-axis data was chosen as the final calibration.

While these criteria provide an estimate of the quality of the calibration, they cannot be used to determine the remaining uncertainty in any individual calibration parameter. In particular, it should be noted that the set of calibration parameters determined by the Fourier analysis method is solely chosen to minimise the signal power at the first and second harmonics of the spin frequency. There may be other sets of values for these parameters which would satisfy this criterion equally well. Accordingly, lacking an independent reference for the individual field components, caution must be used when discussing how these calibration parameters relate to the true gains and alignment angles of the instrument. For example, within a time period as short as an orbit, it is reasonable to assume that the alignment angles of the instrument are independent of the operating range. However, the angles output by the Fourier analysis procedure do vary slightly between ranges, again because the software implementation

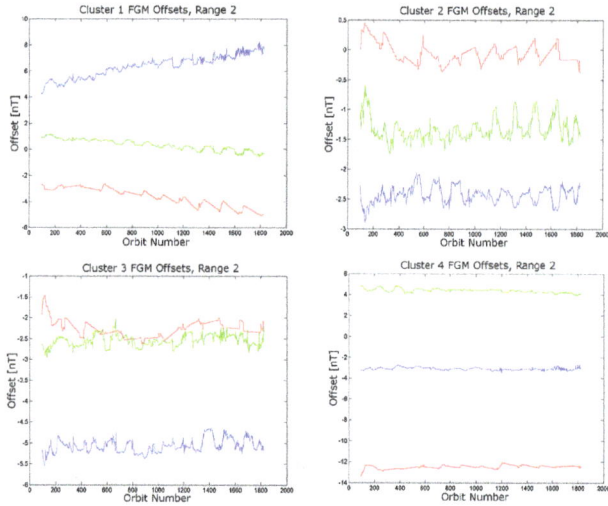

Figure 4. Range 2 spin-axis (O_1, red) and spin-plane (O_2 and O_3, blue and green) offsets in nanoteslas for orbits 93–1825 (February 2001–February 2012).

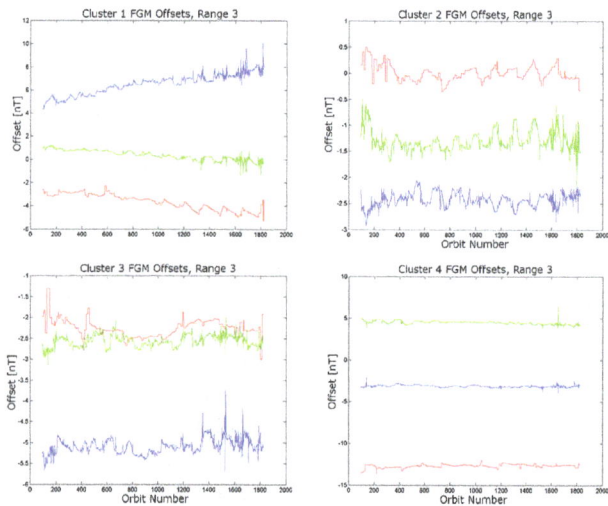

Figure 6. Range 4 spin-axis (O_1, red) and spin-plane (O_2 and O_3, blue and green) offsets in nanoteslas for orbits 93–1825 (February 2001–February 2012).

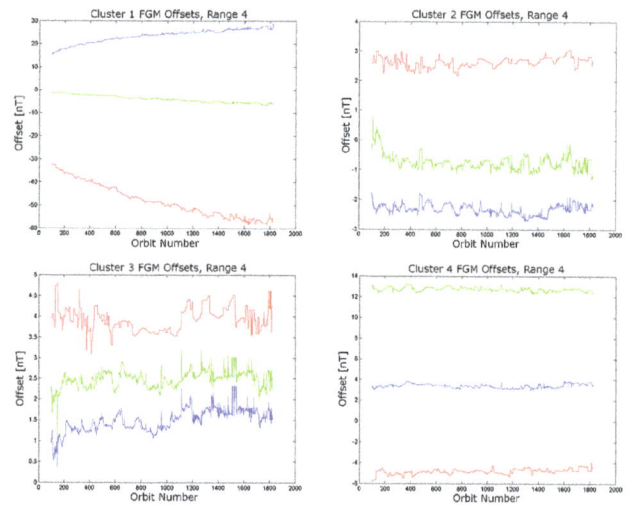

Figure 5. Range 3 spin-axis (O_1, red) and spin-plane (O_2 and O_3, blue and green) offsets in nanoteslas for orbits 93–1825 (February 2001–February 2012).

The original Cluster mission has been extended several times and utilised manoeuvres to configure a range of different spacecraft constellations. Trajectories bringing the spacecraft closer to Earth than originally foreseen necessitated the use of the full instrument ranging capability. From November 2000 to October 2006, ranges 2–4 (see Table 1) were in regular use. Starting in November 2006, range 5 entered routine use. Starting in May 2008, range 6 entered routine use. Starting in December 2009, range 7 entered routine use. Neither range 6 nor range 7 was originally intended for use during the nominal mission hence these ranges were not fully calibrated on the ground. The entry of the spacecraft into the inner magnetosphere and auroral acceleration zone in the extended mission phases meant that the total field magnitude exceeded the capacity of range 5. The calibration parameters for range 6 and range 7 are tied to those of range 5, as only partial ground calibration information was available for them.

3 Long-term trends in FGM parameters

Having applied the above described calibration methods to the Cluster FGM data set over a period of 11 years, we considered it valuable to begin an examination of the long-term behaviour of the FGM calibration parameters. Such a survey serves several purposes:

- It allows us to examine the long-term measurement stability of the FGM instrument. Such stability has been observed in other space-based fluxgate magnetometers such as those aboard the CHAMP (Challenging Minisatellite Payload) and THEMIS (Thermal Emission

only selects values which best reduce the spin-frequency signal power. No attempt is made to harmonise the alignment angles across different ranges. This discussion also underlines the importance of monitoring the output of the calibration process, as it is possible for unrealistic parameter values to be produced by the automated calibration routines.

The accuracy of the recovered parameters is strongly dependent upon the quality of the data available. Excessive signal noise, data gaps, etc. can all affect the efficacy of the calibration procedures. Periods of unavoidably poor calibration are flagged in FGM's CAA caveat files.

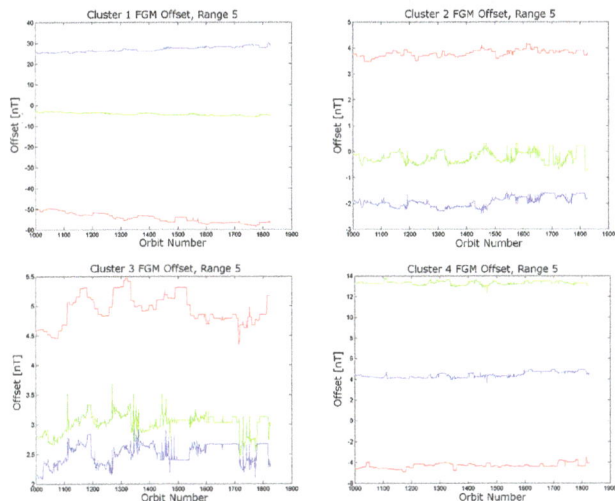

Figure 7. Range 5 spin-axis (O_1, red) and spin-plane (O_2 and O_3, blue and green) offsets in nanoteslas for orbits 1000–1812 (December 2006–February 2012).

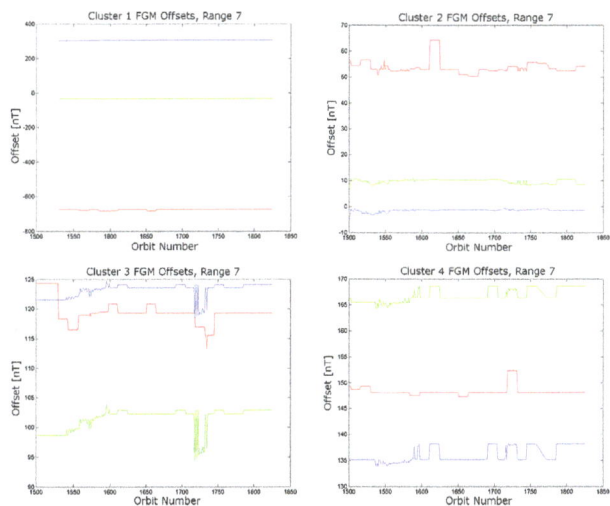

Figure 8. Range 6 spin-axis (O_1, red) and spin-plane (O_2 and O_3, blue and green) offsets in nanoteslas for orbits 1300–1825 (December 2008–February 2012).

Imaging System) satellites (Auster et al., 2008; Yin and Luehr, 2011).

– It allows us to quickly identify periods where the calibration parameters have anomalous values, flagging data that may need to be revisited to see if the calibration can be improved. The most egregious cases have already been corrected as a result of the survey, and the data resubmitted to the CAA.

– It allows us to examine whether or not it is possible to correlate variations in calibration parameters with instrument and spacecraft events, particularly FGM instrument housekeeping telemetry.

– It gives some indication of the validity of interpolating the spin-axis offsets across the tail season, by comparing the change in offset over the tail seasons with overall variation throughout the mission.

More generally, a time history of FGM calibration on board the four Cluster spacecraft represents a unique and valuable body of knowledge in the field of space magnetometry, which should serve to inform the planning of any similar future missions where accurate magnetometer data are important. The 11 years of data discussed in this paper represent an opportunity to examine the results of a calibration campaign of unprecedented duration.

It should be noted that this paper aims to provide a primarily descriptive account of the results of the Cluster FGM calibration campaign over the period 2001–2012. No attempt has been made to quantify the remaining uncertainty in the FGM parameters after calibration, nor to incorporate these results into a comprehensive error analysis of the FGM instrument. Such an analysis lies outside the scope of this paper.

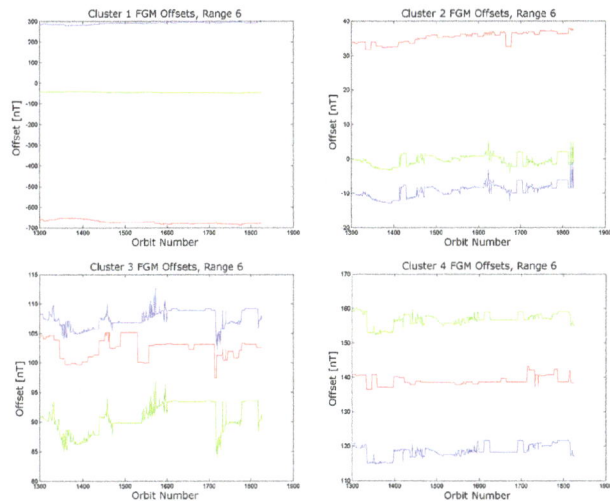

Figure 9. Range 7 spin-axis (O_1, red) and spin-plane (O_2 and O_3, blue and green) offsets in nanoteslas for orbits 1445–1825 (December 2009–February 2012).

The remainder of this paper consists of several parts. The entire data set consists of the time series for each calibration parameter, covering the period from the start of the nominal mission at orbit 93 (February 2001) to orbit 1825 (February 2012). Presentation and discussion of the time series for each parameter is impractical, given that the complete data set encompasses 12 parameters for each of the 6 ranges for each of the 4 Cluster spacecraft. Therefore, only a representative subset of the calibration parameters is discussed, highlighting what we consider the most significant features of the data set.

Table 2. Mean and standard deviation of the offsets (O_i) for the mission segment February 2001 to February 2012 for each coordinate in every range on all spacecraft. Standard deviations for ranges that are at least twice as large as those for all the other spacecraft in that range are highlighted in italic.

Offset (nT)

Coordinate 1 (spin axis) — Mean

	Range 2	Range 3	Range 4	Range 5	Range 6	Range 7
Cluster 1	−3.6472	−3.6363	−47.7804	−54.3680	−673.5475	−674.3962
Cluster 2	−0.0500	0.0035	2.6492	3.7953	35.2965	53.7835
Cluster 3	−2.2598	−2.2513	3.9292	4.9389	102.5378	119.4613
Cluster 4	−12.4734	−12.6754	−4.8020	−4.3008	139.0158	148.3612

Coordinate 1 (spin axis) — Standard deviation

	Range 2	Range 3	Range 4	Range 5	Range 6	Range 7
Cluster 1	*0.6766*	*0.6634*	*7.1659*	*2.4167*	*9.1031*	3.0133
Cluster 2	0.1584	0.1542	0.1725	0.1306	1.5800	2.6557
Cluster 3	0.1901	0.2174	0.2859	0.2473	1.6885	1.9154
Cluster 4	0.1601	0.2154	0.2733	0.2348	1.3228	0.9242

Coordinate 2 (spin plane) — Mean

	Range 2	Range 3	Range 4	Range 5	Range 6	Range 7
Cluster 1	6.3769	6.5161	23.5516	27.1099	290.7744	306.5910
Cluster 2	−2.4250	−2.4272	−2.3178	−1.9328	−9.1467	−1.4474
Cluster 3	−5.0619	−5.0892	1.4972	2.5119	107.5918	123.1910
Cluster 4	−3.0896	−3.0821	3.4234	4.4197	118.6630	136.1260

Coordinate 2 (spin plane) — Standard deviation

	Range 2	Range 3	Range 4	Range 5	Range 6	Range 7
Cluster 1	*0.8277*	*0.8612*	*2.9593*	*1.1104*	*4.6518*	0.7906
Cluster 2	0.1352	0.1352	0.1721	0.1973	1.8764	0.4969
Cluster 3	0.1608	0.1805	0.2780	0.1713	1.4813	1.1673
Cluster 4	0.1226	0.1301	0.1916	0.2587	1.8519	1.5174

Coordinate 3 (spin plane) — Mean

	Range 2	Range 3	Range 4	Range 5	Range 6	Range 7
Cluster 1	0.3404	0.3902	−3.7819	−4.2530	−45.3345	−30.1651
Cluster 2	−1.3266	−1.2950	−0.7289	−0.2210	−0.3983	9.7329
Cluster 3	−2.5431	−2.5589	2.4849	3.0053	90.8152	101.5091
Cluster 4	4.4047	4.5053	12.7467	13.3045	156.8103	166.8933

Coordinate 3 (spin plane) — Standard deviation

	Range 2	Range 3	Range 4	Range 5	Range 6	Range 7
Cluster 1	*0.4079*	*0.4161*	*1.4741*	*0.7083*	2.2685	0.4807
Cluster 2	0.1898	0.1944	0.2567	0.2281	1.5767	0.8230
Cluster 3	0.1376	0.1408	0.2026	0.1633	2.4599	1.8600
Cluster 4	0.1780	0.1866	0.1862	0.1681	1.8072	1.2829

Table 3. Mean and standard deviation of gains (G_1 and ΔG_{32}) for the mission segment from February 2001 to February 2012 for each coordinate in every range on all spacecraft. Standard deviations for ranges that are at least twice as large as those for all the other spacecraft in that range are highlighted in italic.

Gain

Coordinate 1 (spin axis) — Mean

	Range 2	Range 3	Range 4	Range 5	Range 6	Range 7
Cluster 1	0.9501	0.9684	0.9795	0.9962	0.9768	0.9978
Cluster 2	0.9589	0.9759	0.9866	1.0034	0.9853	1.0012
Cluster 3	0.9601	0.9756	0.9954	1.0110	0.9954	1.0106
Cluster 4	0.9595	0.9783	0.9954	1.0130	0.9925	1.0108

Coordinate 1 (spin axis) — Standard deviation

	Range 2	Range 3	Range 4	Range 5	Range 6	Range 7
Cluster 1	0.0037	*0.0024*	*0.0027*	0.0001	0.0046	0.0030
Cluster 2	0.0017	0.0007	0.0010	0.0001	0.0027	0.0007
Cluster 3	0.0030	0.0010	0.0010	0.0001	0.0068	0.0008
Cluster 4	0.0032	0.0008	0.0002	0.0001	0.0010	0.0019

Coordinate 3 – Coordinate 2 (spin plane) — Mean

	Range 2	Range 3	Range 4	Range 5	Range 6	Range 7
Cluster 1	0.0157	0.0151	0.0108	0.0101	0.0084	0.0071
Cluster 2	−0.0057	−0.0066	0.0041	0.0031	0.0049	0.0040
Cluster 3	−0.0180	−0.0163	−0.0130	−0.0112	−0.0104	−0.0085
Cluster 4	0.0246	0.0240	0.0266	0.0259	0.0237	0.0230

Coordinate 3 – Coordinate 2 (spin plane) — Standard deviation

	Range 2	Range 3	Range 4	Range 5	Range 6	Range 7
Cluster 1	0.0008	0.0003	0.0001	0.0000	0.0000	0.0014
Cluster 2	0.0007	0.0003	0.0002	0.0000	0.0000	0.0000
Cluster 3	0.0004	0.0003	0.0001	0.0001	0.0001	0.0001
Cluster 4	0.0004	0.0003	0.0001	0.0001	0.0002	0.0002

Additionally, a preliminary attempt has been made to correlate variation in the calibration parameters with instrument housekeeping telemetry. The instrument housekeeping consists of the following quantities: the FGM electronics box temperature located within the body of the spacecraft, the FGM outboard and inboard sensor temperatures on the boom, and the currents and the voltages of the electronics inside the electronics box. Only the temperatures are discussed here.

3.1 Cross-spacecraft comparisons of instrument housekeeping telemetry values

The data displayed in Figs. 2 and 3 are for the outboard sensors on all of the spacecraft. They cover the period from February 2001 (orbit 93) to August 2012 (orbit 1889).

3.1.1 FGM outboard sensor temperature

The FGM sensors each contain a thermistor, which is independent of the FGM electronics and which is monitored by the spacecraft. Each sensor also contains a heater which can be operated independently from the remainder of the

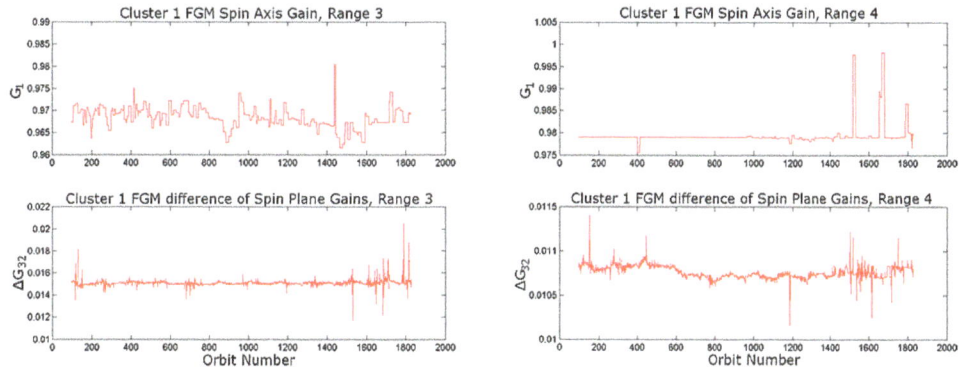

Figure 10. Gain plots (G_1 and ΔG_{32}) from February 2001 to February 2012 for exceptional cases in C1.

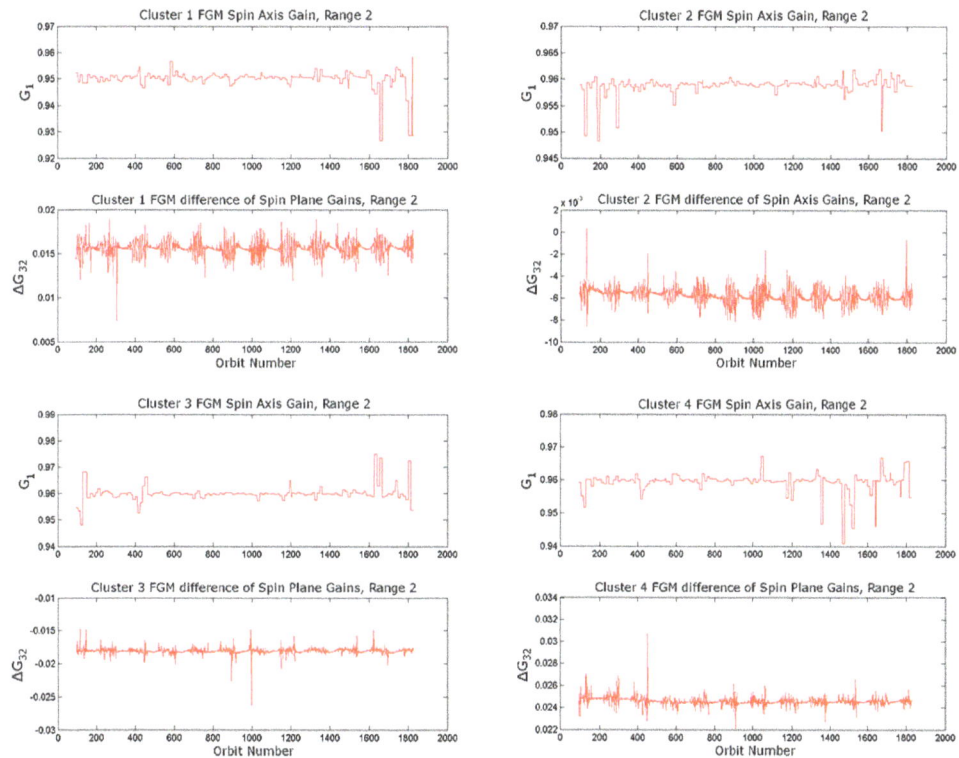

Figure 11. Gain plots (G_1 and ΔG_{32}) from February 2001 to February 2012 showing periodic behaviour of spin-plane gain difference for range 2 in all spacecraft.

spacecraft electronics. Both heaters are powered through a single switch which is controlled by the spacecraft. The sensors have thermal insulation and their temperature can be expected to change at a maximum of $20\,°C\,h^{-1}$. The sensors are monitored at intervals of the order of 30 min (FGM instrument users manual). Although each of these values is monitored more frequently, a single averaged value has been shown for each orbit (51–57 h) in Fig. 2.

The outboard sensor temperatures for all four spacecraft show a cyclical fluctuation over the course of nightside-to-dayside transitions, becoming around $5\,°C$ warmer during the peak of the dayside season. The spikes are due to long

eclipse periods, during which the FGM is off. All sensors have undergone a warming trend over the course of the mission as shown in Fig. 2. Since the outboard FGM sensors are located on the ends of 5 m booms, the warming and cooling cycle is most likely related to the spacecrafts' positions relative to the Sun during dayside and nightside seasons. The overall warming trend is likely related to the spacecrafts' positions relative to the Earth, as both periapsis and apoapsis have become lower over the course of the mission.

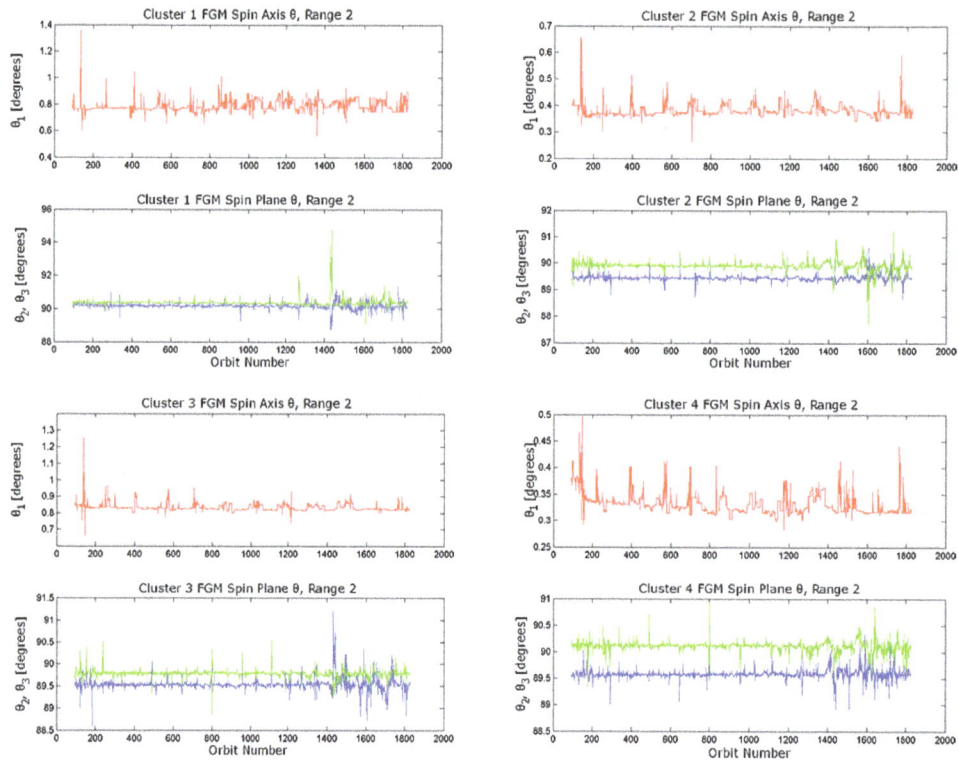

Figure 12. Elevation angle, θ_i, plots from February 2001 to February 2012 for range 2 for all spacecraft.

3.1.2 FGM electronics box temperature

The electronics box temperature is monitored from a thermistor located on the DC–DC converter card. This temperature can be expected to follow the temperature of the main equipment platform. The upper and lower operational limits for the box temperature are $+60\,°C$ and $-25\,°C$. If these limits are exceeded, the FGM instrument is powered off (Brown et al., 2000). Although each of these values is monitored more frequently, a single averaged value has been shown for each orbit (51–57 h) in Fig. 3.

On all four spacecraft, the box temperature is about $21\,°C$, with a declining trend until around orbit 600 (late May 2004), when it becomes cyclical as seen in the outboard sensor temperature. The boxes undergo less dramatic warming than the outboard sensors of around $3\,°C$ during the peak of the dayside season. This is likely due to their less exposed positions on the spacecraft body. The electronics boxes appear to be cooling over the course of the mission, with C1 and C3 cooling less dramatically than C2 and C4. The electronics boxes, due to their position on the spacecraft platform, are coupled to the spacecraft temperature much more strongly than the boom-mounted units. Changes in spacecraft heating strategy over the course of the mission, with more portions of the spacecraft being turned off during eclipses, are reflected in the electronics box temperature behaviour.

3.2 Inter-spacecraft calibration parameter comparisons by range

The data displayed are for the outboard sensors on all of the spacecraft and run from February 2001 (orbit 93) to February 2012 (orbit 1825).

3.2.1 Offsets

In Figs. 4 through 9, the individual offsets have been plotted for each range and spacecraft. The offsets in red have been applied to the spin-axis component of the magnetic field vector, while those in green and blue have been applied to the spin-plane components of the magnetic field vector. One offset value is applied across all data for a given range in an orbit. The spin-plane offsets are adjusted on a per-orbit basis no matter the phase of the mission. The spin-axis offsets, as mentioned in the Introduction, are adjusted on a biweekly or monthly basis during the dayside season when the spacecraft are in the solar wind and then interpolated between the end of one dayside season and the start of the next. The biweekly/monthly adjustment of the spin-axis offset gives a short, step-like appearance to the offset lines, while the interpolation method gives longer sloping steps for the 7 or so months (around 100 orbits) that the spacecraft spend on the nightside portion of their tours.

Comparison of the smooth slopes of the interpolated offsets with the variability of the solar wind-adjusted offsets

Table 4. Mean and standard deviation of elevation angles (θ_i) for the mission segment from February 2001 to February 2012 in every range on all spacecraft. Standard deviations for ranges that are at least twice as large as those for all the other spacecraft in that range are highlighted.

	Theta (°)												
	Mean							Standard deviation					
	Coordinate 1 (spin axis)							Coordinate 1 (spin axis)					
	Range 2	Range 3	Range 4	Range 5	Range 6	Range 7		Range 2	Range 3	Range 4	Range 5	Range 6	Range 7
Cluster 1	0.7878	0.7711	0.7696	0.7736	0.7585	0.7527	Cluster 1	0.0436	0.0123	0.0075	0.0064	0.0053	*0.0142*
Cluster 2	0.3806	0.3669	0.3674	0.3709	0.3645	0.3582	Cluster 2	0.0238	0.0059	0.0049	0.0029	0.0038	0.0035
Cluster 3	0.8343	0.8232	0.8242	0.8223	0.8270	0.8268	Cluster 3	0.0226	0.0073	0.0036	0.0026	0.0019	0.0017
Cluster 4	0.3319	0.3216	0.3352	0.3301	0.3564	0.3445	Cluster 4	0.0195	0.0102	0.0083	0.0048	0.0084	0.0029
	Coordinate 2 (spin plane)							Coordinate 2 (spin plane)					
	Range 2	Range 3	Range 4	Range 5	Range 6	Range 7		Range 2	Range 3	Range 4	Range 5	Range 6	Range 7
Cluster 1	90.1757	90.1731	90.1919	90.1666	90.1940	90.2105	Cluster 1	0.1831	0.2995	*0.0522*	0.0160	0.0144	0.0044
Cluster 2	89.4633	89.4650	89.4575	89.4573	89.4710	89.4661	Cluster 2	0.1162	0.1332	0.0180	0.0080	0.0232	0.0059
Cluster 3	89.5477	89.5165	89.5268	89.5276	89.5205	89.5174	Cluster 3	0.1426	0.1646	0.0147	0.0060	0.0178	*0.0122*
Cluster 4	89.5849	89.5746	89.5695	89.5703	89.5654	89.5660	Cluster 4	0.0920	0.1468	0.0139	0.0088	0.0173	0.0056
	Coordinate 3 (spin plane)							Coordinate 3 (spin plane)					
	Range 2	Range 3	Range 4	Range 5	Range 6	Range 7		Range 2	Range 3	Range 4	Range 5	Range 6	Range 7
Cluster 1	90.3697	90.3718	90.3484	90.3567	90.3604	90.3438	Cluster 1	0.2282	0.3353	0.0355	0.0096	0.0087	0.0103
Cluster 2	89.9023	89.8828	89.9163	89.9048	89.9416	89.9405	Cluster 2	0.1784	0.2616	0.0290	0.0122	0.0226	0.0071
Cluster 3	89.7856	89.7890	89.7895	89.7908	89.7961	89.7948	Cluster 3	0.0852	0.0936	0.0147	0.0046	0.0204	0.0143
Cluster 4	90.1174	90.1101	90.1471	90.1448	90.1573	90.1641	Cluster 4	0.1012	0.1981	0.0190	0.0093	0.0157	0.0032

Table 5. Mean and standard deviation of azimuthal angles (φ_1 and $\Delta\varphi_{32}$) for the segment from February 2001 to February 2012 in every range on all spacecraft. Standard deviations for ranges that are at least twice as large as those for all the other spacecraft in that range are highlighted in italic.

	Phi (°)												
	Mean							Standard deviation					
	Coordinate 1 (spin axis)							Coordinate 1 (spin axis)					
	Range 2	Range 3	Range 4	Range 5	Range 6	Range 7		Range 2	Range 3	Range 4	Range 5	Range 6	Range 7
Cluster 1	−119.1486	−119.7610	−119.9031	−119.3920	−120.2991	−120.1796	Cluster 1	3.0070	1.1482	0.7069	0.3514	0.4094	0.8647
Cluster 2	−64.0441	−64.6107	−64.7980	−64.0815	−63.5844	−62.9304	Cluster 2	3.1663	1.5004	1.0614	0.5710	0.6295	0.5848
Cluster 3	167.4990	167.3238	167.4863	167.3277	167.6714	167.6328	Cluster 3	1.0820	0.4848	0.4424	0.2880	0.1726	0.0572
Cluster 4	−89.3487	−89.6470	−89.6600	−88.7984	−88.5335	−88.1003	Cluster 4	3.0770	2.2709	1.3110	1.0266	0.5088	0.3284
	Coordinate 3 − Coordinate 2 (spin plane)							Coordinate 3 − Coordinate 2 (spin plane)					
	Range 2	Range 3	Range 4	Range 5	Range 6	Range 7		Range 2	Range 3	Range 4	Range 5	Range 6	Range 7
Cluster 1	89.7586	89.7502	89.5435	89.5716	89.5199	89.5177	Cluster 1	*0.0953*	*0.0869*	0.0312	0.0083	0.0027	0.0009
Cluster 2	89.3710	89.3698	89.3617	89.3659	89.3692	89.3635	Cluster 2	0.0301	0.0208	0.0075	0.0059	0.0072	0.0039
Cluster 3	89.2561	89.2549	89.3181	89.3183	89.3300	89.3261	Cluster 3	0.0287	0.0101	0.0033	0.0032	0.0031	0.0018
Cluster 4	88.8318	88.8325	88.8201	88.8252	88.8267	88.8236	Cluster 4	0.0232	0.0157	0.0086	0.0054	0.0052	0.0045

shows that the interpolation is probably masking the natural variability during the tail season. Orbits for which no FGM data were taken (and hence no calibration) have been omitted.

Over the course of the mission, offset drift is negligible in all components. The largest drifts in offsets take place on C1. C4 is the only other spacecraft with offsets that are comparable in magnitude to C1 in ranges 2 through 5, but offset drift is still insignificant. Offset variation with temperature is about $0.2\,\mathrm{nT}\,°\mathrm{C}^{-1}$ on C1 and $0.1\,\mathrm{nT}\,\mathrm{C}^{-1}$ on C2, C3 and C4.

For C1 in range 2 (Fig. 4) there is a clear decreasing trend in the spin-axis offset O_1, which is visible even with the steps introduced by the interpolated values over the course of the

mission. A decreasing trend is also seen in O_3. The total change is about 2 nT in O_1 and 1 nT in O_3. In O_2, the offset appears to increase by about 2 nT. On the other spacecraft, C2–C4, there is no overall drift, although some cyclical behaviour that may be related to instrument parameter cycles, particularly in the outboard sensor temperature, can be seen in C2 and C3.

Range 3 offset trends are very similar in magnitude and type to those observed in range 2 (Fig. 5). This is not too surprising since the range change is achieved by switching a single feedback resistor. On C1, the decreases of 2 nT in the spin-axis offset O_1 and 1 nT in O_3 are observed, as is the 2 nT increase in O_2. On the other spacecraft, C2–C4, there is

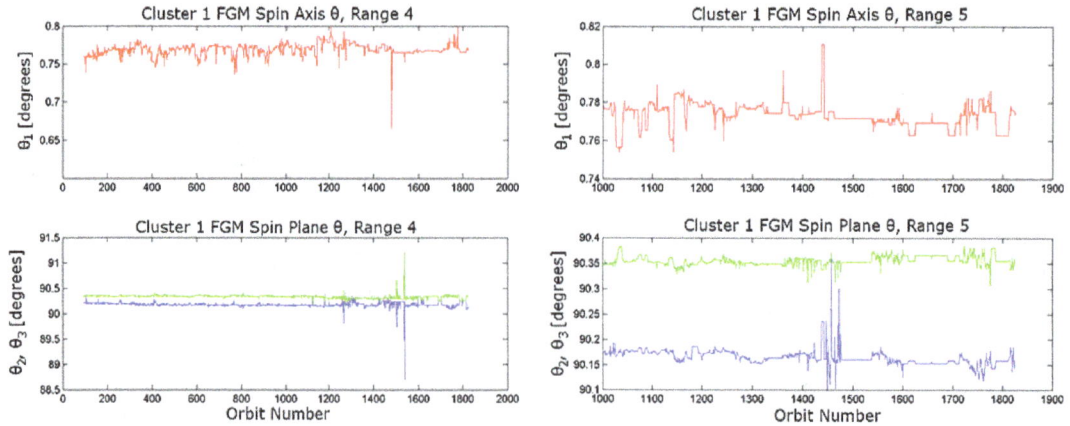

Figure 13. Elevation angle, θ_i, plots for C1, ranges 4 and 5, from February 2001 to February 2012.

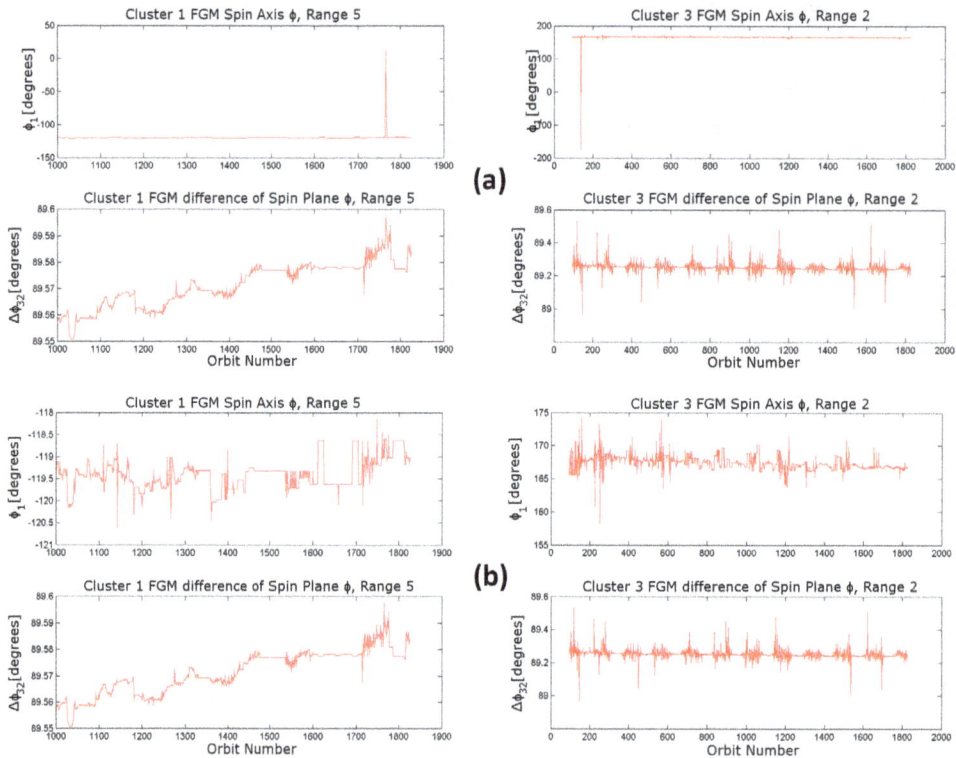

Figure 14. Azimuthal angles, φ_1 and $\Delta\varphi_{32}$, for C1, range 5 and C3, range 2. The top four plots show the angles before the unphysical calibration parameters were corrected and the lower four plots show them after correction.

no overall drift, although some cyclical behaviour that may be related to instrument parameter cycles, particularly in the outboard sensor temperature, can be seen in C2 and C3. The offsets also appear to fluctuate more, particularly the spin-axis offset, early in the mission compared to range 2. Some outlying values in the spin-axis offset for C2 will require further investigation.

Range 4 offset trends are similar in type to those observed in range 2 (Fig. 6). The C1 drifts have increased by approximately an order of magnitude. The decrease of 2 nT has become 20 nT in the spin-axis offset O_1 and 1 nT in O_3 has become 8 nT. The 2 nT increase in O_2 has become 15 nT. On the other spacecraft, C2–C4, there is no overall drift, although some cyclical behaviour that may be related to instrument parameter cycles, particularly in the outboard sensor temperature, can be seen on C2 and C3. The offsets fluctuate less, particularly the spin-axis offset, early in the mission compared to range 3.

Since range 5 did not enter routine use until late November 2006, Fig. 7 covers 800 orbits, or just over 5 years. C1

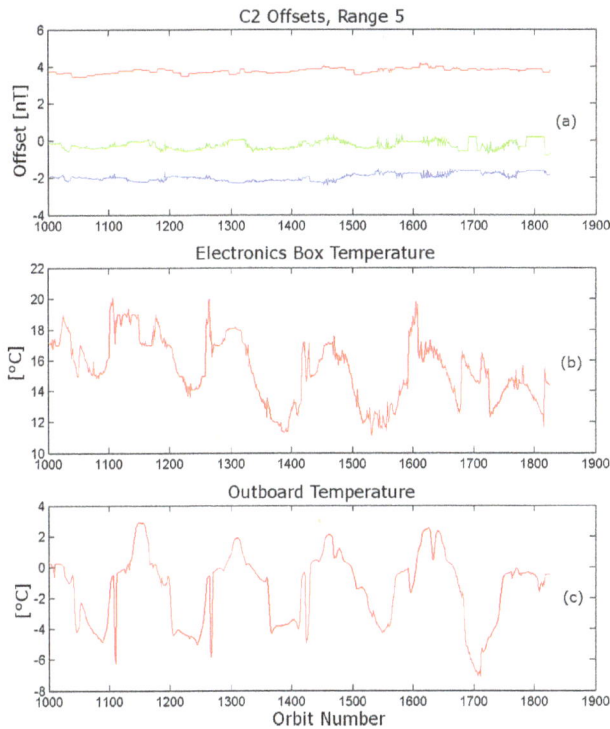

Figure 15. (a) C2 range 5 offsets (O_1 in red, O_2 in blue and O_3 in green), (b) C2 electronics box temperature and (c) C2 outboard sensor temperature. Offsets shown in nanoteslas and temperatures in degrees Celsius for orbits 1000–1812 (December 2006–February 2012).

Figure 16. (a) C3 range 5 offsets (O_1 in red, O_2 in blue and O_3 in green), (b) C3 electronics box temperature and (c) C3 outboard sensor temperature. Offsets shown in nanoteslas and temperatures in degrees Celsius for orbits 1000–1825 (December 2006–February 2012).

follows the trend seen in the lower ranges, where O_1 and O_3 are slowly decreasing and O_2 is slowly increasing. C2 shows slight signs of a cyclical trend like that observed in the outboard sensor and box temperatures. C3 shows strong signs of such a cyclical trend, while C4 shows the same stability and independence of instrument parameter trends exhibited previously. The potential correlation with instrument parameters in C2 and C3 merits further investigation in another section of the Analysis.

Despite the limited data available in range 6, the offset trends mirror those seen in the range 5 data (Fig. 8). At present there are 500 orbits' worth, or just over 3 years, of data. As discussed in the Introduction, limited ground calibration information was available for this range since it was not originally intended for scientific investigation. Changes in the range 6 parameters, including the offsets, are tied to changes in the range 5 parameters, which were used during the initial calibration of range 6 to help discover consistent values. It is therefore sensible that any potential correlation with other instrument parameters that were seen in range 5 should also be observed in range 6. As the mission continues and range 6 is employed more regularly for lower periapsis passes, the consistency between range 5 and range 6 offsets should become clearer.

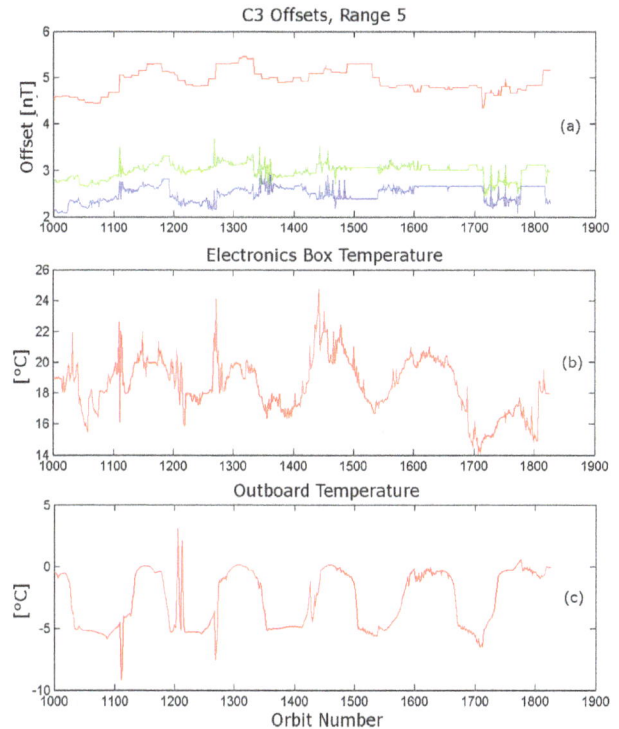

Insufficient data exists in range 7 to distinguish many trends (Fig. 9). One exception is that large month-long deviations in a parameter, such as the one seen in the spin-axis offset around orbit 1600 in C3, are paralleled by similar deviations in range 6. Adjustments to the spin-axis offsets during range jump corrections are primarily responsible for such shifts, since a large change in the range 6 spin-axis offset to eliminate R56 jumps is likely to result in the need for a large change in the range 7 spin-axis offset to eliminate the R67 jumps.

At most (on C2 and C4) there are 60 orbits' worth, or 5 months, of data and at least (on C1) there are 36 orbits' worth, or 3 months, of data. Since the spacecraft are now off due to power-sharing issues during the lower periapsis passes that necessitated the use of range 7, it is unlikely that this limited set will be expanded much. It will therefore not be possible to determine whether range 7 follows the same trends as observed in the lower ranges for each spacecraft. Further discussion of this range has been omitted from the remainder of this article.

In Table 2, the mean value and standard deviations for the spin-axis and spin-plane offsets on each spacecraft over orbits 93–1825 (February 2001–February 2012) have been calculated. The standard deviations in this, and in Tables 3–

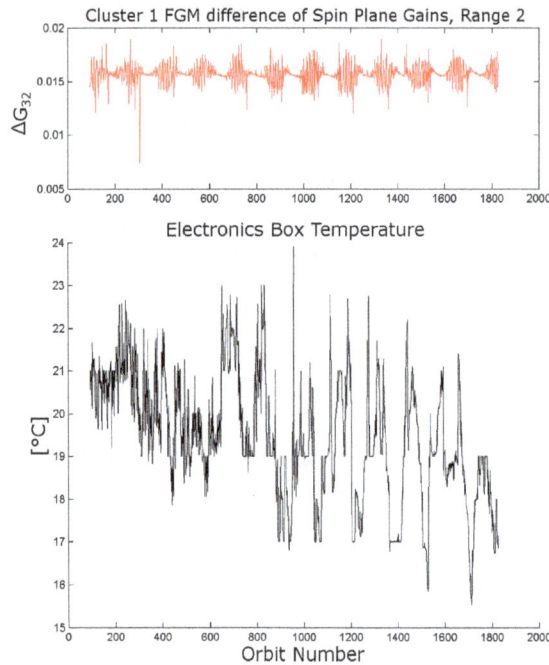

Figure 17. Spin-plane gain difference (ΔG_{32}) for C1, range 2 and electronics box temperatures in degrees Celsius for orbits 93–1825 (February 2001–February 2012).

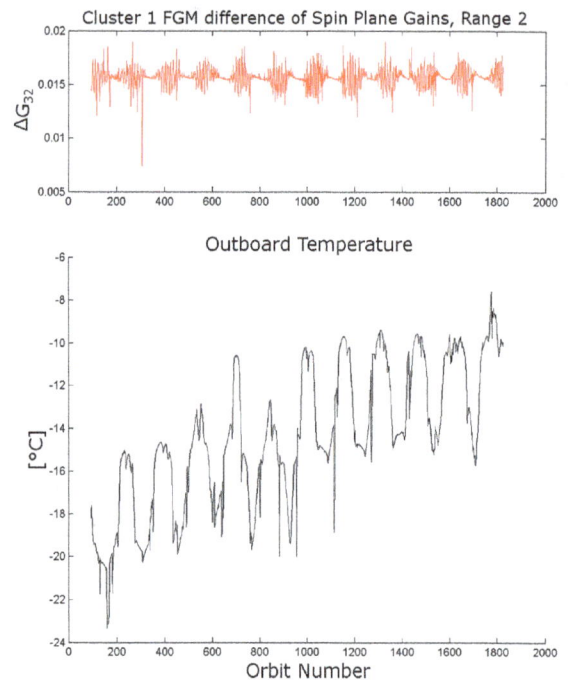

Figure 18. Spin-plane gain difference (ΔG_{32}) for C1, range 2 and sensor temperatures in degrees Celsius for orbits 93–1825 (February 2001–February 2012).

5, should be viewed as a measure of the variability in the output of the calibration procedures and not necessarily as a measure of physical variability in the sensor itself. The standard deviations are fairly consistent between coordinates and across all ranges for C2, C3 and C4. With the exception of range 7, the standard deviation for C1 is significantly larger, from two up to thirty times greater than the other spacecraft. This tallies with the observation of greater long-term drift in the offset parameters for C1 than in the other spacecraft.

3.2.2 Gains and angles

For most of the remaining calibration parameters, the fluctuations across the mission show no visible correlation with instrument parameters and no long-term trends. Mission averages for the parameters, which are the gains (G_i) and angles (θ_i and φ_i) will therefore be discussed in tabular form, with plots shown for exceptional cases.

In Table 3, the mean value and standard deviations for the spin-axis gain (G_1) and the difference of the spin-plane gains (ΔG_{32}) have been calculated. The change in the difference between the spin-plane gains is used as a calibration parameter. The final spin-plane gain values are therefore interdependent, which is why the difference is evaluated here. With two exceptions, notably the spin-axis gains for ranges 3 and 4 on C1, there is little fluctuation in these parameters. The noted gains are shown in Fig. 10.

The two exceptional cases show similar behaviour. In ranges 3 and 4 of C1, the fluctuations begin in orbits cor-

responding to late 2010, with no obvious correlation to other behaviour, through to February 2012. The calibration for the orbits where large, probably non-physical fluctuation in the gain is seen will need to be revised.

One interesting behaviour not reflected in the gains table of averages and standard deviations occurs in range 2 for all spacecraft. The spin-axis gain difference appears to undergo periodic increases in fluctuation, seemingly corresponding with the warming/cooling cycles observed in the instrument housekeeping sensor temperature values, as shown in Fig. 11. However, both the gain difference and the azimuthal angle difference in the spin plane contribute to the second harmonic of the spin frequency used in the Fourier analysis. None of the calculated absolute angles exhibit this behaviour, which would suggest that the link to the temperature cycling may be coincidental. It is also possible that the phenomenon is noise-related. The data are much cleaner in the tail season, which might lead to reduced fluctuation in the calculation of these parameters, for example. The potential causes cannot be distinguished easily and are beyond the scope of this initial investigation.

In Table 4, the mean value and standard deviations for the elevation angles, theta, have been calculated. The levels of fluctuation in theta are consistent across ranges and spacecraft, with the exception of the spin-axis theta on C1 for all except range 6 in which the levels are elevated. This might lead to the assumption that the C1 values are simply consistently elevated. The assumption is borne out when observing the theta values for range 2 in all spacecraft as shown in Fig. 12. However, as shown in Fig. 13, an examination of all of the plots reveals spikes of up to 1.5° occur in ranges 4 and 5 respectively for C1. This highlights the importance of examining the long-term trends in a number of ways. The variability of the spin-plane thetas has increased later in the mission.

In Table 5, mean value and standard deviations for the azimuthal angles, phi, have been calculated. The levels of fluctuation in phi are consistent across ranges and spacecraft, although initially exceptionally large values were seen in range 5 in C1 and range 2 on C3 in the spin axis and range 6 on C3 in the y coordinate of the spin plane. As shown in the top four plots of Fig. 14, these were caused by spikes in the values for single orbits. The calibration for these orbits was reviewed and corrected, resulting in the improved lower four plots of Fig. 14 and more consistent values in Table 5.

3.2.3 Individual calibration and instrument housekeeping parameter comparisons

The calibrated offsets for C2 and C3 exhibited cyclical trends in some ranges that merit individual visual comparison with instrument parameters extracted from telemetry. The cyclical trends become more obvious in the latter half of the mission, so range 5 has been chosen as the primary example. On C2, O_1 and O_2 appear to track the electronics box and outboard sensor temperatures, rising and falling in the same cycle (Fig. 15). O_3 shows an inverted trend. On C3, all three offsets appear to track the electronics box and outboard sensor temperatures, rising and falling in the same cycle (Fig. 16).

As mentioned previously, the calibrated spin-plane gains for all spacecraft exhibited cyclical trends in range 2 that merit individual visual comparison with instrument parameters extracted from telemetry. The spin-plane gains for C1 are the least affected by exceptional single-orbit fluctuations and have thus been chosen for comparison with the electronics box and outboard sensor temperatures as shown in Figs. 17 and 18.

Visual inspection of the plots indicates that there may be a correlation between the warming/cooling sensor temperature cycles and cycling of the spin-plane gain fluctuation. As mentioned above however, it is possible that other factors contribute to the cyclical fluctuation in these calibration parameters, such as the manner in which the gain and azimuthal angle differences are used during the calibration procedure or the noise level in the data. In future, it might be desirable to perform a more thorough data correlation between calibration parameters and temperatures in order to try and discover a temperature coefficient which could be compared with ground data. This was deemed beyond the scope of the present work as an initial survey of parameter comparisons.

4   Conclusions and future work

The Cluster mission marks the first time that the magnetometer data from four spacecraft have been calibrated simultaneously in flight. The FGM measurements, and the parameters determined by the FGM post-launch support team for calibrating the outboard magnetometer sensor, span over 11 years. The offsets on C1 show a steady drift in all ranges (for which there is sufficient data) at the resolution of spacecraft orbits over the course of the Cluster mission to February 2012. The offsets on C2, C3 and C4 remain fairly constant across all ranges. Cyclical trends in the calibration parameters that may be correlated with instrument housekeeping parameters have been identified. Examination of the tabulated means and standard deviations for the gains, elevation and azimuthal angles, has helped to identify cases in which the calibration of certain archived orbits may need to be revisited. However, in general the stability of the outboard sensor calibration parameters over the course of the mission is excellent. Hence, confidence can be placed in the accuracy of the Cluster magnetic field data. In future papers, the features observed in the instrument housekeeping and calibration parameters will be explored further.

Acknowledgements. The Cluster FGM team would like to acknowledge the ESA and the CAA for the ongoing operations and archiving support to fund this work. We acknowledge the STFC for support until UK funding ceased in 2010. We thank the IGEP TU-BS for provision of the FGM data processing software. Finally, we express our regret for the loss of our valuable and respected colleague, Edita Georgescu.

Edited by: V. Korepanov

References

Auster, H. U., Glassmeier, K. H., Magnes, W., Aydogar, O., Baumjohann, W., Constantinescu, D., Fischer, D., Fornacon, K. H., Georgescu, E., Harvey, P., Hillenmaier, O., Kroth, R., Ludlam, M., Narita, Y., Nakamura, R., Okrafka, K., Plaschke, F., Richter, I., Schwarzl, H., Stoll, B., Valavanoglou, A., and Wiedemann, M.: The THEMIS fluxgate magnetometer, Space Sci. Rev., 141, 235–264, 2008.

Balogh, A., Dunlop, M. W., Cowley, S. W. H., Southwood, D. J., Thomlinson, J. G., Glassmeier, K. H., Musmann, G., Lühr, H., Buchert, S., Acuña, M. H., Fairfield, D. H., Slavin, J. A., Riedler, W., Schwingenschuh, K., and Kivelson, M. G.: The Cluster Magnetic Field Investigation, Space Sci. Rev., 79, 65–91, 1997.

Escoubet, C. P., Russell, C. T., and Schmidt, R. (Eds.): The Cluster and Phoenix Missions, Kluwer Academic Publishers, Dordrecht, 1997.

Brown, P., Carr, C. M., Balogh, A., and Oddy, T. M.: FGM Instrument Users Manual, European Space Agency, 2000.

Gloag, J. M., Lucek, E. A., Alconcel, L. N., Balogh, A., Brown, P., Carr, C. M., Dunford, C. N., Oddy, T., and Soucek, J.: FGM Data Products in the CAA, Cluster Active Archive: Studying the Earth's space plasma environment, Springer, New York, 2010, 109–128, 2010.

Hedgecock, P. C.: A correlation technique for magnetometer zero level determination, Space Sci. Inst., 1, 83–90, 1975.

Kepko, E. L., Khurana, K. K., Kivelson, M. G., Elphic, R. C., and Russell, C. T.: Accurate Determination of Magnetic Field Gradients from Four Point Vector Measurements – Part I: Use of Natural Constraints on Vector Data Obtained From a Single Spinning Spacecraft, IEEE Transactions on Magnetics, 32, 377–385, 1996, 377–385, 1996.

Laakso, H., Taylor, M. G. T., and Escoubet, C. P.: Cluster Active Archive: Studying the Earth's Space Plasma Environment, Springer, New York, 2010.

Walsh, A. P., Forsyth, C., Fazakerley, A. N., Chen, C. H. K., Lucek, E. A., Davies, J. A., Perry, C. H., Walker, S. N., and Balikhin, M. A.: 10 years of the Cluster mission, Astron. Geophys., 51, 33–36, 2010.

Yin, F. and Luehr, H.: Recalibration of the CHAMP satellite magnetic field measurements, Measure. Sci. Technol., 22, 055101, doi:10.1088/0957-0233/22/5/055101, 2011.

6

In-flight calibration of the Hot Ion Analyser on board Cluster

A. Blagau[1]**, I. Dandouras**[2,3]**, A. Barthe**[2,3,4]**, S. Brunato**[2,3,5]**, G. Facskó**[6,7,8]**, and V. Constantinescu**[1]

[1]Institute for Space Sciences, Bucharest, Romania
[2]Institut de Recherche en Astrophysique et Planétologie, Université de Toulouse, Toulouse, France
[3]CNRS, Institut de Recherche en Astrophysique et Planétologie, Toulouse, France
[4]AKKA Technologies, Toulouse, France
[5]Noveltis, Toulouse, France
[6]Laboratoire de Physique et Chimie de l'Environnement et de l'Espace, Orléans, France
[7]Geodetic and Geophysical Institute, Research Centre for Astronomy and Earth Sciences, HAS, Sopron, Hungary
[8]now at: Finnish Meteorological Institute, Helsinki, Finland

Correspondence to: A. Blagau (blagau@spacescience.ro)

Abstract. The Hot Ion Analyser (HIA), part of the Cluster Ion Spectrometry experiment, has the objective to measure the three-dimensional velocity distributions of ions. Due to a variety of factors (exposure to radiation, detector fatigue and aging, changes in the operating parameters, etc.), the particles' detection efficiency changes over time, prompting for continuous in-flight calibration. This is achieved by comparing the HIA data with the data provided by the WHISPER (Waves of HIgh frequency and Sounder for Probing of Electron density by Relaxation) experiment on magnetosheath intervals, for the high-sensitivity section of the instrument, or solar wind intervals, for the low-sensitivity section. The paper presents in detail the in-flight calibration methodology, reports on the work carried out for calibrating HIA and discusses plans to extend this activity in order to ensure the instrument's highest data accuracy.

1 Introduction

The Hot Ion Analyser (HIA) and the COmposition and DIstribution Function (CODIF) analyzer are the two sensors of the Cluster Ion Spectrometry (CIS) experiment (Rème et al., 2001) on board Cluster, having the objective to measure the three-dimensional velocity distributions of ions. As a major difference from CODIF, the HIA instrument does not provide mass resolution; however, HIA offers other important advantages, like higher detection efficiency, better angular and en-

ergy resolution, faster electronics capable to handle higher count rates, etc.

The HIA detection system is based on micro-channel plate (MCP) technology. The instrument efficiency has been determined on ground through extensive preflight calibrations. However, due to various reasons, like MCP gain fatigue and aging (Prince and Cross, 1971) or because of the penetrating radiation (in the radiation belts or from cosmic ray bombardment), the detector efficiency changes in the course of the mission, requiring periodic in-flight calibration. An in-flight calibration is needed as well whenever the HIA operating point is changed by commands from ground. The multipoint character of Cluster and its complex payload made it possible to asses the HIA's in-flight performance at an unprecedented level of accuracy.

The calibration methodology to be presented in this paper was developed by taking advantage of the large cross-calibration effort carried on in the framework of ESA's Cluster Active Archive (CAA) program. Before the CAA initiative, the resources allocated to this activity were relatively small when compared with the complexity of the work; also, the less accurate Cluster Prime Parameter data set has been used as a reference for the total electron density (Vallat, 2001).

The CIS experiment is prepared by an international consortium, under the principal responsibility of Institut de Recherche en Astrophysique et Planétologie (IRAP) in Toulouse (formerly Centre d'Etude Spatiale des

Rayonnements). Since 2009, the Institute for Space Sciences in Bucharest assumed a key role in HIA's in-flight calibration, in close collaboration with IRAP.

The CIS data sets available through the CAA interface are described in Dandouras et al. (2010). An updated report on the CIS calibration activities can be found on the CAA web page (current version: 1.4, see Dandouras et al., 2012).

The paper is organized as follows: in Sect. 2 the instrument and its specific parameters are presented. Section 3 discusses the HIA operation modes and data caveats. Section 4 provides details on the calibration tasks and illustrates the calibration methodology by two examples. The next section presents two statistical studies carried out for validating HIA's in-flight calibration. In Sect. 6 the results of HIA's in-flight calibration are summarized and plans to extend this activity are discussed.

2   Instrument presentation

Figure 1 presents the HIA operational principles. The instrument employs a 360° angle imaging "top-hat" (Carlson et al., 1982) toroidal electrostatic analyzer (EA) and a fast detection system, based on MCP electron multipliers. The ions moving along different directions in the plane of the instrument entrance aperture (different polar angle in the upper panel of Fig. 1) are deflected and focused by the EA (middle panel) on the exit plane, where they are recorded by a system of position encoding discrete anodes (lower panel). Ion energies from 5 eV to 32 keV are sequentially measured by rapidly varying (sweeping), in logarithmically spaced steps, the voltage across the hemispherical EA plates. In the detection plane, the MCP plates are arranged in chevron-pair configuration in order to achieve a higher gain of secondary electrons emission. For a better detection efficiency, the ions are post-accelerated by a ~ 2300–2500 V potential applied between the front of the MCP and a high-transparency grid located ~ 1 mm above. The MCP gain can be checked by occasionally stepping this high voltage and by adjusting the discrimination level of the collecting charge amplifiers. Coverage in azimuthal angle is achieved by using the satellite spin.

To accommodate the large dynamic range of ion fluxes that occur in different regions sampled by Cluster, the entrance aperture consists of two narrow fans, each covering 180° in polar angle and having sensitivities that differ by a factor of ~ 25. The high-sensitivity (HS or "high G") section (entrance aperture on the right in the upper panel of Fig. 1), selects ions with the appropriate energy per charge (E/Q) and concentrates them on 16 anodes, 11.25° each, located in the exit plane (on the left in the bottom panel of the figure). This section is designed for analyzing magnetospheric ions. Similarly, the low-sensitivity (LS or "low g") section (entrance aperture on the left in the upper panel of Fig. 1) is tuned for the detection of solar wind ions, i.e., for high ion fluxes with

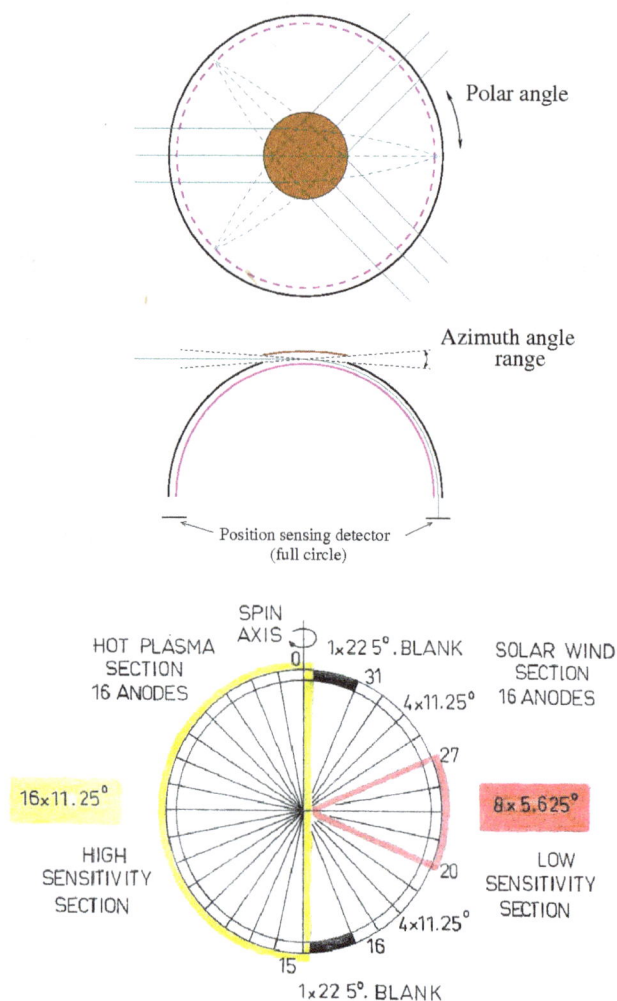

Fig. 1. The upper panels present the top view and cross-sectional view of the HIA instrument. In the bottom panel, the principles of HIA's anode sectoring are shown. In the upper and bottom panels of the figure, the spin axis is in the plane of the paper, along the vertical direction, while in the middle panel it points into the paper. See text for more details. Figure adapted from Klumpar et al. (2001) and Rème et al. (2001).

narrow energy and angular range. The required high angular resolution is achieved through the use of $8° \times 5.625°$ central anodes in the exit plane, the remaining 8 sectors having in principle 11.25° resolution. In the solar wind mode, the HIA voltage sweep is truncated when the "high G" section is facing the Sun, in order to avoid the solar wind detection and to protect the MCP lifetime. The two sections of the instrument can supply data simultaneously in the solar wind mode.

The HIA and CODIF sensors complement each other in terms of sensitivity, mass resolution, and detection efficiency. For CODIF, an additional time-of-flight (TOF) section is present, following the E/Q selection by the EA, allowing thus the separation of ion species. However, for the HIA detector the efficiency is much larger, primarily because this

Table 1. The HIA specific parameters according to Rème et al. (2001) and Dandouras and Barthe (2012).

Energy range (eV)		$\sim$5–32 $\times 10^3$
Energy resolution $\Delta E/E$ (FWHM, %)		~ 16
No. of E microsteps		124
Time resolution (s)	2-D	62.5×10^{-3}
	3-D	4
Angular resolution ($^\circ$)	LS	$\sim 5.6 \times 5.6$
	HS	$\sim 12.5 \times 5.6$
Geom. factor (cm^2 sr keV keV^{-1})	LS	1.9×10^{-4}
	HS	4.9×10^{-3}
Dynamic range (cm^2 s sr)$^{-1}$	LS	$10^6 - 2 \times 10^{10}$
	HS	$10^4 - 2 \times 10^8$
Lower limit density value (cm^{-3})		$0.01 \div 0.02$

sensor has no TOF section. In addition, HIA provides a higher angular resolution (up to $5.6^\circ \times 5.6^\circ$, to be compared with $11.25^\circ \times 22.5^\circ$ for CODIF) and faster electronics, capable to handle higher count rates. This makes the HIA more suitable for the study of the solar wind environment. Table 1 summarizes the HIA specific parameters.

3 HIA operation mode and data caveats

There are 16 operating modes for the CIS instrument, which can be roughly grouped in 2 classes, i.e., "magnetospheric" and "solar wind" modes. In the magnetospheric modes the full energy-angle ranges are covered and the different data products are based on the counts accumulated on the "high G" section. In the solar wind modes the plasma moments are based on data accumulated on the "low g" section when this side is facing the solar wind direction. In each mode, HIA and CODIF share the telemetry bit rate allocated to CIS for transmitting scientific products (on-board computed moments, one-, two- and three-dimensional distributions and pitch-angle distributions) to the ground.

The HIA instrument involves extensive on-board data processing, including the computation of the moments of the velocity-distribution functions (density, bulk velocity vector, pressure tensor, and heat flux vector). The moments are transmitted to the ground every spin period, i.e., about 4 s. The computation uses a table of efficiency coefficients (values dependent on the energy and angular sector θ) based on the ground calibration performed before launch. Assuming the same energy dependence and symmetric anode efficiency evolution in time, the values of these moments are periodically adjusted on ground through the so called absolute calibration (see Sect. 4).

The transmission to the ground of the complete 3-D distribution function (i.e., at full angular and energy resolution) is not possible due to the limited telemetry rates allocated

to the CIS experiment. For example, the nominal operation of HIA's HS section would require the transmission, every 4 s, of a matrix having 62 (or 31) energy channels $\times$ 16 elevation angles $\times$ 32 azimuth angles $= 31.744$ (or 15.872) elements. Therefore a reduced distribution function (particle counts typically binned in 31 energy channels and 88 angular directions) is computed on board and transmitted to the ground with a time resolution of multiple spin periods. Based on the reduced distribution function, the so-called "ground" plasma moments can be computed where, in principle, correction for the efficiency energy dependence and asymmetric anode efficiency evolution can be made.

Since in the solar wind spectrogram the He^{++} trace is clearly separated, appearing as an ion beam at roughly twice the mean proton energy, the HIA is providing the plasma moments for this ion species as well. So far this data product has not been calibrated.

There are a number of data caveats of particular importance for the instrument calibration as well as for regular exploitation of HIA data. These aspects, briefly summarized below, are closely checked when selecting the calibration intervals. For a detailed discussion the reader is referred to the instrument web page http://cluster.irap.omp.eu/ (see also Rème et al., 2001; Dandouras and Barthe, 2012).

- The accuracy of computed moments is affected by the instrument's finite energy and angle resolution, and by its finite energy range. Also, a reliable plasma moment computation requires that enough counts (minimum 100) are accumulated over the spin period.

- Inappropriate operational mode adversely affects the data accuracy. For example, when HIA is in solar wind mode, while the measurements are taken in the magnetosphere, a large portion of the ion distribution is excluded. Similarly, when HIA is in the magnetospheric mode but measures in the solar wind, detector saturation may occur, leading to underestimated values for the plasma density.

- Due to the penetrating particles from the radiation belts, the HIA measures a high background around perigee passes. A similar effect may also occur during some intense solar particle events.

- The detection of low-energy ions may be affected by the spacecraft charging to a positive floating potential that repels these ions.

- On some occasions, instrument artifacts (wrong time tagging, sudden density drops, high-voltage discharges, wrong discriminator levels, etc.) may occur. These events are listed on the CIS Data Caveats list, available on the instrument's web page.

Since the beginning of the mission, the HIA sensors were operational only on C1 (Cluster 1) and C3. Since November

2009, after almost 10 years of very good performance, the CIS experiment on board C3 is no longer operational. Also, since June 2011 the HIA operations on C1 are restricted to magnetospheric modes only, with the instrument switched into a safe stand-by mode when the satellite samples the solar wind region.

4 The HIA in-flight calibration

The HIA detection efficiency as a function of position (polar angle θ) and particle energy E is given by the formula (see Bosqued, 2000, reporting on HIA's ground calibration)

$$\mathrm{Eff}(\theta, E)^{-1} = \mathrm{Norm\_\theta} \cdot \mathrm{Cheff}(\theta) \cdot \frac{A \cdot E + B}{T_0 + T_1 \cdot E_\mathrm{t} + T_2 \cdot E_\mathrm{t}^2}$$

The first part of the RHS describes the position (anode)-dependent efficiency, with $\mathrm{Norm\_\theta}$ designating the anode normalization coefficients (one for each sensitivity side) and $\mathrm{Cheff}(\theta)$ the relative anode-dependent efficiency coefficients. The second part of RHS describes the efficiency energy dependence, with A, B and $T_{0,1,2}$ being the calibration coefficients and $E_\mathrm{t} = E + E_\mathrm{g}$ the total energy (sum of the particle energy E and the MCP – grid acceleration energy E_g) employed for describing the MCP energy-dependent efficiency. In total there are $2 + 2 \cdot 16 + 2 \cdot 2 + 3 = 41$ (39 independent) efficiency calibration coefficients for each validity period and spacecraft. Their values are specified in the calibration files that are constantly provided as the mission progresses.

The efficiency coefficients of the HIA instruments have been determined on ground through extensive preflight calibrations at IRAP vacuum-test facilities in Toulouse. Using ion beams of energies from a few 10 eV up to 30 keV, detailed studies of MCPs gain levels, MCP matching, and angular-energy resolution for each sector (each θ) were performed. Based on these tests, a table of efficiency coefficients is stored on-board in the nonvolatile memory and used by the processing software to compute the on-board moments from the full angular and energy resolution 3-D ion-distribution function.

However, the detector efficiencies change with time due to various reasons presented in Sect. 1. Therefore, as the missions progresses, the on-board calculations are based on out-of-date/in-accurate efficiencies. These unavoidable changes with time of the channel-plate detectors require continuous in-flight calibration. Also, the MCP high voltage is periodically increased by ground commands to compensate for the MCP gain fatigue. Since the procedure has a direct impact on detector efficiency, an in-flight calibration is subsequently required. So far, this operation has been performed five times (see Fig. 8).

The standard procedure for HIA's in-flight calibration, called the absolute calibration, relies on comparing HIA's ion number density with the electron number density pro-

vided by the WHISPER (Décréau et al., 2001) experiment on board Cluster . While the HIA detects individual particles to measure the ion-distribution function, WHISPER is based on a different method to determine the plasma density, i.e., by analyzing, both actively and passively, the electric signals in the neighboring plasma. In active mode, WHISPER measures the total electron density, while in passive mode it provides a survey of natural emissions from about 2 to 80 kHz, which covers the electron plasma resonance frequency.

The following are a number of assumptions involved in the absolute calibration procedure.

– There is a symmetric anode-dependent efficiency evolution with time, and therefore the relative anode-dependent efficiency coefficients $\mathrm{Cheff}(\theta)$, determined in the preflight tests, have not changed.

– The coefficients A, B and $T_{0,1,2}$ describing the efficiency energy dependence do not change as well and assume the values determined in the preflight tests.

– The WHISPER data are well-calibrated and free of errors (at least in the statistical sense). The traces in the WHISPER spectrograms are correctly assigned to some characteristic frequencies in plasma, from where electron density can be inferred. In these circumstances, the WHISPER data can be taken as reference.

The first two assumptions greatly simplify the calibration task, basically implying that only two coefficients (one for each sensitivity side) are needed to correct the HIA efficiency. It also means that the on-board moments are accurate up to a multiplication factor determined through calibration, thus allowing to take advantage of the HIA's highest temporal, directional and energy resolution. Indeed, e.g., in magnetospheric modes, the on-board moments are computed every spin period based on uncompressed data accumulated in 32 energy channels and 16 elevation × 32 azimuth solid angles (Di Lelis and Formisano, 2000), whereas the ground-computed moments are based on the reduced distribution function transmitted to the ground, having typically 31 energy × 88 solid angle bins and poorer time resolution.

Figure 2, based on C1 data from 17 October 2007, presents the individual anode response of the HIA high-sensitivity section in the plasma-sheet environment, where the plasma-distribution function is expected to be highly isotropic. Each of the panels corresponds to one sector in elevation angle (θ angle) for the arriving particles. The relatively homogeneous response from all eight angular sectors qualitatively supports the assumption of a symmetric anode-dependent efficiency evolution with time. The same situation is observed for the HIA instrument on C3 as well. Note that there are only 8 panels, although according to bottom panel in Fig. 1 there should be 16 angular sectors for the HS side of the HIA. This is because the distribution function sent to the ground has

CIS—HIA   Cluster 1   17/Oct/2007

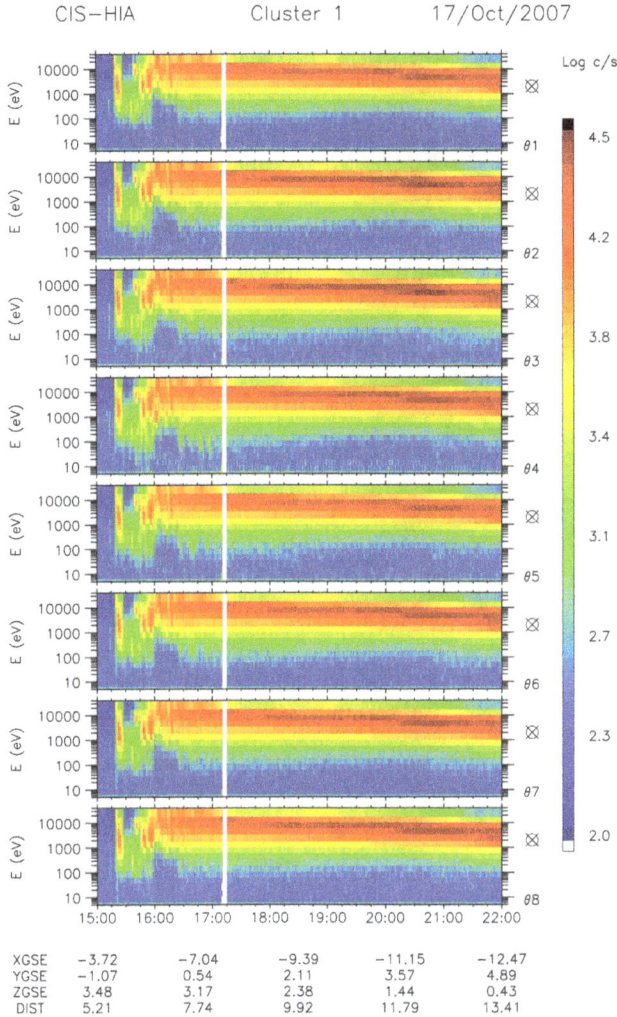

Fig. 2. Individual anode response of HIA's high-sensitivity section in the plasma sheet, an environment where the plasma-distribution function is highly isotropic. Only C1 data are shown. Each of the panels corresponds to one sector in elevation angle for the arriving particles.

been reduced by the on-board processing software in order to comply with the limited capacity of the telemetry; in that process the counts registered by individual anodes are binned in eight angular sectors.

Regarding the second assumption, i.e., constancy of the coefficients describing the efficiency energy dependence, there are some indications that this might not be completely valid (see e.g., the discussion about the HIA measurements in the plasma-sheet environment in Sect. 6) but so far no careful study addressing this problem has been carried out. However, it seems that a change in the efficiency energy dependence is a second-order effect, at least in the plasma environments where the two HIA sides are calibrated (see the statistical studies presented in Sect. 5).

One particular aspect that might be of concern is the role of ion composition in the HIA–WHISPER data comparison. The prevalent minor ions in solar wind and magnetosheath plasma (the environments where the two HIA sides are calibrated; see next sections) are the α particles. If one considers a mixture of protons (number density N_p, mass m_p, and electric charge q_p) and α particles (with corresponding parameters N_α, m_α, and q_α), then for a detector like HIA, unable to discriminate between the ion species, the number density reported by the instrument will be (see e.g., Paschmann et al., 1998) $N_{HIA} = N_p + \sqrt{m_p/m_\alpha}\,N_\alpha = N_p + N_\alpha/2$. However, the WHISPER instrument will report a number density of $N_{WHI} = N_p + (q_\alpha/q_p)N_\alpha = N_p + 2N_\alpha$. Typically, the α particles' abundance in the solar wind and magnetosheath plasma is around few percents of the proton number density. Therefore the discrepancy between the readings of the two instruments is of the same magnitude and consequently can be neglected in the first approximation. In addition, the procedure used to select the final set of calibration intervals (see Sect. 4.1) tends to exclude intervals with lower (than expected) values of the N_{HIA} / N_{WHI} ratio.

In the case of the HIA operating in solar wind mode, where the protons and α particles are clearly separated in the energy/charge channels, the on-board software automatically computes the plasma moments separately for the two ion species. Therefore here one compares $N_{HIA} = N_p$ with $N_{WHI} = N_p + 2N_\alpha$. However, the clear separation of the He^{++} trace in the energy spectrogram allows us to select from the beginning intervals with low presence of α, as will be described in Sect. 4.2.

4.1 Calibration of HIA's high-sensitivity section

For calibrating HIA's high-sensitivity section, magnetosheath (MS) intervals are used since the characteristic values of the plasma parameters in that environment (like density, temperature, and energy spectrum within the energy domain covered by HIA) allow for an optimum instrument performance.

Each part of the year in which Cluster samples the MS environment, i.e., roughly between November and June the following year, is analyzed to obtain one set of calibration coefficients. The number of values in the set depends on the efficiency evolution. Typically, two values, each obtained by combining data from several intervals, are inferred. Nevertheless, when an increase in the MCP HV is commanded from ground, a sudden increase followed by a relatively rapid adjustment in efficiency is expected, which requires the determination of additional values. The last value in the set is assumed to be valid until the beginning of the next MS season.

Intervals for calibration are carefully selected to meet several criteria listed below.

– The HIA energy spectrogram suggests that, presumably, the vast majority of ions are detected, i.e., no indication of a significant ion population below or above the detector energy range exists.

- The evolution of HIA's density is regular, useful for revealing potential instrument artifacts. This condition is also important because the HIA and WHISPER instruments could provide different values when a steep boundary is encountered (due to Larmor radius effects). It is desirable as well to select intervals where the density values span over a wide range, for a better comparison.

- Preferably the same intervals for both C1 and C3 are used to allow for an interspacecraft calibration and to detect potential instrument artifacts.

- Intervals are not on the Data Caveat list (see Sect. 3).

For the selected intervals, the WHISPER density data are requested from the instrument team, which either decides to regenerate it for the purpose of HIA calibration or to revalidate the already available CAA data set.

Figure 3 illustrates one example of an interval selected for the HIA–WHISPER density comparison. The four panels at the top represent the type of plot routinely produced for the identification of calibration intervals. First, the HIA energy spectrograms in three ranges, i.e., above 2 keV, the entire energy range, and below 100 eV, are shown in order to identify intervals that better comply with the requirement that virtually all particles are detected by the instrument. In the fourth panel the HIA (red) and WHISPER (black) raw density data are shown. Here a few sudden HIA density drops can be seen; these signatures are not present in the WHISPER data and are interpreted as instrument artifacts. The next panel compares the two densities after the data processing has been performed, which includes the removal of instrument artifacts, discarding of short intervals with rapid variation in density, data filtering, averaging and interpolation, etc. The sixth panel checks the plasma gyrotropy as measured by the quantity $(p_{\perp 2} - p_{\perp 1})/[(p_{\perp 2} + p_{\perp 1})/2]$, with $p_{\perp 1}$ and $p_{\perp 2}$ being the plasma thermal pressure along two orthogonal directions in the plane perpendicular to the magnetic field. The relatively small deviations from gyrotropy, e.g., around or below 5%, provides supporting evidence of the HIA's symmetric anode response for this interval. The bottom panel shows the N_{HIA} / N_{WHI} density ratio (blue) and its average value for this interval (magenta straight line).

The result of comparison can also be shown in the form of a HIA vs. WHISPER density plot, presented in Fig. 4, where the left panel refers to the event described above. The points are clearly scattered along the regression line (in red) forced to cross the origin; its slope is used to estimate the calibration factor inferred from this interval. The average value of several calibration factors, obtained from different intervals, is then employed to update the calibration files used in processing the HIA data.

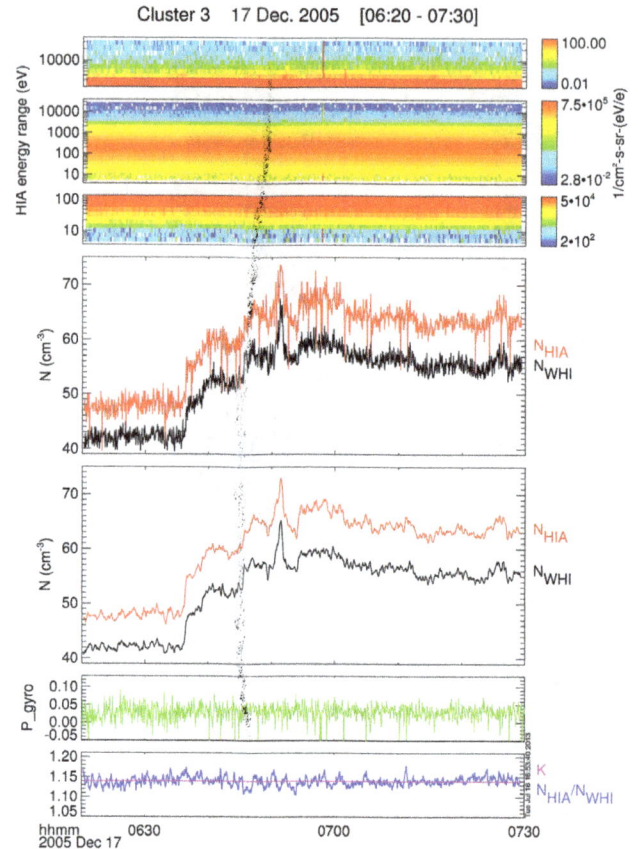

Fig. 3. Calibration of HIA's HS section. The top panels show ion energy spectrograms in three ranges, i.e., above 2 keV, the entire HIA energy range, and below 100 eV. The next panel presents HIA (red) and WHISPER (black) raw density data while in the fifth panel, the data from the two instruments are processed. The sixth panel presents the deviation in gyrotropy, based on the relative differences between the perpendicular pressure components. The bottom panel shows the ratio between the two densities (blue) and its average value K (magenta straight line). More explanations are provided in the text.

In spite of the careful selection, it can happen that the HIA–WHISPER data comparison brings inconsistent results on some intervals. Typically a lower (than expected) value of the N_{HIA} / N_{WHI} ratio is attributed to the presence of plasma population outside the HIA detection range, to the spacecraft charging or to events with relatively high abundance of α particles. Therefore, the final set of intervals to be used in the calibration is established after an additional interspacecraft comparison and/or checking the data provided by other instruments like CODIF, ASPOC, and EFW (spacecraft potential).

4.2 Calibration of HIA's LS section

With some specific differences, the calibration of the HIA low-sensitivity section follows a similar procedure. One uses

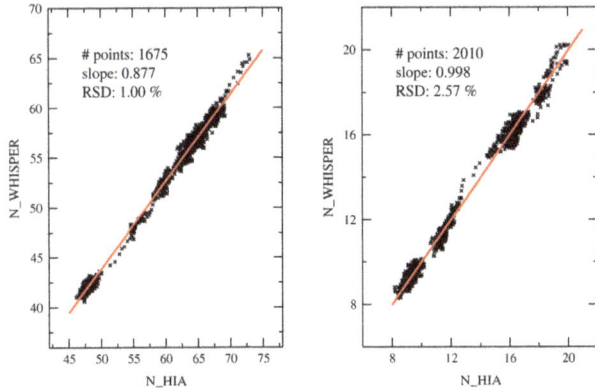

Fig. 4. HIA vs. WHISPER density comparison for the events presented in Fig. 3 (left) and in Fig. 5 (right), employed to calibrate the HIA HS and LS sections, respectively. The slope of the regression line crossing the origin (in red) is used to estimate the calibration factor for the corresponding interval.

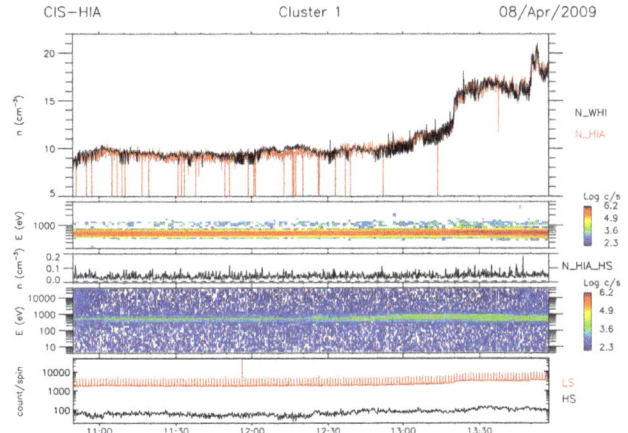

Fig. 5. Standard plot used for identification of suitable HIA LS calibration intervals. From top to bottom, the individual panels present HIA (red) and WHISPER (black) raw density data, the energy spectrogram provided by the LS section, the number density based on data collected by the HS section, the energy spectrogram provided by the HS section, and the number of counts detected by LS (red) and HS (black) sections.

solar wind (SW) data and compares the HIA and WHISPER density on carefully selected intervals.

The selection is based on the type of plot presented in Fig. 5. The top panel shows HIA (red) and WHISPER (black) raw density data, A number of sudden HIA density drops (instrument artifacts) can be seen here as well. In the second panel the energy spectrogram based on data accumulated on the LS section is shown. For the purpose of calibration, it is desirable to select events with low presence of α particles, (seen in the second panel as the faint green line, at around twice the peak proton energy). Since an accurate calibration of the LS section requires no significant counts accumulated on the HS section (blocked when facing the SW direction, see Sect. 2), the third and the fourth panel present the plasma density and the energy spectrogram based on data accumulated on this side. The bottom panel compares the LS (red) and HS (black) count rates.

After the selection of SW intervals suitable for calibration, the raw HIA and WHISPER data are processed in a manner similar to that described in the previous section. For the event presented above, the result of the comparison is shown in the right panel of Fig. 4 in the form of a HIA vs. WHISPER density plot. The points are aligned along the regression line (in red) forced to cross the origin; its slope is taken as the value of the calibration factor corresponding to this interval. By averaging several calibration factors, inferred from different intervals, a value is obtained that will be used to update the calibration files.

5 Statistic comparison between HIA and WHISPER data

To validate the results of calibration methodology presented in Sect. 4, two statistical studies have been carried out, each based on data provided by one HIA sensitivity side.

All MS and SW intervals observed in Cluster 1 data between 15 December 2006 and 15 February 2007 have been analyzed in order to compare calibrated HIA and WHISPER density data. Stable detector efficiency is expected in that period since the previous MCP high-voltage increase occurred in January 2006, the next one being performed on 16 February 2007, i.e., just after the chosen interval. Using selection criteria similar to those presented in Sect. 4, a total of 54 MS intervals sampled by the HIA HS section have been identified, comprising around 55 h of data. For the LS section, 64 SW intervals have been identified, covering around 96 h of data.

The results of the analysis are presented in Fig. 6 for the HS section and in Fig. 7 for the LS section. In the former case, the regression line (in red) forced to cross the origin corresponds to a proportionality factor of ~ 1.01; the relative standard deviation (RSD) of calibrated HIA data with respect to WHISPER data is 8.3 %. For the LS side, the regression line corresponds to a proportionality factor of ~ 1.02, with the RSD of points of 5.7 %.

The following are a number of conclusions that can be drawn from these comparisons.

- The dependence between N_{HIA} and N_{WHI} is linear. The proportionality factor is close to the ideal value of 1, a result that validates the calibration methodology outlined in Sect. 4.

- At least in the two plasma environments used for HIA's in-flight calibration, the assumptions of symmetric anode-dependent efficiency evolution and constancy of the coefficients describing the efficiency

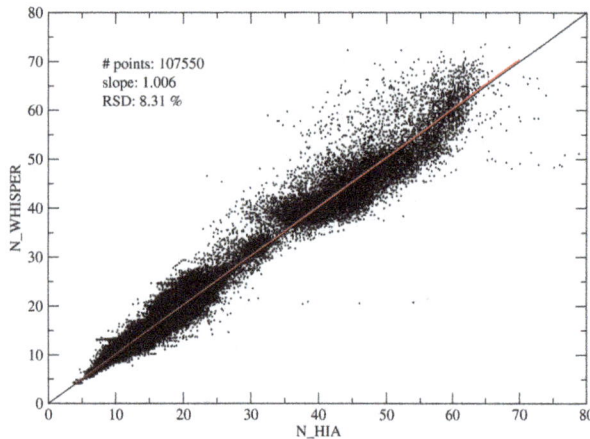

Fig. 6. Comparison between calibrated HIA and WHISPER data on Cluster 1. All magnetosheath intervals sampled by the HIA HS side between 15 December 2006 and 15 February 2007 were selected for comparison using selection criteria similar to those presented in Sect. 4. The regression line crossing the origin is shown in red.

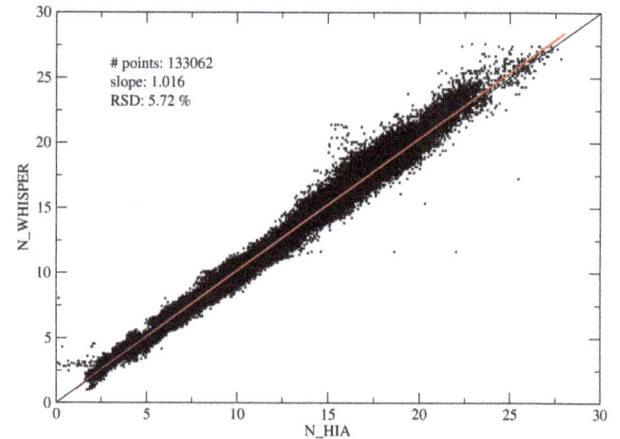

Fig. 7. Comparison between calibrated HIA and WHISPER data on Cluster 1. All solar wind intervals sampled by the HIA LS side between 15 December 2006 and 15 February 2007 were selected for comparison using selection criteria similar to those presented in Sect. 4. The regression line crossing the origin is shown in red.

energy dependence are verified in the statistical sense, any deviation being of a second-order importance.

- Since the instrument's artifacts have not been removed before comparison, the results indicate that they have no significant influence on the data, in a statistical sense.

6 Summary and future work

The first two panels in Fig. 8, showing the evolution of HS and LS detection efficiency in the course of the mission, summarize the results of HIA's in-flight calibration activity. Our comments below will refer mainly to the period starting from October 2005, subject to the calibration methodology presented in this paper.

The top and middle panels in Fig. 8 refer to the HS and LS sections, respectively. The detector efficiencies shown, obtained by comparing the HIA and WHISPER data densities, are relative to the beginning of the mission. Black lines present the evolution of C1 while the red lines the evolution for C3. Each value of the efficiency is shown by a horizontal segment, with the length indicating its validity period. There are no HIA data on C3 after November 2009 as well as on the C1 LS side after June 2011 (see the last paragraph of Sect. 3).

It is worth noting that the HIA detection efficiency stayed at a reasonable level in the course of the mission. Taking, for example, the values corresponding to 2009, the relative efficiency on C1 was around 1 for the HS side and 0.93 for the LS side, whereas in the case of C3, the relative efficiency was around 1.25 for the HS side and 1.38 for the LS side.

The high voltage applied to the MCPs is presented in the bottom panel; the vertical dashed lines indicate the dates when this HIA operating parameter was raised by ground commands for the purpose of increasing the detection efficiency. With some exceptions (e.g., see the LS side after the last change, on 17 February 2007) such an increase has been noticed. Note also that sometimes the efficiency has slightly increased without raising the MCP HV, like at the end of 2007. This unexpected behavior has been observed on both spacecraft and on both sensitivities, i.e., both on MS and SW intervals, which argues for a real effect. An MCP efficiency recovery has also been reported by the PEACE (Plasma Electron And Current Experiment) team in that period.

The HIA in-flight calibration is a complex task that requires considerable effort. This activity will continue and expand in order to ensure the highest data accuracy. Below we present a list of topics to be addressed in the future.

- Applying the calibration methodology described in this paper to data provided by HIA in the first years of the Cluster mission, i.e., before October 2005.

- Investigating more closely whether or not the anode-dependent efficiency factors evolved symmetrically in the course of the mission. Although Fig. 2 indicates a relatively homogeneous response from all HIA angular sectors, no quantitative assessment of this assumption was made so far and only few data intervals have been qualitatively evaluated. In the case of CODIF such an investigation is regularly performed, bringing significant improvements to data quality.

- Investigating possible changes in the efficiency energy dependence in the course of the mission.

- Calibrating the He^{++} data provided by HIA in the SW.

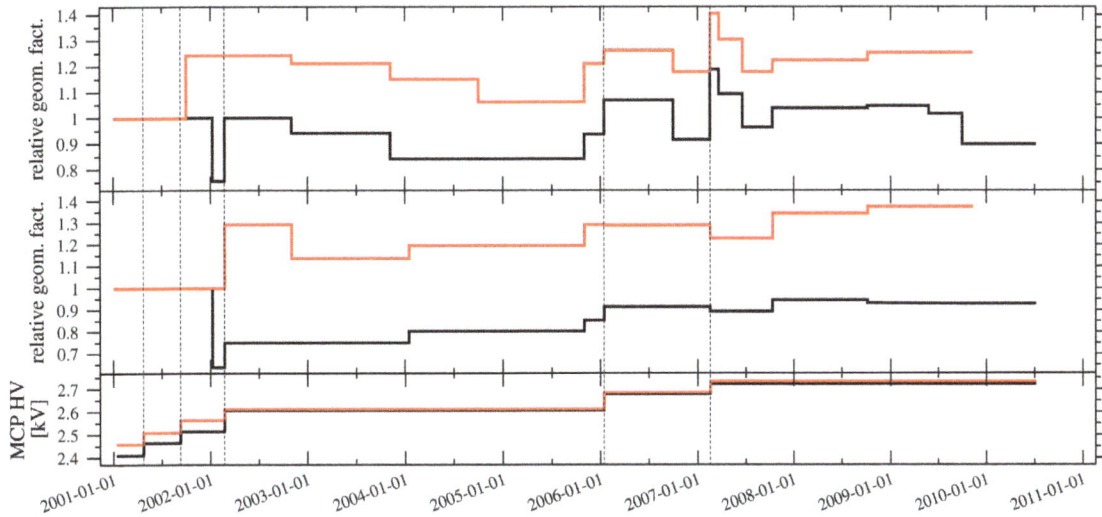

Fig. 8. Evolution of the HIA HS and LS efficiency (top and middle panels, respectively), and of the high voltage applied to the MCPs (bottom panel). Black lines refer to C1 and red lines to C3. The time axis spans from 1 January 2001 (around the start of Cluster's operational phase) to the beginning of 2011. The dates of MCP HV increase are also shown by vertical dashed lines.

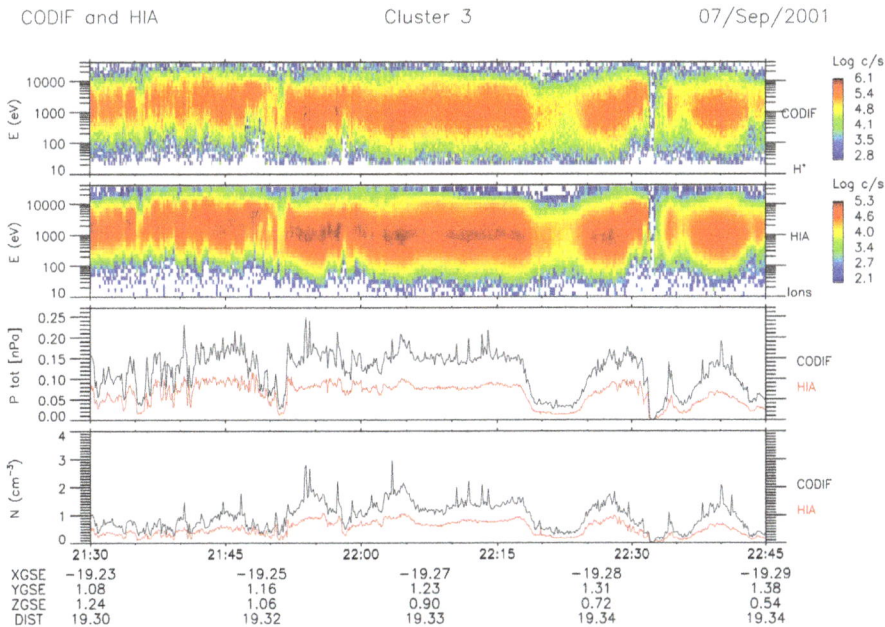

Fig. 9. CODIF and HIA data comparison in the plasma-sheet region. From top to bottom, the first two panels show the differential energy spectrogram from CODIF and HIA, respectively. The next two panels show the plasma pressure and plasma density as provided by CODIF (black line) and HIA (red line).

Related to the third item above, there is evidence suggesting the need for correction of the efficiency energy dependence. Figure 9 shows a comparison between HIA and CODIF data in the plasma-sheet environment on 7 September 2001. The energy spectrograms (CODIF first panel and HIA the second panel) support the idea that most of the particles are detected by the two instruments. However, the plasma pressure and density (next panels) are different, with HIA (red lines) providing values around two times lower that

CODIF (black lines). The two instruments are calibrated and there is convincing evidence that CODIF provides the correct measurements.

The situation presented in Fig. 9 is typical for the plasma-sheet. The HIA–CODIF data discrepancy in this environment could be explained by changes in the efficiency energy dependence, since the average plasma energy in the plasma-sheet is higher than in the magnetosheath, where the instrument was calibrated. It means that a correction that results in lower efficiency for the high-energy channels would provide a plasma density correction in the right direction (i.e., an increase in density and, mostly, in pressure). The same type of correction has been suggested by the instrument response to the penetrating radiation during perigee passes (see Ganushkina et al., 2011).

Acknowledgements. The authors would like to thank the WHISPER team (Principal Investigator, J. L. Rauch; former Principal Investigators, P. M. .E. Décréau and Jean Gabriel Trotignon) for their effort to ensure reliable electron density data for the HIA calibration.

The calibration-routine package is based on the *cl* (credit to E. Penou) and CCAT (credit to J. McFadden, P. Schroeder and C. Mouikis) programs for CIS data analysis.

The work has been supported through the project VALS (contract 45/19.11.2012) of the Romanian Space Agency STAR program, and by the OTKA grant K75640 of the Hungarian Scientific Research Fund.

Edited by: H. Laakso

References

Bosqued, J. M.: Cluster-2: CIS-2 Instrument: Summary of Calibration Tables, CESR report, Toulouse, 2000.

Carlson, C. W., Curtis, D. W., Paschmann, G., and Michel, W.: An instrument for rapidly measuring plasma distribution functions with high resolution, Adv. Space Res., 2, 67–70, doi:10.1016/0273-1177(82)90151-X, 1982.

Dandouras, I. and Barthe, A.: User Guide to the CIS measurements in the Cluster Active Archive (CAA), http://caa.estec.esa.int/caa/ug_cr_icd.xml (last access: 4 July 2013), 2012.

Dandouras, I., Barthe, A., Kistler, L. M., and Blagau, A.: Calibration Report of the CIS measurements in the Cluster Active Archive (CAA), http://caa.estec.esa.int/caa/ugcricd.xml (last access: 15 April 2014), 2012.

Dandouras, I., Barthe, A., Penou, E., Brunato, S., Rème, H., Kistler, L. M., Bavassano-Cattaneo, M. B., and Blagau, A.: Cluster Ion Spectrometry (CIS) Data in the Cluster Active Archive (CAA), in: The Cluster Active Archive, Studying the Earth's Space Plasma Environment, edited by: Laakso, H., Taylor, M., and Escoubet, C. P., 51–72, Springer, Berlin, 2010.

Décréau, P. M. E., Fergeau, P., Krasnoselskikh, V., Le Guirriec, E., Lévêque, M., Martin, Ph., Randriamboarison, O., Rauch, J. L., Sené, F. X., Séran, H. C., Trotignon, J. G., Canu, P., Cornilleau, N., de Féraudy, H., Alleyne, H., Yearby, K., Mögensen, P. B., Gustafsson, G., André, M., Gurnett, D. C., Darrouzet, F., Lemaire, J., Harvey, C. C., Travnicek, P., and Whisper experimenters (Table 1): Early results from the Whisper instrument on Cluster: an overview, Ann. Geophys., 19, 1241–1258, doi:10.5194/angeo-19-1241-2001, 2001.

Di Lelis, A. M. and Formisano, V.: Cluster CIS-2 Instrument, FM Normal Operation Software, CNR IFSI report, Frascati, 2000.

Ganushkina, N. Y., Dandouras, I., Shprits, Y. Y., and Cao, J.: Locations of boundaries of outer and inner radiation belts as observed by Cluster and Double Star, J. Geophys. Res.-Space, 116, A09234, doi:10.1029/2010JA016376, 2011.

Klumpar, D. M., Möbius, E., Kistler, L. M., Popecki, M., Hertzberg, E., Crocker, K., Granoff, M., Tang, L., Carlson, C. W., McFadden, J., Klecker, B., Eberl, F., Künneth, E., Kästle, H., Ertl, M., Peterson, W. K., Shelly, E. G., and Hovestadt, D.: The Time-of-Flight Energy, Angle, Mass Spectrograph (Teams) Experiment for Fast, Space Sci. Rev., 98, 197–219, 2001.

Paschmann, G., Fazakerley, A. N., and Schwartz, S. J.: Moments of plasma velocity distributions, in: Analysis Methods for Multi-Spacecraft Data, edited by Paschmann, G. and Daly, P. W., ISSI Scientific Reports, 125–158, ESA Publications Division, Noordwijk, 1998.

Prince, R. H. and Cross, J. A.: Gain Fatigue Mechanism in Channel Electron Multipliers, Rev. Sci. Instrum., 42, 66–71, doi:10.1063/1.1684879, 1971.

Rème, H., Aoustin, C., Bosqued, J. M., Dandouras, I., Lavraud, B., Sauvaud, J. A., Barthe, A., Bouyssou, J., Camus, Th., Coeur-Joly, O., Cros, A., Cuvilo, J., Ducay, F., Garbarowitz, Y., Medale, J. L., Penou, E., Perrier, H., Romefort, D., Rouzaud, J., Vallat, C., Alcaydé, D., Jacquey, C., Mazelle, C., d'Uston, C., Möbius, E., Kistler, L. M., Crocker, K., Granoff, M., Mouikis, C., Popecki, M., Vosbury, M., Klecker, B., Hovestadt, D., Kucharek, H., Kuenneth, E., Paschmann, G., Scholer, M., Sckopke, N., Seidenschwang, E., Carlson, C. W., Curtis, D. W., Ingraham, C., Lin, R. P., McFadden, J. P., Parks, G. K., Phan, T., Formisano, V., Amata, E., Bavassano-Cattaneo, M. B., Baldetti, P., Bruno, R., Chionchio, G., Di Lellis, A., Marcucci, M. F., Pallocchia, G., Korth, A., Daly, P. W., Graeve, B., Rosenbauer, H., Vasyliunas, V., McCarthy, M., Wilber, M., Eliasson, L., Lundin, R., Olsen, S., Shelley, E. G., Fuselier, S., Ghielmetti, A. G., Lennartsson, W., Escoubet, C. P., Balsiger, H., Friedel, R., Cao, J.-B., Kovrazhkin, R. A., Papamastorakis, I., Pellat, R., Scudder, J., and Sonnerup, B.: First multispacecraft ion measurements in and near the Earth's magnetosphere with the identical Cluster ion spectrometry (CIS) experiment, Ann. Geophys., 19, 1303–1354, doi:10.5194/angeo-19-1303-2001, 2001.

Vallat, C.: Evaluation des performances en vol des détecteurs de l'expérience "Cluster Ion Spectrometry", CNRS report, Toulouse, 2001.

7

Interinstrument calibration using magnetic field data from the flux-gate magnetometer (FGM) and electron drift instrument (EDI) onboard Cluster**

R. Nakamura[1], F. Plaschke[1], R. Teubenbacher[1,*], L. Giner[2], W. Baumjohann[1], W. Magnes[1], M. Steller[1], R. B. Torbert[3], H. Vaith[3], M. Chutter[3], K.-H. Fornaçon[4], K.-H. Glassmeier[4], and C. Carr[5]

[1] Space Research Institute, Austrian Academy of Sciences, 8042 Graz, Austria
[2] Graz University of Technology, 8010 Graz, Austria
[3] University of New Hampshire, Durham, NH 03824, USA
[4] Institut für Geophysik und extraterrestrische Physik, Technische Universität Braunschweig, 38106 Braunschweig, Germany
[5] Blackett Laboratory, Imperial College London, London, UK
* now at: Materials Center Leoben Forschung GmbH, Leoben, Austria

Correspondence to: R. Nakamura (rumi.nakamura@oeaw.ac.at)

*** This paper is dedicated to the memory of Edita Georgescu.*

Abstract. We compare the magnetic field data obtained from the flux-gate magnetometer (FGM) and the magnetic field data deduced from the gyration time of electrons measured by the electron drift instrument (EDI) onboard Cluster to determine the spin-axis offset of the FGM measurements. Data are used from orbits with their apogees in the magnetotail, when the magnetic field magnitude was between about 20 and 500 nT. Offset determination with the EDI–FGM comparison method is of particular interest for these orbits, because no data from solar wind are available in such orbits to apply the usual calibration methods using the Alfvén waves. In this paper, we examine the effects of the different measurement conditions, such as direction of the magnetic field relative to the spin plane and field magnitude in determining the FGM spin-axis offset, and also take into account the time-of-flight offset of the EDI measurements. It is shown that the method works best when the magnetic field magnitude is less than about 128 nT and when the magnetic field is aligned near the spin-axis direction. A remaining spin-axis offset of about 0.4 ~ 0.6 nT was observed for Cluster 1 between July and October 2003. Using multipoint multi-instrument measurements by Cluster we further demonstrate the importance of the accurate determination of the spin-axis offset when estimating the magnetic field gradient.

1 Introduction

Magnetic field and plasma environments of the Earth and other bodies in the solar system have been studied in situ since decades (Balogh, 2010). Therefore, magnetic field experiments onboard of spacecraft are of primary importance. Most commonly, flux-gate magnetometers (FGMs) are used due to their high accuracy, measurement range, resolution, and stability, paired with reasonable mass, power consumption, level of complexity, and overall costs (Acuña, 2002).

A FGM that is able to measure the strength and direction of the ambient magnetic field (B) with high precision, requires extensive pre-flight (ground-based) and in-flight calibration (e.g., Glassmeier et al., 2007; Auster et al., 2008). The aim of the calibration is to determine 12 parameters needed to convert raw measurements (B_{raw}) into components of a magnetic field vector (B_{cal}) in a usable coordinate system (e.g., Kepko et al., 1996). The calibration parameters are six angles describing the orientation of the sensor axes in, e.g., a spacecraft-fixed frame of reference (constituting matrix $\mathbf{M}$), three gain values (elements of a diagonal matrix $\mathbf{G}$), and three zero level offset values (elements of vector O). Therewith, the conversion of B_{raw} into B_{cal} is given by (e.g., Kepko et al., 1996; Acuña, 2002; Auster et al., 2008)

$$B_{\mathrm{cal}} = \mathbf{G} \cdot \mathbf{M} \cdot B_{\mathrm{raw}} - O. \tag{1}$$

Despite pre-flight calibration under a variety of conditions (magnetic fields, temperatures), in-flight calibration remains necessary to account for slight changes of the calibration parameters during launch, instrument drifts over time while the mission proceeds, and, most importantly, spacecraft-caused disturbances which are beyond the scope of ground-based tests.

Variations in ambient magnetic field strengths and temperatures may have a minor influence on gain levels ($\mathbf{G}$) and orientations ($\mathbf{M}$) of the sensor axes relative to the spacecraft body. Spacecraft generated fields (e.g., due to electrical currents or magnetic materials) strongly contribute to the zero level offsets ($\boldsymbol{O}$), as these offsets represent the field values measured under the absence of an external magnetic field. Influence of the spacecraft on the magnetic field measurements can be reduced either by placing the FGM sensor on a long boom (e.g., Dougherty et al., 2004), hence, furthest possible away from the spacecraft's main structure, or by implementation of a magnetic cleanliness program (e.g., Ludlam et al., 2008). Unfortunately, both measures tend to be extremely expensive.

Spin stabilization of the spacecraft greatly supports the in-flight calibration process, as the presence and content of spin tone and/or higher harmonics in the magnitude and/or spin-axis component of $\boldsymbol{B}_{\mathrm{cal}}$ is influenced by 8 of the 12 calibration parameters (see, Auster et al., 2002), namely the spin-plane components of $\boldsymbol{O}$ (which shall be O_1 and O_2), the ratio of the spin-plane components of $\mathbf{G}$ (i.e., G_{11}/G_{22}), and five elements of $\mathbf{M}$ (all but the angle defining the absolute orientation of the two spin-plane axes within that plane).

The in-flight determination of the spin-axis component of $\boldsymbol{O}$ (which we denote with O_3) is often dependent on the availability of prolonged solar wind observations, where Alfvénic fluctuations are prevalent. These fluctuations are characterized by rotations in the magnetic field while the field strength ($|\boldsymbol{B}|$) remains constant. Hence, O_3 can be determined by minimization of variance of $|\boldsymbol{B}_{\mathrm{cal}}|$ while observing Alfvénic fluctuations, as proposed in Hedgecock (1975). Improvements to his method are discussed in Leinweber et al. (2008) and, more recently, in Pudney et al. (2012).

If solar wind measurements are not available, O_3 may be determined with the help of complementary magnetic field observations, for instance from an electron drift instrument (EDI), which is the main subject of this paper. The EDI (Paschmann et al., 1997, 2001) onboard Cluster consists of two electron gun/detector units placed on opposite sides of the spacecraft, similar to that flown on the Equator-S spacecraft (Paschmann et al., 1999). Amplitude-modulated electron beams are fired by the two guns in specific directions. They perform one (or more) gyrations due to the ambient magnetic field and are eventually collected by the detectors after times T_1 and T_2. The primary objective of the EDI is to measure the drift of the electrons caused by electric fields or magnetic field gradients.

The drift step, $\boldsymbol{d} = \boldsymbol{v}_{\mathrm{d}} T_{\mathrm{g}}$, during the gyration time T_{g} (drift velocity: $\boldsymbol{v}_{\mathrm{d}}$) is a direct result from EDI measurements: small d can be determined by triangulation, based on the two beam-firing directions (for a detailed description see, Paschmann et al., 1997; Quinn et al., 1999). Large d are more accurately determined by time-of-flight observations of the two beams (Paschmann et al., 1997; Vaith et al., 1998). These times are different for electron release in parallel or anti-parallel directions to $\boldsymbol{v}_{\mathrm{d}}$: $T_{1,2} = T_{\mathrm{g}}(1 \pm |\boldsymbol{v}_{\mathrm{d}}|/|\boldsymbol{v}_{\mathrm{e}}|)$, where $\boldsymbol{v}_{\mathrm{e}}$ is the electron velocity dependent on their (known) kinetic energy: the sum of T_1 and T_2 yields twice the gyration time T_{g}, their difference is proportional to d (Paschmann et al., 1999). The use of different electron energies further allows one to distinguish drifts caused by electric fields or magnetic field gradients (see, Paschmann et al., 1997).

Since the gyration time T_{g} is inversely proportional to the magnetic field strength $|\boldsymbol{B}|$, EDI measurements allow for a determination of ambient $|\boldsymbol{B}|$:

$$|\boldsymbol{B}| = \frac{2\pi m_{\mathrm{e}}}{e T_{\mathrm{g}}}, \tag{2}$$

where m_{e} is the electron mass and e the elementary charge. These values are practically not influenced by spacecraft fields, as electrons perform most of their gyration at sufficient distances from the spacecraft. Hence, they are ideally suited as a reference for FGM measurements. Comparison of EDI and FGM magnetic field data yields FGM zero level offset vectors $\boldsymbol{O}$ and, in particular, their spin-axis components O_3, as shown by Georgescu et al. (2006).

Their methods were developed further by Leinweber et al. (2012) in order to obtain absolute spin-plane and spin-axis FGM gains (i.e., G_{11} and G_{22} with constant ratio G_{11}/G_{22}, and G_{33}), in addition to O_3, with the help of EDI time-of-flight $|\boldsymbol{B}|$ values. Note that the spacecraft spin does not support calibration of any of these three parameters, as they do not influence the content of spin tone or higher harmonics in $\boldsymbol{B}_{\mathrm{cal}}$.

Both studies (Georgescu et al., 2006; Leinweber et al., 2012), however, do not take into account that the time-of-flight measurements themselves are known to be subject to offsets (Georgescu et al., 2012). T_1 and T_2 values differ systematically from the respective true values; and deviations depend on instrument mode as we will show later.

Accurate calibration of FGM gains and zero level offsets with EDI $|\boldsymbol{B}|$ measurements is only possible if electron time-of-flight offsets are previously determined and corrected for. In this paper, we show how this can be achieved by using Cluster data from the EDI and FGM (Balogh et al., 2001) and present the possible schemes of interinstrument calibration. We further examine the characteristics of the FGM spin-axis offsets in the low field region and demonstrate the importance of accurate calibration when determining magnetic field gradient using multipoint Cluster measurements.

2   Method of analysis: interinstrument calibration

Since our main interest is to determine the spin-axis offset component, we use the flux-gate spin reference (FSR) coordinates, where Z points along the spin axis and X and Y are the spin-plane components. Here we assume that except for some residual spin-axis offset, $\Delta B_{Z\,\mathrm{fgm}}$, all the calibration parameters have been accurately determined. Since the time-of-flight data provide the magnitude of the magnetic field, B_{edi}, from Eq. (2) we use the spin-plane components of the FGM data to deduce the spin-axis component, $B_{Z\,\mathrm{edi}}$:

$$B_{Z\,\mathrm{edi}}{}^2 = B_{\mathrm{edi}}^2 - B_{X\,\mathrm{fgm}}^2 - B_{Y\,\mathrm{fgm}}{}^2. \tag{3}$$

The spin-axis offset, $\Delta B_{Z\,\mathrm{fgm}}$, can then be obtained from

$$|B_{Z\,\mathrm{edi}}| = |B_{Z\,\mathrm{fgm}} + \Delta B_{Z\,\mathrm{fgm}}|, \tag{4}$$

if the spin-axis component of the magnetic field deduced from the EDI time-of-flight measurements and the spin-plane component of the FGM magnetic field are obtained with sufficient quality.

For determining B_{edi}, we have simply used all the time-of-flight data from the two gun-detector units, GDU1 and GDU2, without identifying the pairs of long and short time of flight to obtain the gyration time from their average, such as described before, based on an assumption that the usage of large numbers of data of both times of flight is equivalent to effectively averaging the measurement pairs. We use the high resolution FGM data (22.4 Hz for normal mode) and match them with the nearest neighbor to the EDI time-of-flight data. The EDI time-of-flight data are irregularly spaced data with a smallest interval of 16 ms, but are sparse compared to the FGM data, since detection of the returning electron beam is required.

In this study we use Cluster data from July to October 2003 and from July to October 2006, when the apogee of Cluster orbit is at night side. The interspacecraft distance was on the order of 200 km in 2003 and 10 000 km in 2006. During these summer seasons, when Cluster stayed in the magnetosphere and no solar-wind data were available, it is of particular interest to determine the FGM offset using the EDI measurements since the Hedgecock method (Hedgecock, 1975) cannot be applied. Furthermore, one of the scientific interests in the tail region is the magnetic reconnection process, for which the magnetic field component normal to the current sheet, corresponding to the spin-axis component, is key in detecting the process. Hence an accurate determination of the spin-axis component is crucial in this region.

Since both the FGM and EDI instruments are designed to obtain optimized field measurements in different regions of space, the digital resolution of the measurements change. In this study we analyzed magnetic field data with magnitudes less than 600 nT. For FGM, within our region of interest, this corresponds to 3 different ranges, i.e., digital resolutions, changing from 7.813 to 0.125 nT depending on the

field magnitude as will be discussed later. The EDI time-of-flight measurement, however, is operated by tracking electron beams that are amplitude-modulated with a pseudonoise (PN) code, with a certain code period, T_{PN}, or alternatively represented as the code repetition frequency (CRF), which is $1/T_{\mathrm{PN}}$. The PN code consists of either a 15-chip or 127-chip code with different code chip lengths, T_{chip}. The accuracy of EDI measurements depend on the T_{chip}, and therefore T_{PN} or CRF, which is usually given in unit of kilohertz. T_{PN} varies between 30 μs and 2 ms for the data set used in this study. The time resolution of EDI is defined by the shift-clock period, which is the shift in the PN code to track small time-of-flight variations, that varies from 1.907 to 0.119 μs depending on the magnetic field; see more details in Georgescu et al. (2006). Further details about these parameters and the EDI operation schemes are given by Vaith et al. (1998) and Paschmann et al. (2001). Here we call the different measurement settings of the EDI "CRF mode" for convenience. As will be discussed later in more detail, these different resolutions/modes need to be taken into account when data are calibrated.

Figure 1 shows FGM and EDI magnetic field magnitude data during a quiet interval of about 3 min from Cluster 3 for different FGM calibration schemes. The FGM data shown in the three left panels a–c use the orbit calibration file provided for the Cluster Active Archive (CAA) data set (Gloag et al., 2006), the three middle panels d–e use the daily calibration file (Fornaçon et al., 2011) used for the Cluster Prime Parameter (PP) and Summary Parameter (SP) data set in the Cluster Science Data System (CSDS), and the three right panels g–i use the fine-tuned calibration file using the daily calibration file as an input. Figure 1a shows the magnetic field magnitude data estimated from EDI and FGM, in which the latter data are time-matched data to EDI using the nearest neighbor data selected from the high-time resolution (22.4 Hz) data shown in Fig. 1b. Although the example shown here is from a period when the numbers of the returning beams are quite evenly distributed all the time, EDI data depend on the availability of the returning beam and can be also sparse in time. Hence it is essential to compare EDI data with the time-matched FGM data. Figure 1c shows 1 Hz averaged data for both FGM and EDI. It can be seen that both data sets have a small standard deviation (about 0.1 nT) during this interval and there exists a clear difference between FGM and EDI magnitudes of about 0.5 nT. The same comparison has been done for data calibrated using the daily calibration file (Fig. 1d–f). The 22.4 Hz data have a slightly larger standard deviation compared to the CAA data, but the difference between EDI and FGM is smaller, about 0.14 nT. The relatively large scatter of the 1 Hz data (Fig. 1f) comes from the spin-tone, which can be more clearly seen in the 22 Hz data (Fig. 1e). Data shown in Fig. 1g–i are using the same daily calibration file, as was used for data in Fig. 1d–f, as input and then further refined the calibration file to reduce the spin tone. This additional procedure, however, has little effect on the average

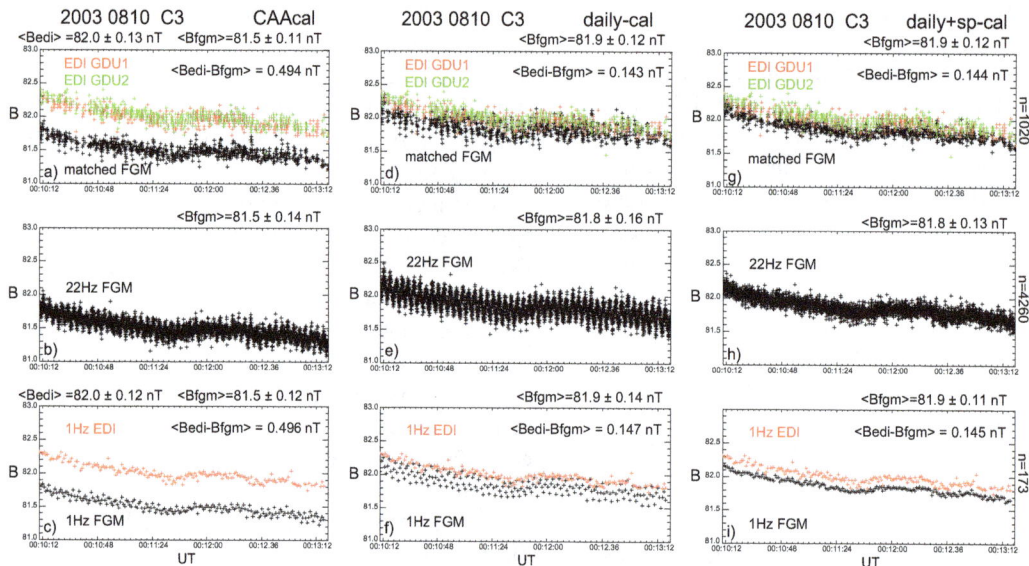

Fig. 1. FGM and EDI magnetic field magnitude data during a quiet interval from Cluster 3 using FGM data with different calibration schemes: the orbit calibration, method used for CAA data set (**a–c**); daily calibration method used for CSDS data set (**d–f**); and refined calibration applied to daily calibration input (**g–i**). The upper three panels (**a, d, g**) show high-resolution EDI and time-matched FGM 22.4 Hz data, the middle three panels (**b, e, h**) show the 22.4 Hz FGM data, and the lower three panels (**c, f, i**) show 1 Hz averaged data for both FGM and EDI.

FGM–EDI difference as can be seen in the numbers obtained for the high-resolution data (Fig. 1d, g) and for the 1 Hz data (Fig. 1f, i). Note that for following discussions on offset calibration procedure we use the daily calibration file, prepared since the Cluster launch by the Technical University of Braunschweig Cluster Co-I team. That is, we use the same data set as shown in Fig. 1d–f. It should be therefore noted that when we write "spin-axis offset" in this paper, we are not speaking about an offset from the raw data as given in the Eq. (1), but about a remaining offset correction from an already in-flight calibrated data set.

Figure 2a shows the number of EDI time-of-flight data points from Cluster 1 in August 2003, when corresponding FGM data were available, binned by the magnitude of the field, B_{fgm}. The size of the bins is 16 nT. The number of points are grouped by different CRF modes. Note that these different CRF modes generally correspond to data from different field magnitude regions, which are marked as R1–R6 next to the legend. More details of the meaning of these different magnitude regions, R1–R6, and the EDI measurement resolution are explained later (Fig. 3). It can be seen in the histogram that for smaller field regions, in particular, the EDI observations have been made with several different CRF modes. Figure 2b shows the differences between the $|B_{Z\,edi}|$ and $|B_{Z\,fgm}|$. The bin averages (dotted line) and medians (solid line) are also depicted in the figure. When both B_Z values are positive, it corresponds to the spin-axis offset. It can be seen that the values are widely scattered, particularly with increasing magnitude of the field. Also, instead of seeing a constant offset value of FGM, the difference is increas-

ing with magnetic field magnitude but not monotonically. As will be discussed below, these variable differences can be due to (i) the effects of different magnetic field angles relative to the spin axis, (ii) the different CRF modes of the instruments and different offsets, and (iii) the effects from variable calibration parameters other than the offsets considered here. In the following we mainly examine the first two effects when obtaining the spin-axis offset of FGM and further discuss the possible effect due to (iii) based on the obtained offsets.

Since we are interested in the spin-axis offset, it is important to use measurements with sufficient magnitude of the spin-axis direction. As mentioned before, a meaningful comparison of the two spin-axis components using Eq. (4) can only be performed when both have the same (positive, for majority of the data used in this study) sign even when the possible offset values are subtracted, because Eq. (3) does not provide the sign of the magnetic field along the spin axis. The unknown sign of the $B_{Z\,edi}$ will lead to miscalculation when the spin-axis offset effect changes the sign of the spin-axis component. This corresponds to cases when the expected spin-axis offset becomes significant compared to the spin-axis component of the magnetic field. Considering that we use an already calibrated data set as an input, a typical offset value is expected to be small, i.e., less than a couple of nanoteslas. For the Cluster data we are examining in this paper, such offset can be more than 10 % of the field magnitude. Hence we need to take into account only data when $|\cos b| \equiv |B_{Z\,fgm}/B_{fgm}|$ is sufficiently large so that the offset subtraction will not make any difference in the change

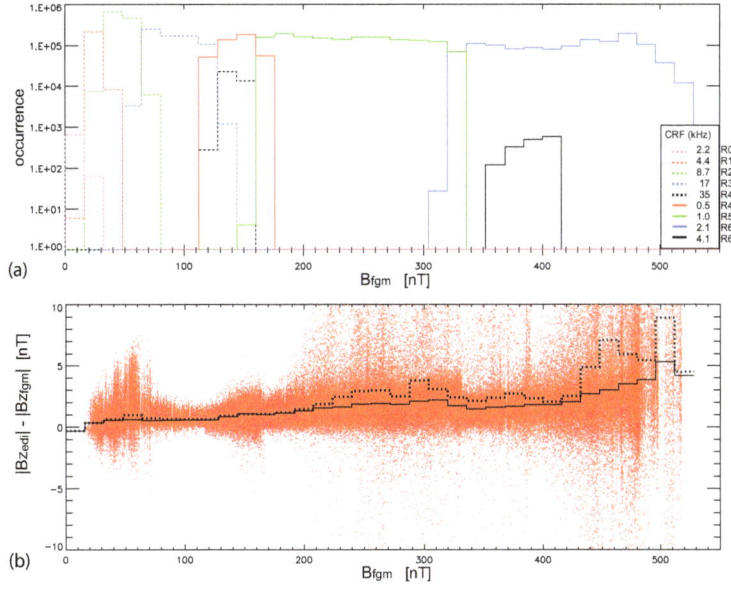

Fig. 2. (a) Number of points for all available Cluster 1 EDI data in August 2003, binned by the magnitude of the field B_{fgm}. The size of the bins is 16 nT. The number of points are grouped for different CRF modes (see details in text). **(b)** Differences between $|B_{Z\,\mathrm{edi}}|$ and $|B_{Z\,\mathrm{fgm}}|$ for the same data set. The solid line shows the median and the dotted line shows the average of the data within each bin. Here every 20th point from the entire data set shown in **(a)** is plotted.

of sign. As we will show later, $|\cos b| \geq 0.4$ would typically work for the analysis.

In this study we consider a time-of-flight offset of EDI, ΔT_{edi}, which is expected to have different values for different CRF modes. For simplicity we assume the same offset value for the time-of-flight measurements from GDU1 and GDU2. That is, when calculating the magnetic field from EDI measurement, we use

$$B_{\mathrm{edi}} = \frac{2\pi m_{\mathrm{e}}}{e(T_{\mathrm{edi}} + \Delta T_{\mathrm{edi}})}, \tag{5}$$

to determine both ΔT_{edi} and $\Delta B_{Z\,\mathrm{fgm}}$ from the data, instead of Eq. (2).

Significance of the EDI and FGM offsets varies for different field magnitudes as is shown in Fig. 3. The four solid curves in Fig. 3a show the effective spin-axis offset value caused by an EDI time-of-flight offset, $\Delta T_{\mathrm{edi}} = 0.5\,\mu$s, that will appear when the EDI and FGM measurements are compared, such as in Fig. 2. They are plotted for different angles of the magnetic field, $\cos b$. Here, the effective EDI magnetic field measurement resolution based on the digital resolution of the EDI measurements discussed by Georgescu et al. (2006) is also given as a dashed curve for the different magnetic field regions, R1–R6, as indicated at the bottom of Fig. 3b. The borders of R0–R6 are shown with the vertical dotted line, which corresponds to 16, 32, 64, 128, 164, and 326 nT. The horizontal brown line indicates the 0.5 nT level, as a typical number for the spin-axis offset of FGM. In a similar way, we plotted the effective time-of-flight offsets caused by a FGM spin-axis offset of $\Delta B_{Z\,\mathrm{fgm}} = 0.5$ nT. The

dashed lines indicate the same EDI digital resolution of the time-of-flight measurement as given in Fig. 3a. The horizontal brown line shows $0.5\,\mu$s as a typical number for the time-of-flight offset of EDI. It can be immediately seen that the time-of-flight offset will have no effect in the small field region regardless of the angle to the magnetic field (brown line located above the curves in Fig. 3). Therefore, these curves show that the different angle of the fields as well as the time-of-flight offset can easily cause the large scatter of points in Fig. 2b. One can also conclude that for determining the offset in B_Z in a given field magnitude, it would be most effective to use data from large $\cos b$, since the relative importance of the EDI time-of-flight offset would be smallest. Furthermore, in the low-field region, a time-of-flight offset of about $0.5\,\mu$s will have only negligible effect in the spin-axis component of the magnetic field, which is a value below the instrument resolution. In the high-field region, however, a 0.5 nT spin-axis offset is a negligible value in the time-of-flight data and comparable to the resolution of the EDI measurement. It is also important to note that when we determine ΔT_{edi}, it is most efficient to use data with low $\cos b$, i.e., when the field direction is mainly along the spin-plane direction. Vice versa, $\Delta B_{Z\,\mathrm{fgm}}$ should be determined for large $\cos b$ as mentioned before. Due to these variable effects over the field magnitude, we need to consider different approaches for different magnetic field magnitudes depending on the importance of the offset. In Sect. 3 we demonstrate an example of a calibration in which all the different offsets are obtained using a large number of points and for different magnetic field magnitude

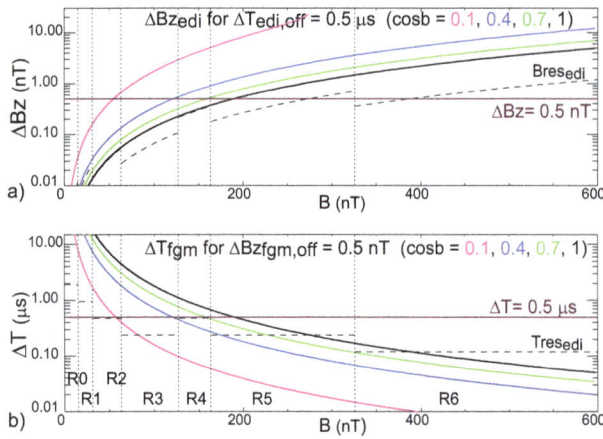

Fig. 3. (a) The effective spin-axis offset value caused by an EDI time-of-flight offset, $\Delta T_{\text{edi}} = 0.5\,\mu\text{s}$, that will appear when the EDI and FGM measurements are compared, plotted for selected angles of the magnetic field, $\cos b$. The dashed lines show the resolution of the EDI magnetic field measurement. The horizontal brown line shows 0.5 nT level, which represents a typical number for the spin-axis offset of FGM. **(b)** The effective time-of-flight offsets caused by a FGM spin-axis offset, $\Delta B_{Z\text{fgm}} = 0.5\,\text{nT}$, plotted for selected values of $\cos b$. The dashed lines indicate the EDI digital resolution of the time-of-flight measurement. The horizontal brown line shows the $0.5\,\mu\text{s}$ level, which represents a typical number for the time-of-flight offset of EDI. The vertical dotted lines indicate the border of different EDI measurement settings, R0–R6. See text for further details.

regions. We also specifically use data from the low-field region to examine the possibility for estimating an offset with a small number of samples.

3   Example of interinstrument calibration

Figure 4a shows the number of EDI measurements from Cluster 1 in August 2003 in the same format as in Fig. 2a, but only for $\cos b > 0.7$. As discussed before, this condition angle allows one to select data when the relative importance of the B_Z offset is higher than the possible time-of flight offset, and additionally to fulfill the condition of the same positive sign of FGM and EDI in spin-axis components. As discussed before, EDI is operated with different CRF modes in different magnetic field regions. For this field angle, data were available only between the regions R2 and R6 (see Fig. 3 for definition of the regions). The FGM range changes at 256 nT, which is a value within R5. Depending on the importance of the offset, we determined ΔT_{edi} or $\Delta B_{Z\text{fgm}}$ in the following way.

- Low-field region (R1–R3), when the effect of $\Delta B_{Z\text{fgm}}$ is important: $\Delta B_{Z\text{fgm}}$ is first determined for $\cos b > 0.7$. ΔT_{edi} is then determined using data obtained for R1–R3 separately.

- Mid-field region (R4), when both effects from EDI time of offset and FGM offset in spin-axis component are comparable: ΔT_{edi} is determined using $\Delta B_{Z\text{fgm}}$ determined for R2–R3. Since there are two different CRF modes used for EDI measurements in this region, we calculated the time-of-flight offsets for each CRF mode separately.

- Mid-field region (R5), when both effects are comparable and FGM range changes within the same EDI CRF mode: same method as R4 is used for data with $B_{\text{fgm}} < 256\,\text{nT}$. Determine $\Delta B_{Z\text{fgm}}$ for $\cos b > 0.7$ using ΔT_{edi} determined for R5 data with $B_{\text{fgm}} < 256\,\text{nT}$.

- High-field region (R6), when the effect of ΔT_{edi} is important: determine ΔT_{edi} taking into account the FGM offset determined for R5. Since the effect of spin-axis offset is not important regardless of $\cos b$ all data are used.

Figure 4b shows the FGM and EDI differences of original calibrated data as shown in Fig. 2 except for $\cos b > 0.7$. The bin averages and median are shown as solid lines, although the difference between the two are hardly recognizable in this plot.

The average profile in Fig. 4b shows some jumps coinciding with CRF-mode change and more monotonic increase in the high-field region within the same CRF mode as expected in the curve shown in Fig. 3a. Figure 4c shows the results of the calibration procedure for August 2003. The points are the differences between the offset-corrected FGM and EDI data. The lines again show the bin average and the median of the differences of the offset-corrected FGM and EDI data. Here again the differences between the two lines are hardly seen. It can be seen that the bin average (or median) runs at almost the zero level except for some fluctuations of $\leq 0.1\,\text{nT}$ in the higher field region. The nearly zero level of the bin's average (or median) profile suggests that the spin-axis component difference between EDI and FGM was well explained due to the spin-axis offset of FGM and time-of-flight offset of EDI.

Table 1 provides the monthly average results of the different offsets between July and October 2003 for Cluster 1: $\Delta B_{Z\text{fgm}}$ for low field range ($< 256\,\text{nT}$) and high field range ($> 256\,\text{nT}$) and ΔT_{edi} for different CRF modes, corresponding to R1–R6 (as given in the legend in Fig. 4a). Although we used all the available data without selecting, for example, quiet time data, it can be seen that $\Delta B_{Z\text{fgm}}$ determined from the low field region (R2–R3), which corresponds to $B \sim 32$–128 nT, stays at about 0.4–0.6 nT with a relatively small standard deviation. The standard deviation is quite large for the FGM offset at the high-field region (R5), while the values stay at a similar value to the low-field region within 0.1 nT during the four months. ΔT_{edi}, however, is stably obtained only in the field region larger than about 128 nT (R4–R6), while the time-of-flight offsets could be poorly

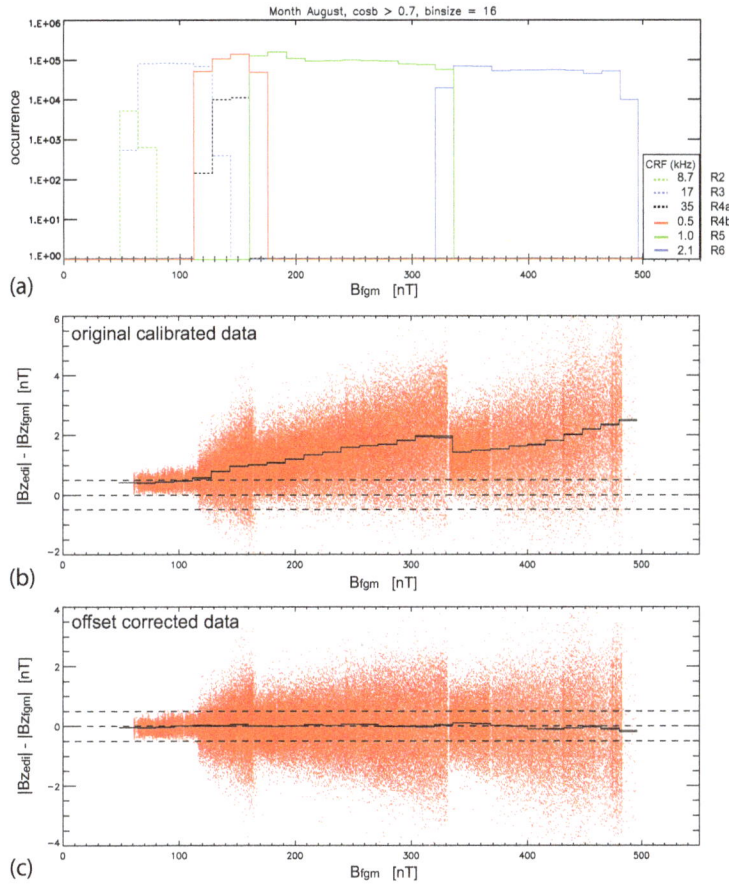

Fig. 4. (a) Number of points for Cluster 1 EDI data in August 2003 binned by the magnitude of the field B_{fgm} as in Fig. 2 except for $\cos b > 0.7$. Difference between spin-axis component EDI and FGM fields for $\cos b > 0.7$ (**b**) for the original calibrated data and (**c**) for the offset-corrected data. Bin average and median for the original and offset-corrected data are shown in each panel. Note that both curves are nearly identical and therefore their differences can hardly be seen. As in Fig. 2, every 20th point from the corresponding data sets given in panel (**a**) is plotted. Dashed lines indicate $-0.5, 0.0$, and 0.5 nT levels.

determined with a large standard deviation only in the low-field region. This behavior can be understood with the characteristics of resolution of the EDI measurements (Fig. 3), i.e., finer B resolution of EDI for the smaller field region, and smaller (larger) effect of ΔT_{edi} in a smaller (larger) field region relative to the effect of $\Delta B_{Z fgm}$. Except for the poorly determined ΔT_{edi} (R1–R3), the values shown in Table 1 were used to calculate the points in Fig. 4c.

We have performed the same procedure for every orbit in August 2003 for Cluster 1 and the results are shown in Fig. 5. $\Delta B_{Z fgm}$ for the low field (< 256 nT) and high field (> 256 nT) and their corresponding numbers of points are shown in Fig. 5a and b, respectively. As described before, low-field data points are from EDI CRF modes for R2 and R3 (see Fig. 4b), while high-field data points are from EDI CRF modes for R5. ΔT_{edi} for each orbit in R2, R5, and R6 and the corresponding numbers of data points are shown in Fig. 5c and d. Note that measurements in low field regions did not take place in every orbit in this month and therefore

Table 1. Average offsets determined for different modes/ranges for Cluster 1.

Parameters	July 2003	August 2003	September 2003	October 2003
$\Delta B_{Z fgm,l}$ [nT]	0.51 ± 0.15	0.46 ± 0.16	0.64 ± 0.17	0.57 ± 0.17
$\Delta B_{Z fgm,h}$ [nT]	0.40 ± 0.99	0.41 ± 0.99	0.57 ± 1.04	1.00 ± 0.19
$\Delta T_{edi,R1}$ [μs]	2.92 ± 5.77	1.90 ± 4.77	2.87 ± 6.42	1.92 ± 4.98
$\Delta T_{edi,R2}$ [μs]	1.81 ± 2.42	1.60 ± 1.96	1.81 ± 1.89	1.85 ± 2.40
$\Delta T_{edi,R3}$ [μs]	0.38 ± 1.15	1.03 ± 1.27	0.70 ± 0.89	1.20 ± 1.04
$\Delta T_{edi,R4a}$ [μs]	0.21 ± 0.25	0.15 ± 0.20	0.19 ± 0.16	0.05 ± 0.23
$\Delta T_{edi,R4b}$ [μs]	0.65 ± 0.97	0.63 ± 0.96	0.50 ± 0.97	0.48 ± 1.00
$\Delta T_{edi,R5}$ [μs]	0.55 ± 0.42	0.55 ± 0.43	0.59 ± 0.46	0.57 ± 0.50
$\Delta T_{edi,R6}$ [μs]	0.28 ± 0.19	0.26 ± 0.19	0.27 ± 0.19	0.26 ± 0.20

values using those data points can only be seen every second or fourth orbit. It can be seen that $\Delta B_{Z fgm}$ obtained from the low-field region is relatively stable compared to that obtained from the high-field region. As for ΔT_{edi}, the values of R6 are most stable among the three offsets. ΔT_{edi} is larger for R2

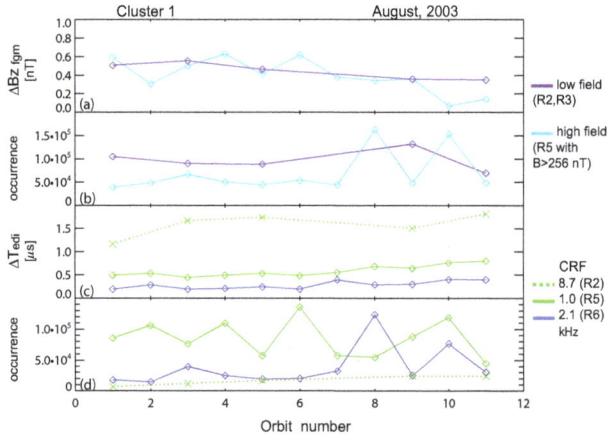

Fig. 5. (a) $\Delta B_{Z\text{fgm}}$ determined for every orbit in the low field (< 256 nT) and high field (> 256 nT) and **(b)** the corresponding numbers of data points from Cluster 1 in August 2003. **(c)** ΔT_{edi} for each orbit in R2, R5, and R6 and **(d)** corresponding numbers of data points.

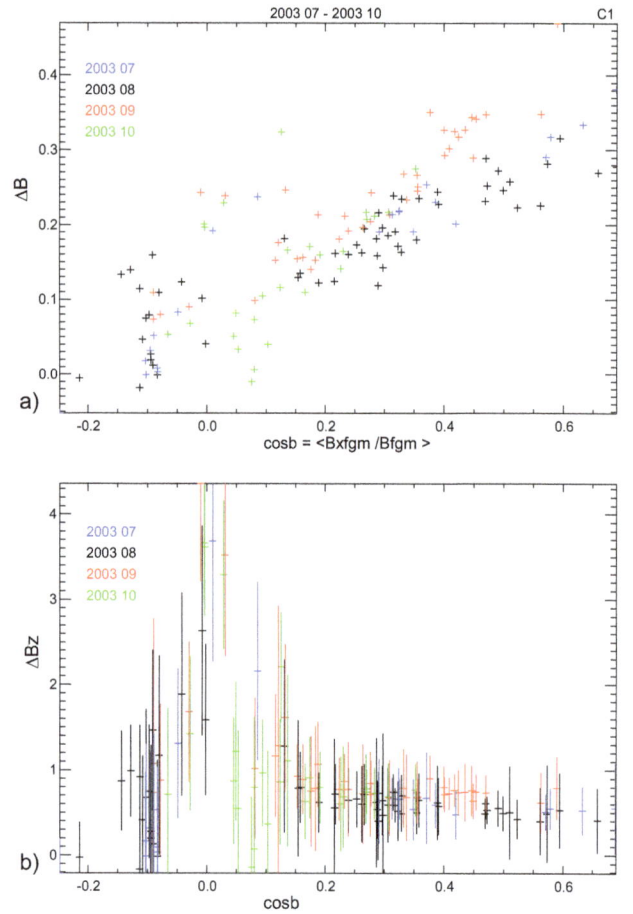

Fig. 6. (a) Average magnitude difference, $\Delta B \equiv B_{\text{edi}} - B_{\text{fgm}}$, and **(b)** average difference in the spin-axis components, $\Delta B_Z \equiv |B_{Z\text{edi}}| - |B_{Z\text{fgm}}|$, plotted vs. the field angle, $\cos b$, obtained using quiet, low field (30–60 nT), short time interval (7 min) data sets in July–October 2003. The vertical bars in **(b)** show the standard deviation.

compared to R5 and R6; yet the effect from $\Delta B_{Z\text{fgm}}$ can still be expected to dominate in R2 for these values (see Fig. 3).

The spin-axis direction, which is approximately the Z direction in geocentric solar ecliptic (GSE) coordinates, is closely aligned to the normal component of the current sheet in the magnetotail, where the apogee is located for Cluster between July and October. This normal component drops to zero when magnetic reconnection occurs, which is an important science target in magnetospheric missions such as Cluster as well as for the upcoming Magnetospheric Multiscale (MMS) mission. Therefore, to detect the process accurately, it is required that the spin-axis offset be corrected. It is therefore desirable that the calibration will take place close to such target intervals, that is, in a relatively small field region when the disturbance of the field is small. Below we use Cluster data for a short interval, i.e., several minutes, in a small field region, such as the example shown in Fig. 1, to examine the effect of the spin-axis component offset in the difference between FGM and EDI magnetic fields. We searched for quiet and constant field intervals using data between July and October 2003 in small field region (R2), corresponding to the magnetic field between about 30 and 60 nT. A quiet field's short time interval is defined as an interval with standard deviation less than 0.1 nT. We chose a time period of 7 min. We obtained 579 such intervals for C1 during the four months. Figure 6a and b show the magnitude difference, $\Delta B \equiv B_{\text{edi}} - B_{\text{fgm}}$, and difference in the spin-axis components, $\Delta B_Z \equiv |B_{Z\text{edi}}| - |B_{Z\text{fgm}}|$, plotted vs. the field angle, $\cos b$. On average, the magnitude difference is small when the magnetic field is nearly aligned to the spin plane (small $|\cos b|$) within an error of about 0.1 nT and justifies our assumption that the main discrepancy between the two data sets are attributed to the spin-axis offset. When the

$|\cos b|$ is small, $|\cos b| < 0.1$, it is not possible to obtain the correct sign of ΔB_Z. In such cases the comparison between the spin-axis components will contain large errors. That is, we may obtain the sums of the two measurements instead of differences, meaning that the ΔB_Z will rather become twice the average of the spin-axis component value ($2B \cos b$). If we assume, for example, that such errors happen in about half of the cases we can expect an average to be estimated as $B \cos b$. For the field magnitude in this data set, i.e., $B = 30$–60 nT, a "wrongly" estimated ΔB_Z of ≤ 3–6 nT can be expected for $\cos b \leq 0.1$, which was in fact the case as shown in Fig. 6b. However, the spin-axis offsets are more stable for a larger $\cos b$, i.e., $\cos b \geq 0.4$, indicating the importance of preselection of the angle of the field when determining the spin-axis offset.

The essential advantage of multipoint measurements such as with Cluster is the ability to determine spatial gradients. We finally examine the possible effect of the offset

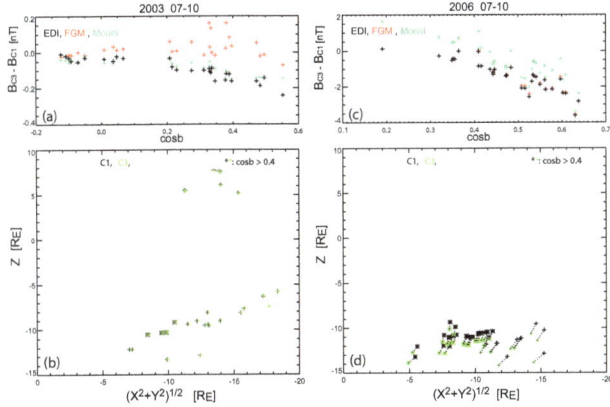

Fig. 7. Average magnetic field differences between C1 and C3 for B_{edi} (black cross) and B_{fgm} (red cross), and a model magnetic field (green cross) during quiet time intervals (standard deviation of $B_{edi} < 0.07$ nT for 5 min interval) plotted vs. $cos b$ for data from **(a)** July to October in 2003, when the interspacecraft distance was about 200 km, and from **(c)** July to October in 2006, when the interspacecraft distance was about 10 000 km. The location of the spacecraft in GSM (geocentric solar magnetospheric) coordinates during these two sets of intervals are shown in **(b)** and **(d)**, respectively.

calibration by comparing the magnetic field gradients (differences between two spacecraft) for B_{edi}, B_{fgm}, and an empirical magnetic field, i.e., combined IGRF and Tsyganenko 89, Kp = 2, as shown in Fig. 7. Here we select again quiet time intervals, when standard deviation of $B_{edi} < 0.07$ nT for 5 min intervals and when data from both Cluster 1 and 3 are available. Cluster data are used from an interval between July and October in 2003, when the interspacecraft distance was about 200 km, and between July and October in 2006, when the interspacecraft distance was about 10 000 km. Figure 7a shows the spacecraft differences, $\Delta B_{edi,C1-C3}$ (black), $\Delta B_{fgm,C1-C3}$ (red), and model (green), plotted again over $cos b$ (of Cluster 1) observed at locations shown in Fig. 7c for the events in 2003.

The model provides a reference value of the magnetic field profile and is constructed based on fitting a number of previous satellite data. Therefore we can expect that the model represents some averages of different randomly distributed "offsets" among the different previous measurements providing an empirical value of the field. The $\Delta B_{edi,C1-C3}$ and model generally agree well. This suggests that $\Delta B_{edi,C1-C3}$ provides closer values to an empirical value of the magnetic field. $\Delta B_{fgm,C1-C3}$ shows a smaller difference in the small $cos b$ region, which corresponds to the magnetic field direction where the spin-axis component does not play a role, suggesting that the spin-plane components are well calibrated. The differences, however, become larger for larger $cos b$ indicating that the effect of the spin-axis offset is apparent and causing these larger differences. Figure 7b and d show the results of the same analysis performed for the data in 2006 for comparison. In contrast to 2003, the gradients obtained from

the two measurements show similar values, while the model values deviate from these two. The interspacecraft distance of 200 km is small enough that the effect of the offset calibration exceeds the magnetic field gradient, while such offset determination plays no difference for the interspacecraft distance of 10 000 km. Hence, depending on the interest of the gradient scales it will become essential to perform special offset calibrations when determining the gradient of the magnetic field.

4   Discussion

Based on a simple comparison between the magnetic field of FGM and the magnetic field deduced from the time of flight of the EDI measurements, we have shown that the remaining spin-axis offset of FGM data can be well determined from the calibrated data set by selecting the appropriate interval, by taking into account the measurement conditions such as the angle of the magnetic field relative to the spin-plane, magnetic field magnitude, and by also considering the effect of the time-of-flight offset of the EDI measurement. While the effect of the time-of-flight offset was unimportant in determining the spin-axis offset in the low field region, it was the major source of the discrepancy between the two data sets in the large field region. Once the effects of these two offsets are taken into account, the difference between the two measurements are reduced to be well below the 0.1 nT level. Note that there is a tendency of somewhat larger fluctuations superposed with a negative trend for the larger field region (R6) in Fig. 4. This might suggest that some additional FGM gain correction needs to be considered. The current offset correction does not take into account any gain correction. If there is a gain error, it should appear as a linear trend if all the other calibration parameters are perfectly determined. Such gain error curve, however, is difficult to differentiate from the EDI time-of-flight profile particularly for a low-resolution measurement. Therefore each EDI range may show a different resultant curve and may not appear a continuous line in Fig. 4c even if there is a gain error. In the low field region, we cannot see any systematic trend, for example. If we take the ~ -0.1 nT deviation in the R6 region (covering an about 200 nT-wide region), as an observed number, it will correspond to a linear gain correction of 0.0005. Such change in the gain may likely happen due to the change in the temperature. Indeed if we use the ground-calibration result from one of the Cluster ground sensors, i.e., 0.00004 K^{-1} (Othmer et al., 2000), this corresponds to a gain drift for a temperature change of about 12°, which would not be an unrealistic variation within an orbit. For an accurate determination of the gain from these comparisons, however, only a statistical approach is possible because in this high-field region. EDI can measure the field only at a resolution of about 1 nT, while the effects expected from gain errors would be less than at a

0.1 nT scale, which is also below the FGM resolution in this range and therefore fluctuations are unavoidable.

While we demonstrate that the simple comparison is overall working, particularly for the spin-axis determination in low field regions, once we are interested to determine also other parameters, such as time-of-flight offsets throughout the EDI CRF modes or FGM offsets and gain factors for high field ranges, further investigations would be necessary. For example, our simplified approach of pre-selecting the data set based on specific conditions in angle and magnitude of the field, as discussed in Sect. 3, limits the number of useful data. Instead one may consider to use all the data from different field magnitudes (and therefore with different EDI CRF modes) and try to determine the EDI and FGM offsets at once by applying appropriate weighting factors that depend on the contribution of the EDI and FGM offsets in the measurement, and by minimizing the differences between the two measurements. Furthermore determining the EDI time-of-flight offset for the two GDUs, separately, may be also important particularly for mid- and high-field regions.

In this study we only used the time-of-flight data of the EDI measurements to compare with the FGM measurements. Another useful approach is to use the direction of the EDI electron beam, $\mathbf{u}_{edi}$, which should be perpendicular to the ambient magnetic field, and use the condition of $\mathbf{u}_{edi} \cdot (\mathbf{B}_{fgm} + \mathbf{O}_{fgm}) = 0$, to determine the offset of the FGM measurement, $\mathbf{O}_{fgm}$. A combination of these two methods will further improve the accuracy of the offset determination.

5 Conclusions

We have shown that the concept of determining the spin-axis offset of a flux-gate magnetometer (FGM) using absolute field magnitude data determined from the electron gyration time data of the electron drift instrument (EDI) works best when the magnetic field magnitude is small; i.e., less than about 128 nT corresponding to the EDI modes for the low field, so that the EDI time-of-flight offset is negligible, and when the spin-axis component becomes the major component ($\cos b > 0.7$). A remaining spin-axis offset of about $0.4 \sim 0.6$ nT was observed for Cluster 1 between July and October 2003, which is important for studies using the magnetic field component normal to the current sheet in the central plasma sheet such as magnetotail reconnection or thin-current sheet dynamics or particle trajectories near the center of the current sheet.

When the effect of time-of-flight offset from EDI is taken into account, it is shown that data from higher fields can be also used for calibration. It is shown that additional determination of the gain factor of the FGM instrument would most likely also also possible.

The EDI–FGM comparison method is of particular interest for the observations, when no solar-wind data are available for calibration. It will play an essential role for accu-

rate determination of the small normal component (and its reversals) in the current sheet required for studying magnetic reconnection, which is the main objective of NASA's Magnetospheric Multiscale (MMS) mission.

Acknowledgements. This paper is dedicated to the memory of E. Georgescu, Max Planck Institute for Solar System Research, Germany. E. Georgescu initially started the EDI–FGM comparison work. Without her inspiring comments and suggestions this study would not have been done. We thank P. Daly, H. Eichelberger, G. Laky, and the CAA team for the valuable suggestions/comments and supporting the data analysis. This research was partly supported by the Austrian Science Fund FWF I429-N16 and I23862-N16. K.-H. Fornaçon and K.-H. Glassmeier were financially supported through grants 50OC1102 and 50OC1001 by the German Bundesministerium für Wirtschaft und Technologie and the Deutsches Zentrum für Luft- und Raumfahrt.

Edited by: H. Laakso

References

Acuña, M. H.: Space-based magnetometers, Rev. Sci. Instrum., 73, 3717–3736, doi:10.1063/1.1510570, 2002.

Auster, H. U., Fornacon, K. H., Georgescu, E., Glassmeier, K. H., and Motschmann, U.: Calibration of flux-gate magnetometers using relative motion, Meas. Sci. Technol., 13, 1124–1131, doi:10.1088/0957-0233/13/7/321, 2002.

Auster, H. U., Glassmeier, K. H., Magnes, W., Aydogar, O., Baumjohann, W., Constantinescu, D., Fischer, D., Fornacon, K. H., Georgescu, E., Harvey, P., Hillenmaier, O., Kroth, R., Ludlam, M., Narita, Y., Nakamura, R., Okrafka, K., Plaschke, F., Richter, I., Schwarzl, H., Stoll, B., Valavanoglou, A., and Wiedemann, M.: The THEMIS Fluxgate Magnetometer, Space Sci. Rev., 141, 235–264, doi:10.1007/s11214-008-9365-9, 2008.

Balogh, A.: Planetary magnetic field measurements: Missions and instrumentation, Space Sci. Rev., 152, 23–97, doi:10.1007/s11214-010-9643-1, 2010.

Balogh, A., Carr, C. M., Acuña, M. H., Dunlop, M. W., Beek, T. J., Brown, P., Fornacon, K.-H., Georgescu, E., Glassmeier, K.-H., Harris, J., Musmann, G., Oddy, T., and Schwingenschuh, K.: The Cluster Magnetic Field Investigation: overview of in-flight performance and initial results, Ann. Geophys., 19, 1207–1217, doi:10.5194/angeo-19-1207-2001, 2001.

Dougherty, M. K., Kellock, S., Southwood, D. J., Balogh, A., Smith, E. J., Tsurutani, B. T., Gerlach, B., Glassmeier, K.-H., Gliem, F., Russell, C. T., Erdos, G., Neubauer, F. M., and Cowley, S. W. H.: The Cassini magnetic field investigation, Space Sci. Rev., 114, 331–383, doi:10.1007/s11214-004-1432-2, 2004.

Fornaçon, K.-H., Georgescu, E., Kempen, R., and Constantinescu, D.: Fluxgate magnetometer data processing for Cluster, Tech. Rep., Institut für Geophysik und extraterrestrische Physik, Technischen Universität Braunschweig, 38106 Braunschweig, Germany, 2011.

Georgescu, E., Vaith, H., Fornaçon, K.-H., Auster, U., Balogh, A., Carr, C., Chutter, M., Dunlop, M., Foerster, M., Glassmeier, K.-H., Gloag, J., Paschmann, G., Quinn, J., and Torbert, R.: Use of EDI time-of-flight data for FGM calibration check on

CLUSTER, in: Cluster and Double Star Symposium, Vol. 598 of ESA Special Publication, ESTEC, Noordwijk, the Netherlands, 2006.

Georgescu, E., Puhl-Quinn, P., and Vaith, H. and Matsui, H.: Cross Calibration Report of the EDI Measurements in the Cluster Active Archive (CAA), http://caa.estec.esa.int/documents/CR/CAA_EST_CR_EDI_v14.pdf (last access: 29 July 2013), 2012.

Glassmeier, K.-H., Richter, I., Diedrich, A., Musmann, G., Auster, U., Motschmann, U., Balogh, A., Carr, C., Cupido, E., Coates, A., Rother, M., Schwingenschuh, K., Szegö, K., and Tsurutani, B.: RPC-MAG The Fluxgate Magnetometer in the ROSETTA Plasma Consortium, Space Sci. Rev., 128, 649–670, doi:10.1007/s11214-006-9114-x, 2007.

Gloag, J. M., Carr, C., Forte, B., and Lucek, E. A.: The status of Cluster FGM data submissions to the CAA, in: Cluster and Double Star Symposium, Vol. 598 of ESA Special Publication, ESTEC, Noordwijk, the Netherlands, 2006.

Hedgecock, P. C.: A correlation technique for magnetometer zero level determination, Space Sci. Instrum., 1, 83–90, 1975.

Kepko, E. L., Khurana, K. K., Kivelson, M. G., Elphic, R. C., and Russell, C. T.: Accurate determination of magnetic field gradients from four point vector measurements. I. Use of natural constraints on vector data obtained from a single spinning spacecraft, IEEE T. Magn., 32, 377–385, doi:10.1109/20.486522, 1996.

Leinweber, H. K., Russell, C. T., Torkar, K., Zhang, T. L., and Angelopoulos, V.: An advanced approach to finding magnetometer zero levels in the interplanetary magnetic field, Meas. Sci. Technol., 19, 055104, doi:10.1088/0957-0233/19/5/055104, 2008.

Leinweber, H. K., Russell, C. T., and Torkar, K.: In-flight calibration of the spin axis offset of a fluxgate magnetometer with an electron drift instrument, Meas. Sci. Technol., 23, 105003, doi:10.1088/0957-0233/23/10/105003, 2012.

Ludlam, M., Angelopoulos, V., Taylor, E., Snare, R. C., Means, J. D., Ge, Y. S., Narvaez, P., Auster, H. U., Le Contel, O., Larson, D., and Moreau, T.: The THEMIS magnetic cleanliness program, Space Sci. Rev., 141, 171–184, doi:10.1007/s11214-008-9423-3, 2008.

Othmer, C., Richter, I., and Fornaçon, K.-H.: Fluxgate magnetometer calibration for Cluster II, Tech. Rep., Institut für Geophysik und Meteologie, Technischen Universität Braunschweig, 38106 Braunschweig, Germany, 2000.

Paschmann, G., Melzner, F., Frenzel, R., Vaith, H., Parigger, P., Pagel, U., Bauer, O. H., Haerendel, G., Baumjohann, W., Scopke, N., Torbert, R. B., Briggs, B., Chan, J., Lynch, K., Morey, K., Quinn, J. M., Simpson, D., Young, C., McIlwain, C. E., Fillius, W., Kerr, S. S., Mahieu, R., and Whipple, E. C.: The Electron Drift Instrument for Cluster, Space Sci. Rev., 79, 233–269, doi:10.1023/A:1004917512774, 1997.

Paschmann, G., Sckopke, N., Vaith, H., Quinn, J. M., Bauer, O. H., Baumjohann, W., Fillius, W., Haerendel, G., Kerr, S. S., Kletzing, C. A., Lynch, K., McIlwain, C. E., Torbert, R. B., and Whipple, E. C.: EDI electron time-of-flight measurements on Equator-S, Ann. Geophys., 17, 1513–1520, doi:10.1007/s00585-999-1513-3, 1999.

Paschmann, G., Quinn, J. M., Torbert, R. B., Vaith, H., McIlwain, C. E., Haerendel, G., Bauer, O. H., Bauer, T., Baumjohann, W., Fillius, W., Förster, M., Frey, S., Georgescu, E., Kerr, S. S., Kletzing, C. A., Matsui, H., Puhl-Quinn, P., and Whipple, E. C.: The Electron Drift Instrument on Cluster: overview of first results, Ann. Geophys., 19, 1273–1288, doi:10.5194/angeo-19-1273-2001, 2001.

Pudney, M. A., Carr, C. M., Schwartz, S. J., and Howarth, S. I.: Automatic parameterization for magnetometer zero offset determination, Geosci. Instrum. Method. Data Syst., 1, 103–109, doi:10.5194/gi-1-103-2012, 2012.

Quinn, J. M., Paschmann, G., Sckopke, N., Jordanova, V. K., Vaith, H., Bauer, O. H., Baumjohann, W., Fillius, W., Haerendel, G., Kerr, S. S., Kletzing, C. A., Lynch, K., McIlwain, C. E., Torbert, R. B., and Whipple, E. C.: EDI convection measurements at 5–6 R_E in the post-midnight region, Ann. Geophys., 17, 1503–1512, doi:10.1007/s00585-999-1503-5, 1999.

Vaith, H., Frenzel, R., Paschmann, G., and Melzner, E.: Electron Gyro Time Measurement Technique for Determining Electric and Magnetic Fields, American Geophysical Union Geophysical Monograph Series, 103, 47–52, 1998.

8

Harmonic quiet-day curves as magnetometer baselines for ionospheric current analyses

M. van de Kamp

Finnish Meteorological Institute, P.O. Box 503, 00101 Helsinki, Finland

Correspondence to: M. van de Kamp (kampmax@fmi.fi)

Abstract. This paper presents a novel method to determine a baseline for magnetometer data. This baseline consists of all magnetic field components not related to ionospheric and magnetospheric disturbances, i.e. all field components due to solar quiet variations and other background variations, such as tidal and secular variations, as well as equipment effects. Extraction of this baseline is useful when the magnetic field variations due to solar disturbances are analysed. This makes magnetometer data suitable, for instance, for the calculation of ionospheric equivalent currents related to geomagnetic storms and substorms.

The full baseline is largely composed of two main constituents: the diurnal baseline and the long-term baseline. For the diurnal baseline, first "templates" are derived, based on the lowest few harmonics of the daily curves from the quietest days. The diurnal variation of the baseline is obtained by linear interpolation between these templates; this method ensures a smooth baseline at all times, avoiding any discontinuities at transitions between days. The long-term baseline is obtained by linear interpolation between the daily median values of the data; this way the baseline is ensured to follow long-term trends, such as seasonal and tidal variations, as well as equipment drift. The daily median values are calculated for all days except the most disturbed ones; a procedure for this selection is included.

The method avoids many problems associated with traditional baseline methods and some of the other recently published methods, and is simpler in procedure than most other recent ones. As far as can be compared, the distribution of the resulting field after removal of the baseline is largely similar to that using other recent baseline methods. However, the main advantage of the method of this paper over others is that it removes equipment drift and other artefacts efficiently without discarding too much data, so that even low-quality data from remote unmanned magnetometers can be made suitable for analysis. This can give valuable contributions to the database of ionospheric equivalent currents, especially in the area near the polar cap boundary.

1 Introduction

1.1 Equivalent currents

Currents flowing in the ionosphere are related to the field-aligned currents in the magnetosphere through ionosphere–magnetosphere coupling (e.g. Kamide and Baumjohann, 1993). The magnetospheric currents are strongly dependent on solar activity, such as solar wind variations and solar storms, and these solar-dependent magnetospheric current variations are also reflected to the ionospheric currents. Since strong disturbances in the ionospheric currents can affect technological systems on the ground (e.g. Boteler et al., 1998), there is a great interest in the dynamics of these currents in relation to solar activity.

The concept of ionospheric "equivalent currents" models the ionospheric currents as present in a thin shell, usually the highly conductive E layer at 100 km height, and representing only the divergence-free, horizontal part of the total currents. Under many circumstances these give valuable information about spatial and temporal characteristics of the actual 3-D ionospheric currents (e.g. Untiedt and Baumjohann, 1993), especially when analysed together with data from other instruments, such as rockets, satellites, or radars. The variations of the equivalent currents can effectively be estimated from magnetometer measurements from a two-dimensional ground-based magnetometer network.

The European Cluster Assimilation Technology (ECLAT) (http://www.space.irfu.se/ECLAT/eclat-web/eclatdetail. html) project funded by the EU FP7 programme provides a selection of useful supporting data sets to the Cluster Active Archive (http://caa.estec.esa.int/caa) (Laakso et al., 2009). The work described in this paper has been used in an ECLAT work package in which ionospheric equivalent current vectors caused by solar disturbances are computed in the Fennoscandia region from data from the ground magnetometer network IMAGE. These source data are described in the next subsection. In the project, equivalent currents are to be analysed for an area over Fennoscandia, over the period 2001–2010. An important motivation for publishing this paper is to provide for the ECLAT database users accurate information about the equivalent current generation procedure used in the service.

For the equivalent current estimate, use is made of the method of spherical elementary current systems (Amm and Viljanen, 1999; Pulkkinen et al., 2003). The currents induced inside the earth are ignored, as they can be assumed to be relatively small compared to the ionospheric currents. For the calculation, only the x (north) and y (east) components of the magnetic field at all ground stations are necessary, as only these are related to the divergence-free part of the ionospheric currents.

1.2 IMAGE magnetometer measurements

The input data for the calculation of the ionospheric currents in the project ECLAT are the magnetometer recordings of the ground magnetometer network IMAGE (http://space.fmi.fi/image/beta/) (Viljanen and Häkkinen, 1997). This network consists of 32 magnetometers over geographic latitudes from 58 to 79° N, which is especially favourable for electrojet studies. The magnetometers return data at a time resolution of 10 s. Figure 1 shows a map of the locations of the magnetometers of the IMAGE network.

1.3 Various magnetic field components: baselines

In addition to the currents caused by solar and magnetospheric disturbances (storms and substorms), which are of interest for the analysis, the ionospheric currents and the ground magnetic field consist of several other components.

The magnetic field measured at the earth's surface is a superposition of the field originating from the inner earth, and that from electromagnetic effects caused by the ionospheric currents. These ionospheric currents, in turn, contain on the one hand components due to the dynamo effect of the revolving magnetic earth inside the conducting ionosphere, whose conductivity varies, depending (among others) on solar radiation, and on the other hand components due to storms and substorms caused by coronal mass ejections and other

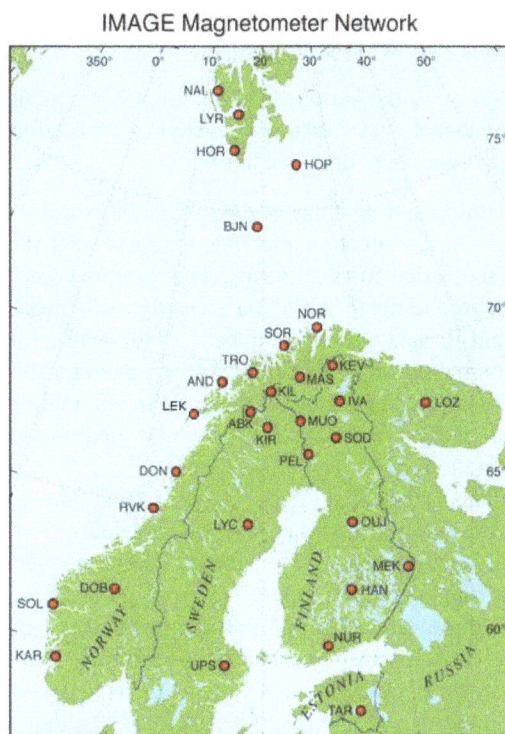

Fig. 1. Map of the IMAGE magnetometer stations.

disturbances in solar activity, and transferred to the ionosphere via ionosphere–magnetosphere coupling. The latter are the components of interest for the ionospheric studies.

As a consequence, in the measured magnetic field the following variations of different timescales can be distinguished:

- the diurnal variations caused by the variation of ionospheric conductivity due to solar radiation during quiet times, usually referred to as "S_q variations" or "solar quiet variations". Although this variation is diurnal, its characteristics can change from day to day.

- lunar variations: variations over a lunar day (24 h 50 m), due to changes in the global distribution of the conducting ionosphere due to lunar attraction. These are usually referred to as "L variations".

- seasonal variations: variations with a period length of one year, due to seasonal changes in the ionospheric conductivity (i.e. slow component of S_q variations), as well as seasonal changes in the ionosphere–thermosphere–magnetosphere interactions.

- secular (long-term) variations, due to very slow variations in the earth's internal magnetic field pattern.

- variations due to solar storms and substorms. These variations are irregular in nature and contain strong fast fluctuations. Still, the nature of these variations can show statistical dependencies on time of day, season, and solar activity.

- in addition to the various field variation components, some magnetometer data also contain system effects such as equipment drift (see later).

Note that within the time range of days, lunar, seasonal and secular variations are much smaller than the solar quiet variations and those due to (sub)storms. The various components of the ground magnetic field are clearly explained by Chapman and Bartels (1940). Campbell (1989) gave an introduction on the physics of S_q and L variations and a short overview of the research history of these, as an editorial to a special issue of *Pure and Applied Geophysics*, dedicated to S_q and L variations.

In the context where the effects of the solar storms and substorms are assessed and analysed by calculating ionospheric equivalent currents, all magnetic field components unrelated to this should be subtracted from ground-based magnetic field measurements, before further analysis. In this context, the earth's main magnetic field, all the S_q and L variations, seasonal variations, and system effects in the output data of any magnetometer are here treated together as "baselines", indicating the components to be removed. The remaining component is assumed to contain only the variations due to solar storms and substorms, and will be referred to in this paper as "disturbance field".

(Note, however, that the term "baseline" has sometimes also been used in the context of studying S_q variations, where the term represented the relatively long-term variations. For instance, when Matsushita (1968) referred to the S_q study by Price and Wilkins (1963), he used the term "base line" for the variations of the midnight field from day to day.)

Although the magnetometer baseline method of this paper is introduced as the basis of calculating ionospheric equivalent currents, this method can also be useful for any other application where either the disturbance magnetic field or the S_q field is to be analysed.

1.4 Baseline removal

A traditional method of removing a baseline from magnetometer data of a particular day is to look for a magnetically quiet day near the day of interest, and calculate the average value of the magnetic field of this day. This constant value is used as the baseline, and subtracted from the data for the day of interest, leaving only the disturbance field. For instance, Davis and Sugiura (1966) proposed to derive the auroral electrojet (AE) index from magnetometer data using this as a baseline method.

Although it has long been known that this way the S_q variations are not included in the baseline and will therefore be considered as part of the disturbances, it was considered that these variations are small compared to the disturbance field, and hence the introduced error is relatively small.

The magnetically "quiet" days used can be more or less frequent, varying from about one to five quiet days per month. They can be determined in various ways (e.g. based on the variations of the data of each day, or using global magnetic indices as Kp or Dst). Also, they can be downloaded directly from a global database (http://www-app3. gfz-potsdam.de/kp_index/definitive.html).

The baseline method using averages of quiet days leads to inaccuracies in the resulting disturbed data in several ways:

1. there may not be any day in the entire month which is completely free from disturbances. In this case, the "quiet" day is not really "quiet" and the data, and even the average value, will still be affected by some disturbance effects.

2. as mentioned above, the diurnal variation of the S_q field is not included in the baseline and will hence be considered as part of the disturbance field.

3. for two consecutive days, the "nearest" quiet day and therefore the baseline value can be different, and hence the baseline may show a discontinuity at midnight between these two days. If this baseline is subtracted from other data for a period around midnight (e.g. to calculate equivalent ionospheric currents), this may cause an artificial jump in the resulting disturbed data.

4. the magnetometer data generally contain slow variations over the course of several days, months and years, caused by long-term magnetic field variations, as well as equipment effects (see Sect. 2.2 for examples of each). As a consequence, the average value of the quiet day may not be representative of that of other days which can be several weeks earlier or later.

In spite of its drawbacks, the above-mentioned baseline procedure is still being used in some applications today, mainly because of its simplicity. Alternatively, many studies have also been performed using more elaborate baseline procedures. In particular, S_q variations (point 2) have been incorporated in baselines by appropriate smoothing of the quiet-day data in one way or the other. Below, a few of the most recent baseline procedures are summarised. It should be emphasised that this is not intended to be a complete review of baseline development, but only to show some examples of the current state of the art.

Janzhura and Troshichev (2008) presented a running automatic method which overcomes some of the problems: point 1 is avoided by looking instead of for quiet days, for any quietest bits of data within a 30-day period; point 2 is avoided by only smoothing the quiet daily curve, but retaining the diurnal variations. Point 4 is only reduced, not entirely removed, by averaging all quiet data over the 30-day period.

However, their method retains point 3, and even introduces more discontinuities. Since the smoothed daily baseline is in general at the end of the day not the same as that at the start of the day, a discontinuity becomes more likely to occur at every midnight. This can be seen, for example, from Fig. 5 in the paper by Janzhura and Troshichev (2008) (the values of the curves at the end of the day do not connect to those at the start of the day).

More recently, Stauning (2011) presented another version of the procedure from Janzhura and Troshichev (2008). The main difference with their method is that his method carefully selects data from similar conditions as the day in question, by giving larger weights to quiet data from days close to the day in question, as well as to data from days approximately 27 days (one solar rotation) before and after. This is because he finds the highest correlation coefficients for such displacement periods. This way, his method further reduces point 4 as long as the slow variations are fluctuations over periods of 27 days. In addition, very slow fluctuations are removed separately in his method. However, his method does not take care of unexpected fluctuations, such as instrument drifts over only a few days (see Sect. 2.2). Also, it is not certain whether any midnight discontinuities (due to point 3, or those as introduced by Janzhura and Troshichev (2008)) remain.

Gjerloev (2012) described the automatic data preprocessing procedure of the worldwide magnetometer network SuperMAG, which also involves a baseline removal technique. Acknowledging the problems associated with the identification of quiet days, he avoids using these altogether, thus avoiding point 1. Instead, he averages the data from several days around the day of interest (3 days in case of relatively quiet periods; longer in disturbed periods), to obtain a daily trend, and applies appropriate smoothing, avoiding point 2. Point 4 is taken care of by separately determining seasonal variations (referred to as "yearly trend"), although it is not clear how well this works to capture relatively fast and irregular equipment drifts. Also here it is not clear whether the midnight discontinuities remain.

In this context it is good to note that all of the above methods were designed to work for rather well-controlled stable instrumentation, in which case point 4 represents only slow and regular variations, which are well taken care of by these methods.

In this paper, however, a method is described which overcomes all problems in the list above, including in the case of unstable instrumentation, and in addition is simpler than the other methods described above. The procedure will be described in the next section.

2 Procedure

The baseline procedure is performed for each magnetometer station separately. The following sections describe the various steps in this procedure.

2.1 Data jumps

Many of the magnetometer time series, especially those from remote unmanned stations, exhibit occasional artificial discontinuities, or "jumps", in the data. These can vary in size from a few nanoteslas to thousands of nanoteslas, and can be both positive and negative. The jumps are probably due to adjustments and resetting of the equipment, and happen mostly at midnight, but not only. These jumps, if not taken care of, will affect the results of the data analysis.

Small discontinuities in the time series need not be a problem in a statistical analysis of the magnetic field. However, they do become important when dynamics aspects of geomagnetic events are analysed. For example, the results of the time derivative of the magnetic field will be affected by artificial jumps. Viljanen et al. (2001) analysed the time derivative of the horizontal magnetic field vector from the IMAGE network, and found significant correlations between this parameter and geomagnetic activity, on various timescales from hours up to years, as well as a significant directional variability of this parameter, which in turn also depends on location and time. Plausible relations were found with, among other things, ionospheric currents, pulsations, and geomagnetically induced currents. Because of this, it is evidently important to avoid the measured time derivative being contaminated by artificial discontinuities. (For the same reason, the artificial jumps in magnetometer data introduced by the baselines, i.e. problem 3 in Sect. 1.4, should also be avoided.)

Furthermore, the data jumps will also affect the baseline determination described in this paper, which involves, among other things, some curve fitting and calculation of standard deviations (see later). Because of this, as a first step in the baseline procedure, these jumps need to be removed from the measured data.

The jumps are removed as follows. A software module has been designed which detects and displays suspected discontinuities in the data. The software lets the user examine each discontinuity and decide whether it is artificial, in which case it is listed in a file. Based on this information, a jump baseline $B_J(t)$ is generated, separately for the x and y components of the field, which contains all the jumps found in the data in the form of a superposition of step functions, and is otherwise constant. This jump baseline is subtracted from the raw data, before any further processing as described in the following sections.

2.2 Long-term baseline

Next, to determine the baseline, first the part is considered which consists of the long-term variations in the magnetic field (i.e. variations over periods longer than 24 h): on the scale of days, months, or years. These variations will be represented in the baseline by determining from every day of data a single daily "background" value. For this value, it might be considered to use the average of the measured data.

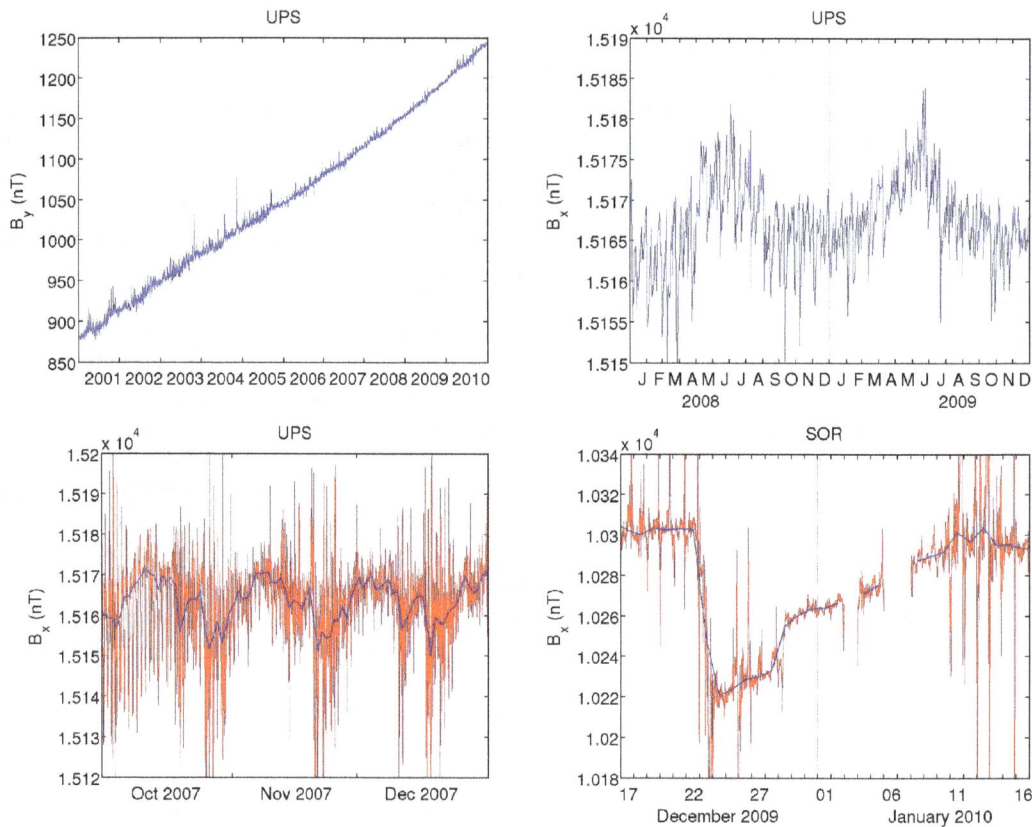

Fig. 2. Daily median values of magnetometer data (blue), demonstrating the long-term variations. Top left panel: secular variation in B_y in Uppsala over the full 10 yr period; top right panel: seasonal variations in B_x in Uppsala over 2008–2009; bottom left panel: tidal effects in B_x in Uppsala in autumn 2007 (raw data included in red); bottom right panel: equipment drift in B_x in Sørøya in December 2009 (raw data in red).

However, rather than this, the median is considered more stable, being less sensitive to extreme values during moderately disturbed days.

Figure 2 shows this daily median calculated over various periods of data, and demonstrates the long-term variations revealed by it. This graph shows that these variations consist of, among others, the following components:

– *Secular variations*: due to very slow variations in the earth's internal magnetic field pattern, the y and z components of the magnetic field at all stations steadily increased, at a rate of slightly over $35\,\mathrm{nT\,yr^{-1}}$, during the period 2001–2010. (The x component did not change significantly.) The upper left-hand panel of Fig. 2 demonstrates this for B_y in Uppsala.

– *Seasonal variations*: the seasonal changes in the ionospheric conductivity and in the ionosphere–thermosphere–magnetosphere interactions result in variations in all magnetic field components with a period length of one year. As an example, the upper right-hand panel of Fig. 2 shows that B_x in Uppsala

oscillated at about 10 nT peak-to-peak within the years of 2008 and 2009.

– *Tidal variations*: variations with periods of about 27 days can be observed, which are the result of interference between S_q and L variations. As an example, the lower left-hand panel of Fig. 2 shows that B_x in Uppsala varied at this frequency at about 15 nT peak-to-peak from October to December 2007. (Note that the raw data, which are also shown (red), include the disturbance field and therefore also show statistical variations due to the period of solar activity changes, which also have a period of 27 days. The median values however should not be much affected by this.)

– *Equipment drift*: some of the magnetometers occasionally exhibit some variations over the course of one or a few days, which do not repeat, and show no correlation with space weather parameters or with any of the other magnetometer results. The lower right-hand panel of Fig. 2 shows an example for Sørøya, where on 23 December 2009, the readings for B_x decreased by 80 nT over two days, and then gradually recovered, over the course of 18 days.

The equipment drifts especially happened at the stations in remote locations, which are not continuously manned and monitored. These drifts should obviously be classified as measurement errors. However, if they can be quantified using the current method, they can be removed along with the above-mentioned long-term variations in the magnetic field, which means that these low-quality data need not be discarded. This is a very useful outcome, since these magnetometer stations in remote places often are some of the most crucial ones. A significant number of them are located in northern Scandinavia and the sea between Norway and Svalbard, which is the area above which the auroral oval or its boundaries are often located, and where therefore much of the magnetic activity occurs. But at the same time, not many magnetometers are present there, due to the difficult accessibility of the area. The resulting relevance of the area is precisely why these particular magnetometers were placed in these locations, despite their inaccessibility. It is therefore very useful if these relatively valuable results, even if not of perfect quality, can still be used in ionospheric analyses.

The median is used for the long-term baseline as follows. The median value is calculated for every day of data. The long-term baseline is considered to be equal to the median at 12:00 UT on the respective day. At any other time, the long-term baseline is linearly interpolated between these values. (Note that linear interpolation is good enough, even though it causes "corners" in the data (discontinuities in the first derivative); these corners are not a problem for the magnetic field data as long as the data themselves stay continuous.)

The resulting time-dependent long-term baseline will be referred to in this paper as $B_T(t)$. The procedure to determine $B_T(t)$ is performed separately not only for each different station but also for each field component.

Obviously, during magnetically very disturbed periods, the median value calculated over a day may not be representative of the long-term baseline. In these cases, the median of these particular days will not be used, but the long-term baseline will be interpolated between other median values. Further on, it will be shown how the classification of these "usable" median values is performed.

2.3 Quiet days

For every separate magnetometer station, a list of the quietest days of each month is generated.

At any magnetometer station, a day is considered "quiet" if the magnetic field variations measured on this day are (almost) entirely caused by S_q variations and not by magnetic disturbances driven by solar activity. Hence, a "quiet" day would mainly contain slow variations; its fast variations should be relatively small.

Because the fast variations due to disturbances are mostly much larger than the S_q variations, it seems logical to calculate simply the standard deviation of every day of data, and look for the smallest standard deviations of each month. Indeed this method works reasonably. However, there are cases where this method is too coarse, as will become clear in the description below of the method of this paper.

In this paper, the quiet-day selection is performed as follows. Each day of data is partitioned into 24 one-hour sections (UT). In each one-hour section, a straight line is fitted to the data of the x and y components of the magnetic field. This straight line is subtracted from the data, and from the remaining data, the hourly standard deviation, σ_H, is calculated. The result of this is 2×24 values (2 components and 24 h) of σ_H. Of these 48 values, the daily maximum is calculated, referred to as $\sigma_{H_{max}}$, for each day and each station. These values of $\sigma_{H_{max}}$ are the indicators for days with and without disturbances: of each month, the day with the lowest $\sigma_{H_{max}}$ is selected as the "quiet day" for that month, at that station.

As an extra requirement, the "quiet day" should contain no data gaps, so only days with 100 % data availability for both magnetic components can be potential candidates for any month's "quiet day".

Figure 3 demonstrates this procedure, showing B_x in Oulujärvi on three different days in March 2002. The day shown in the bottom two panels, 28 March, resulted from this procedure as the quietest day of this month (note that the actual procedure also takes into account the y component of the magnetic field, which is not shown in the figure).

Figure 4 demonstrates why this procedure works better than calculating the overall standard deviation of the whole day. On 28 March (the quiet day at the bottom of Fig. 3) the full-day standard deviation of B_x is 19.9 nT, while on 7 March (Fig. 4), it is smaller: 10.0 nT. However, the latter day contains some fast variations, which make it less suitable as a "quiet" day. This property is revealed by the maximum hourly standard deviation $\sigma_{H_{max}}$, which is larger on 7 March: 6.7 nT rather than 1.4 nT.

Of course, even though according to this procedure a "quietest" day of each month can always be found (as long as at least one full day of data is available), it may be that this quietest day still contains too much magnetic disturbance to be used for a baseline. Especially during and around the solar active year of 2003, disturbances can be so frequent that not necessarily within every month a full day can be found where these disturbances are insignificantly small. In other words: the "quietest" day of the month may not be really "quiet".

Because of this, an extra criterion is applied: if for any month, the value of $\sigma_{H_{max}}$ of the quiet day is above a certain threshold value, then this quiet day is discarded and no quiet day is assigned for this month. Later it will become clear that these cases do not cause a problem in the rest of the procedure. The optimum threshold value of $\sigma_{H_{max}}$ to be used for this depends on the typical level of both slow and fast variations in the magnetic field. The threshold value was empirically chosen by visual inspection of many data sets from

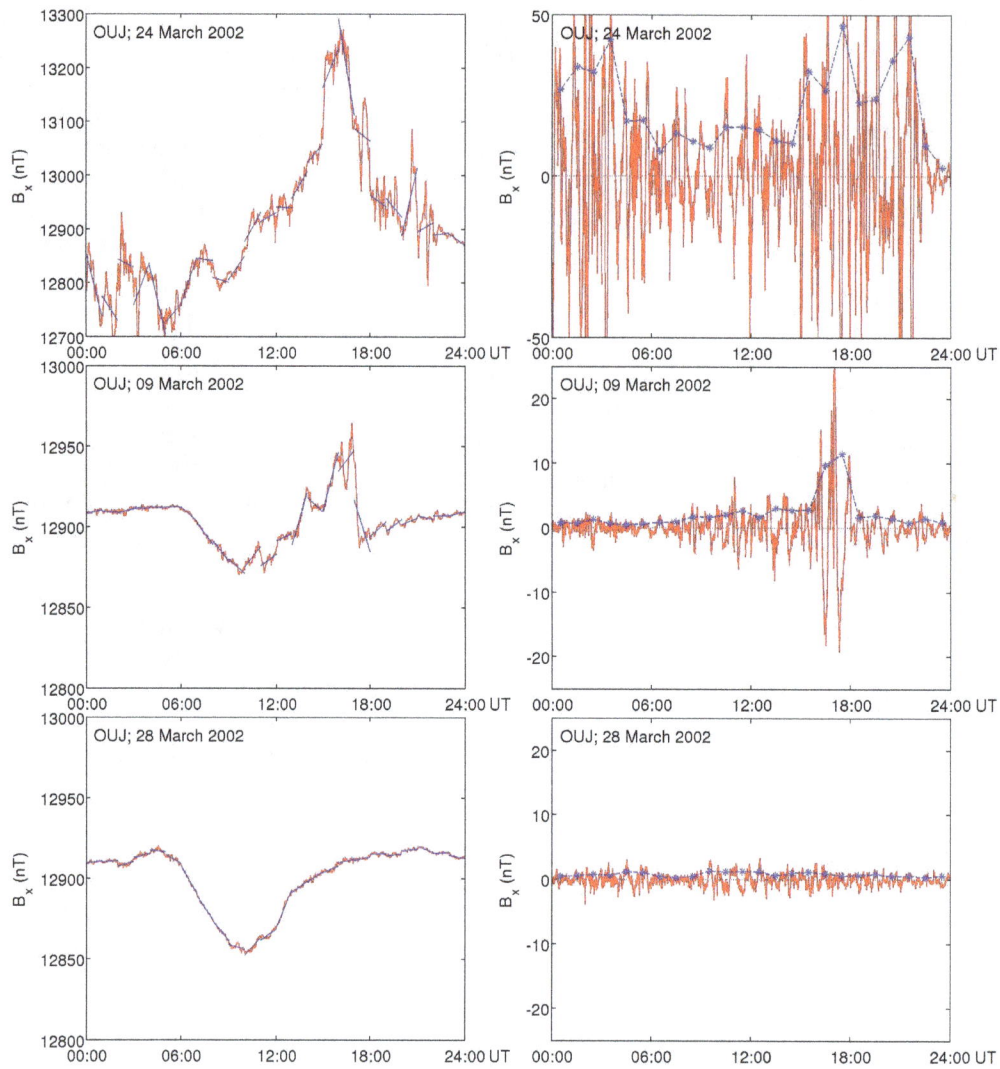

Fig. 3. The quiet-day selection procedure, demonstrated by the B_x field in Oulujärvi in March 2002. Left-hand panels: raw data and the hourly lines fitted to it. Right-hand panels: the data after subtraction of the fitted lines, and the calculated hourly standard deviations. Top panels: on a disturbed day; centre panels: on a mostly quiet day with some disturbance; bottom panels: on a quiet day. The maximum σ_H values (from B_x alone) are 46.7 11.5, and 1.4 nT, respectively.

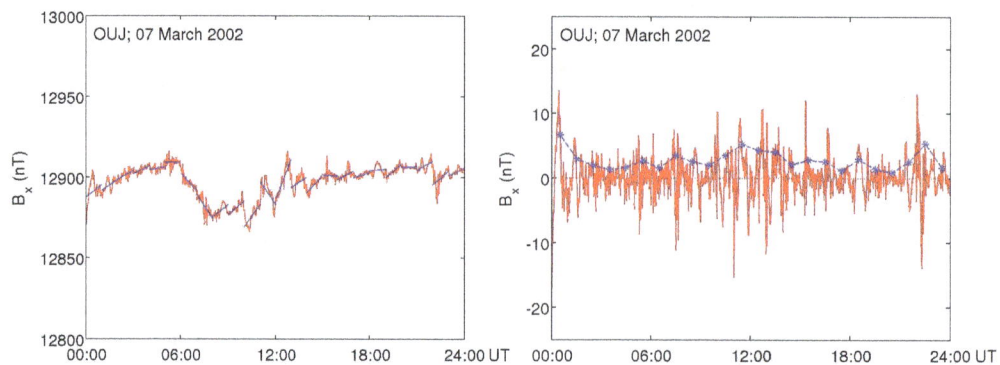

Fig. 4. Similar panels as in Fig. 3, for 7 March 2002, which, although the overall variation is quite small, contains some disturbances. The maximum value of σ_H (from B_x) is 6.7 nT.

Table 1. Threshold values of $\sigma_{H_{max}}$ and $\sigma_{H_{med}}$ for each IMAGE station.

Station	Name	Latitude (°)	Longitude (°)	Threshold $\sigma_{H_{max}}$ (nT)	Threshold $\sigma_{H_{med}}$, B_x (nT)	Threshold $\sigma_{H_{med}}$, B_y (nT)
NAL	Ny-Ålesund	78.92	11.95	18	14	13
LYR	Longyearbyen	78.20	15.82	21	16	15
HOR	Hornsund	77.00	15.60	25	20	17
HOP	Hopen Island	76.51	25.01	29	22	14
BJN	Bjørnøya	74.50	19.20	28	21	15
NOR	Nordkapp	71.09	25.79	19	25	12
SOR	Sørøya	70.54	22.22	20	20	12
KEV	Kevo	69.76	27.01	14	18	9
TRO	Tromsø	69.66	18.94	17	20	11
MAS	Masi	69.46	23.70	13	18	11
AND	Andenes	69.30	16.03	17	20	11
KIL	Kilpisjärvi	69.06	20.77	13	18	10
IVA	Ivalo	68.56	27.29	10	16	8
ABK	Abisko	68.35	18.82	11	17	10
LEK	Leknes	68.13	13.54	10	15	9
MUO	Muonio	68.02	23.53	9	15	8
LOZ	Lovozero	67.97	35.08	8	12	6
KIR	Kiruna	67.84	20.42	9	14	8
SOD	Sodankylä	67.37	26.63	8	13	7
PEL	Pello	66.90	24.08	7	12	7
DON	Dønna	66.11	12.50	7	11	6
RVK	Rørvik	64.94	10.98	6	9	5
LYC	Lycksele	64.61	18.75	8	9	7
OUJ	Oulujärvi	64.52	27.23	6	7	5
MEK	Mekrijärvi	62.77	30.97	5	4	4
HAN	Hankasalmi	62.25	26.60	5	4	4
DOB	Dombås	62.07	9.11	4	5	4
SOL	Solund	61.08	4.84	4	4	4
NUR	Nurmijärvi	60.50	24.65	4	5	3
UPS	Uppsala	59.90	17.35	4	4	3
KAR	Karmøy	59.21	5.24	4	3	3
TAR	Tartu	58.26	26.46	4	3	3

the station Abisko, which shows a continuous set of good-quality data. The optimum value (for this station) was found to be 11 nT. This value for $\sigma_{H_{max}}$, which coincides with daily median values of Kp of around 3, is in Abisko exceeded for 8 % of all quiet days, and for a maximum of 3 consecutive monthly quiet days (this is in the particularly active time of April–June 2003).

However, because the optimum threshold value of $\sigma_{H_{max}}$ depends on the typical level of both slow and fast variations in the magnetic field, it varies from station to station. The threshold values to be used for all other stations were empirically adjusted, to make sure all different threshold values represent the same level of irregularities in the respective data sets. To do this, for each station, all values of $\sigma_{H_{max}}$ were selected which coincided with $\sigma_{H_{max}}$ in Abisko being within 11 ± 1 nT, and the median of these values was taken as the threshold for that station. The resulting values are listed for all IMAGE stations in Table 1 (5th column). Using these

threshold values, generally no more than three consecutive months without quiet days were encountered for any station in the entire IMAGE database for 2001–2010.

2.4 Very disturbed days

The information of the hourly standard deviations σ_H, calculated as described in the previous section, will also be used to classify certain days as being too disturbed for calculation of the median value, used in the derivation of the long-term baseline $B_T(t)$ (see Sect. 2.2).

In the top panel of Fig. 5, three days of B_x data from Abisko in November 2001 are shown, along with their daily median values (green stars). On 4 November, conditions are mostly quiet, and the median value is a good representative of the long-term baseline. On 5 November, some disturbances start late in the day, but the median value is not significantly affected by them. However, on 6 November, conditions are disturbed all day, and the median value is dominated by these

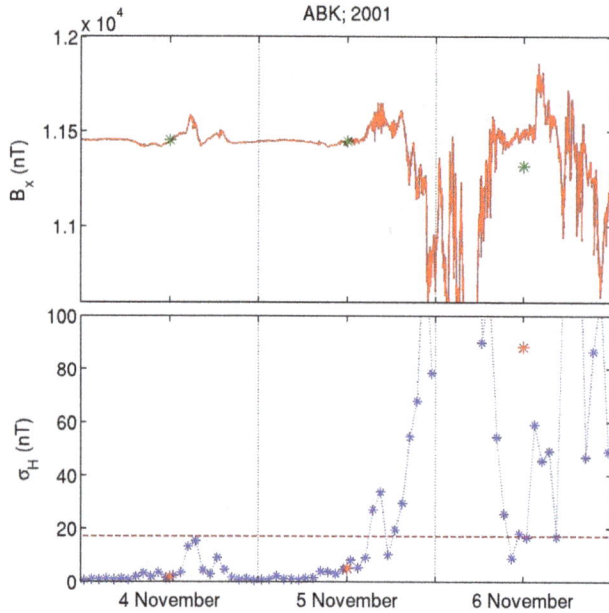

Fig. 5. Top panel: B_x in Abisko on 4–6 November 2001, and the daily median values calculated from it (green stars). Bottom panel: hourly standard deviations σ_H of these data (blue stars), daily medians $\sigma_{H_{med}}$ of these (red stars), and suggested threshold value for $\sigma_{H_{med}}$ of B_x in Abisko (red dashed line).

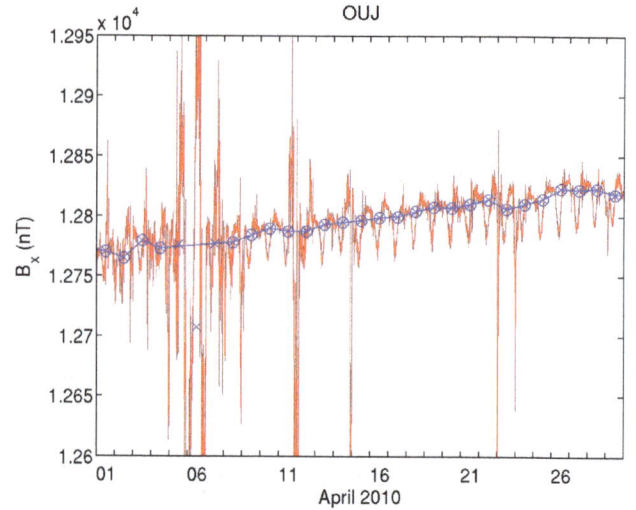

Fig. 6. B_x field in Oulujärvi in April 2010 (red), the daily median values (blue x), the ones that are qualified to be used for the long-term baseline (circles), and this baseline $B_T(t)$ (blue solid line).

disturbances, and unsuitable to be used for the long-term baseline.

A rule of thumb can be described as follows: a median value is relatively insensitive to irregularities in data as long as these irregularities consist of less than half of the data. This can be seen on 5 November in Fig. 5: since the irregularities consist of less than half the day, the median is not significantly affected by them.

Because of this rule, the distribution of hourly standard deviations σ_H can serve as a useful indicator for the stability of the daily median value. If more than half of the day's σ_H values indicate disturbed data during their hours (as in the previous section), it can be assumed that more than half of the data of the day are disturbed, and the daily median value will be unreliable. On the other hand, if more than half of the σ_H values are low, then more than half of the day's data will be quiet, and the median value will be relatively unaffected by any disturbances. These different cases are clearly illustrated in the bottom panel of Fig. 5, where the σ_H values of the data in the upper panel are shown (blue stars).

This criterion is easily represented by the median value of the day's hourly standard deviation, which will be referred to as $\sigma_{H_{med}}$. If $\sigma_{H_{med}}$ is above a certain threshold, this means that at least half the σ_H values are above this threshold. This can be seen in the bottom panel of Fig. 5, where $\sigma_{H_{med}}$ values are the red stars. As a threshold value of B_x for Abisko, 17 nT was empirically chosen by visual inspection of many data series. This value, which coincides with daily median values

of Kp of about 10, is exceeded for 11 % of all days, and for a maximum of 10 consecutive days (in May 2003). The value of 17 nT is indicated as a dashed line in the bottom panel of Fig. 5.

The decision of whether a median value is used for the long-term baseline $B_T(t)$ is made separately for B_x and B_y. Because typical variations in B_y are often different from those in B_x, also different threshold values for these two must be used. Furthermore, the threshold values are dependent on station location, just as is the case for the threshold of $\sigma_{H_{max}}$ (see previous section). Because of this, the thresholds for B_y in Abisko and for both B_x and B_y for all other stations were all empirically adjusted in the same way as $\sigma_{H_{max}}$ in Sect. 2.3: each time, the median of all values of $\sigma_{H_{med}}$ coinciding with $\sigma_{H_{med}}$ for B_x in Abisko being within 17 ± 1 nT was taken as the threshold for that component and that station. The resulting values are included in Table 1 (rightmost two columns).

Figure 6 shows an example of the result of this procedure, by the B_x field in Oulujärvi in April 2010. These data contain some long-term baseline variations, as well as some irregularities. Using the procedure described in this section, the median values of the entire month except 5–7 April were considered suitable for the long-term baseline. The long-term baseline $B_T(t)$, interpolated between the suitable median values, follows the long-term behaviour well, and is unaffected by the disturbance on 5–7 April.

It may be considered in which cases this procedure might give a wrong result: if, in the case of a solar storm which lasts several days, there is a day with a significant magnetic field deviation due to the storm but very few irregularities, $\sigma_{H_{med}}$ might be below the threshold, so that the daily median value, while still affected by the storm, is used for the long-term baseline. In this case this "quiet" part of the storm will be

Table 2. Frequencies of the 7 lowest harmonics of data of 1 day.

h	f (Hz)	period
0	0	inf
1	1.1574×10^{-5}	1 day
2	2.3148×10^{-5}	12 h
3	3.4722×10^{-5}	8 h
4	4.6296×10^{-5}	6 h
5	5.7870×10^{-5}	4 h 48 m
6	6.9444×10^{-5}	4 h

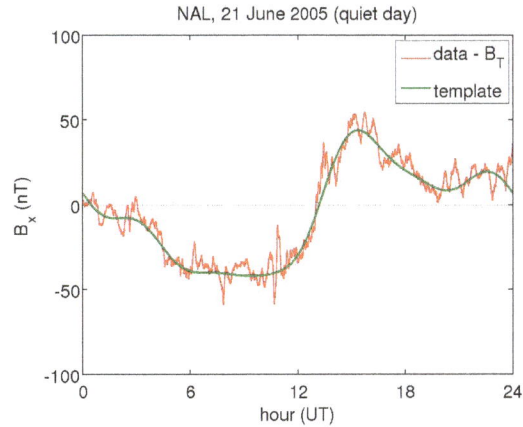

Fig. 7. Red: B_x field measured by the magnetometer in Ny-Ålesund (NAL) on 21 June 2005, one of the assigned quiet days, with the long-term baseline subtracted, leaving only the diurnal variation. Green: the template derived from this quiet day.

included in the baseline variation, and removed from the data as if it were a slow S_q variation. However, these cases will not be frequent, give only a small contribution to the magnetic field, and, most importantly – they are very hard to tell apart from real S_q variations, which *should* be removed from the data. Because of this, it is considered that the procedure described above works well enough, if no other information is used to tell quiet storm parts apart from S_q variations.

2.5 Templates composed of harmonics

Coming back to the subject of quiet days, defined in Sect. 2.3, this section describes how the diurnal variations are derived from these quiet days.

The diurnal S_q variations of the magnetic field can be approximated by harmonic components, as various researchers have already done in the past. Overviews of results of such harmonic analyses performed since 1889 are given by Matsushita (1968) and Campbell (1989). More recently, another harmonic analysis of S_q currents was performed by Pedatella et al. (2011). Following similar principles, in this section the diurnal baseline of the two horizontal components of the ground-measured magnetic field is derived by harmonic analysis of the measured data from quiet days, separately for the x and y components.

For all of the quiet days, and each component, the long-term baseline $B_T(t)$ is subtracted from the data of the entire day. On the residual data, a fast Fourier transform (FFT) is performed. From the result of this, only the lowest few frequency components are used. The exact number of frequency components to be used is somewhat arbitrary; in the method of this paper up to the 6th component is used, similarly as in the S_q study by Pedatella et al. (2011), and in the determination of the Dst index by Sugiura (1964). This means that only the first 7 (the 0th through the 6th) values resulting from the FFT are used. These complex values represent the amplitudes and phases of harmonics with frequencies which are all multiples of the inverse of 1 day. The frequencies of the first 7 harmonics are given in Table 2. Effectively, this means that S_q variations are assumed to be confined to periods of 4 h and longer, while the disturbance field will contain all faster components.

The curve, composed of these 7 lowest harmonics of the quiet day, is equivalent to a low-pass-filtered version of the quiet-day data. The resulting curve, which will here be called a "template", will be used as a basis of the baseline construction. The template is described as follows:

$$T(t_d) = \sum_{h=0}^{6} |X_h| \cos\left(\frac{2\pi h t_d}{86\,400} + \arg(X_h)\right), \qquad (1)$$

where t_d = "time of day", time elapsed since midnight (s); h = index number of harmonic (0...6), and X_h = (complex) coefficient of harmonic h (nT).

There will be one set of harmonic coefficients X_h, and therefore one template $T(t)$, defined for each quiet day (and each field component). One example day of data, with the long-term baseline $B_T(t)$ subtracted, as well as the template derived from this, is shown in Fig. 7.

It is worth noting that since all the cosine arguments in Eq. (1) cover a whole number of cycles over the length of one day (86 400 s), the template value at midnight at the end of the day (i.e. $T(86\,400\,\text{s})$) will always be equal to that at midnight at the start ($T(0\,\text{s})$), thus ensuring continuity at midnight if the template were to be used on consecutive days. However, the templates are not used directly as such for the baselines, which will be shown in the next subsection.

2.6 Diurnal baseline

As the next step, a curve representing the diurnal S_q variations (i.e. the diurnal variation of the background magnetic field) is derived from the templates, separately for the x and y components. This curve will be referred to as the "diurnal baseline", and will be expressed as $B_S(t)$.

To obtain the diurnal baseline, the templates are linearly interpolated continuously between midday on the previous

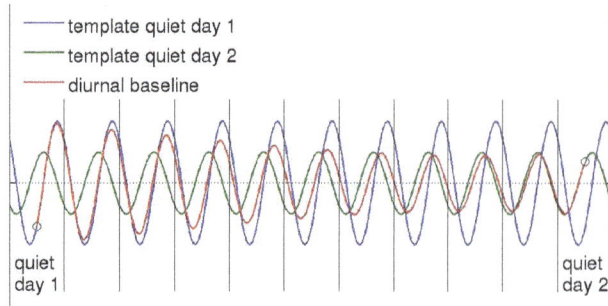

Fig. 8. The principle of interpolation between templates. Templates derived from two consecutive quiet days, and the diurnal baseline, linearly interpolated between them. The vertical lines mark midnights (separation between days).

assigned quiet day and midday on the next assigned quiet day. This can be expressed as follows:

$$B_S(t) = T_1(t_d) + (T_2(t_d) - T_1(t_d)) \frac{t - t_1}{t_2 - t_1}, \qquad (2)$$

where t = the time point of interest (s), t_d = "time of day" as in Eq. (1), $T_{1,2}$ = the template as a function of time of day on the previous and next quiet day, respectively, and $t_{1,2}$ = the time point of midday on the previous and next quiet day, respectively (s).

It should be noted that, for the sake of consistency, also on the quiet days themselves the templates are interpolated. Consequently, only at noon, the diurnal baseline of a quiet day is exactly equal to the template of the same day. After noon, it is linearly interpolated between this template and the next template, and before noon, it is interpolated with the previous template.

Figure 8 shows a schematic example, using imaginary templates consisting of only 1st harmonics. The blue curve is the template derived from quiet day 1 (the day on the left side); the green curve is the template from quiet day 2 (on the right side). The diurnal baseline (red curve) is linearly interpolated between these two from midday on the first quiet day to midday on the second quiet day. Only at noon on the quiet days (marked as "o" in the graph), the baseline is exactly equal to a template.

Note that the cases of months without any quiet day, which can occur during some of the most disturbed periods in solar maximum years (see Sect. 2.3), do not cause a serious problem in this procedure; they merely mean a longer period between two quiet days over which the template is interpolated. To give an indication of how long these periods can be: in the current database, the largest amount of data between any two quiet days was 66 days of data, over a period of 99 days (Masi station; between 6 February and 17 May 2003).

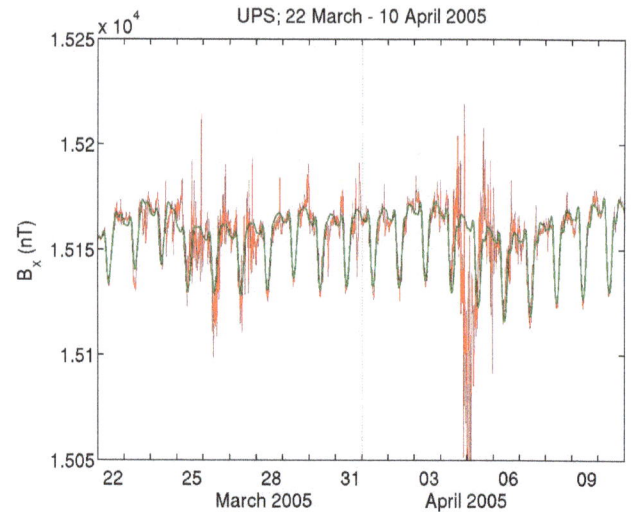

Fig. 9. B_x field at Uppsala from 22 March to 10 April 2005 (red), and the corresponding baseline derived as described in this paper (green).

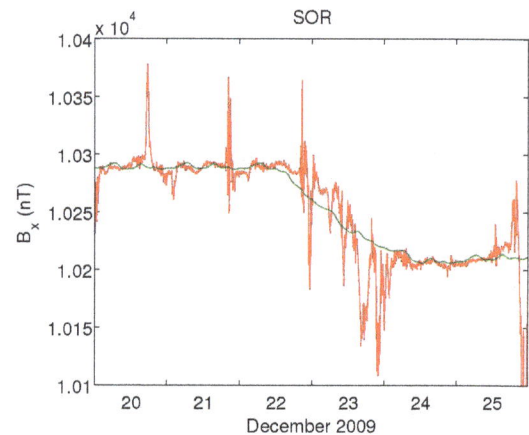

Fig. 10. B_x field at Sørøya from 20 to 25 December 2009, and the corresponding baseline derived as described in this paper.

2.7 Full baseline

To obtain the full baseline, separately for the x and y components, the diurnal baseline B_S is added to the long-term baseline B_T and the jump baseline B_J:

$$B_B(t) = B_S(t) + B_T(t) + B_J(t), \qquad (3)$$

where B_B = full baseline; B_S = diurnal baseline, derived from quiet days as described in Sects. 2.3, 2.5 and 2.6; B_T = long-term baseline, derived from suitable median values as described in Sects. 2.2 and 2.4; and B_J = jump baseline, containing only the data jumps as described in Sect. 2.1.

Figure 9 presents an example of the result of the procedure described in this section. It shows the B_x field in Uppsala from 22 March to 10 April in 2005 (two quiet days), and the corresponding baseline. In this example, the diurnal

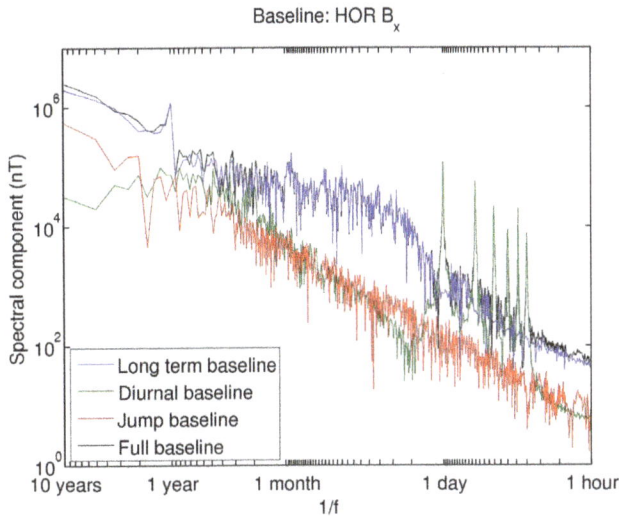

Fig. 11. Spectra of the different baseline components, of the B_x field in Hornsund: blue: long-term baseline; green: diurnal baseline; red: jump baseline; black: full baseline. The x axis labels show the inverse of the frequency f.

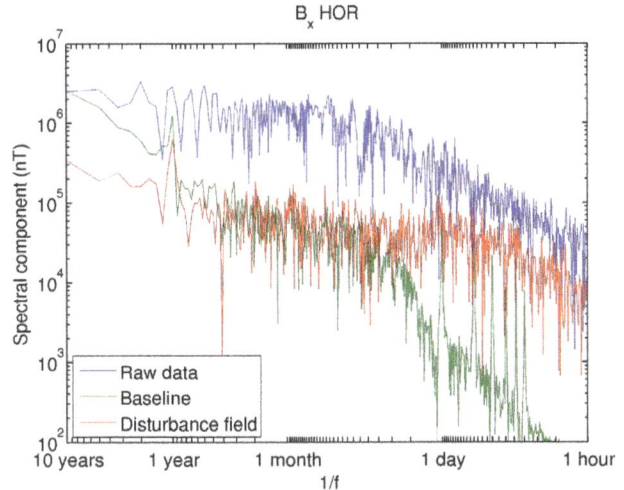

Fig. 12. Spectra of the the B_x field in Hornsund (blue), the full baseline (green), and the magnetic field data after removal of the baseline (i.e. the disturbance field; red).

variation of the baseline follows that of the data not only on the assigned quiet days but also on relatively quiet intervals in the middle of this period (e.g. 1 April). Furthermore, the long-term (mostly tidal) variation of the data is well followed by the baseline.

Figure 10 shows an example for the B_x field in December 2009 in Sørøya, which experienced some significant equipment drift in this period (also shown in Fig. 2). The figure shows that the baseline derived for this period follows the equipment drift well, making this data reasonably usable for equivalent current calculations.

3 General results

The usefulness of the baseline procedure described in this paper is mainly demonstrated by the fact that it avoids all the problems which some other baseline procedures have (as listed in Sect. 1.4), and so makes the resulting data suitable for the calculation of ionospheric equivalent currents. However, it can also be demonstrated by showing some general characteristics of the baselines and the resulting magnetic disturbance field data.

3.1 Spectral analysis

As explained in Sect. 2.5, the baseline of this paper is designed to remove only components of periods of 4 h or longer. This means effectively that the S_q variations are assumed to be confined to those periods, while the disturbance field retains all faster components (but not necessarily only those). The effect of this should be reflected in the signal spectrum of each component.

In this section, the different components of the baseline (long-term, diurnal and jump baseline), as well as the full baseline, and the magnetic field measurements are spectrally analysed. This analysis reveals which frequencies are contained in which baseline components and which frequencies remain present in the magnetic field data after removal of the baselines. For this purpose, the 10 yr data (2001–2010) of all stations were read with a time resolution of 30 min, which allows a spectral analysis of frequencies from 1/(10 yr) up to 1/(1 h).

Figure 11 shows the spectrum for the three baseline components for the B_x field in Hornsund. This case is a good representation of the general result of the spectral analysis for the different magnetometers. The graph shows that the long-term baseline (blue) has its strongest frequency contributions in the low-frequency area (periods of months/years). The diurnal baseline (green) generally has a lower spectrum than the long-term baseline, except, as expected, for 6 strong peaks at the frequencies of the harmonic components which make up the template (e.g. see Table 2). The spectrum of the jump baseline (red) is also lower than of the long-term baseline. The spectrum of the full baseline (black) is hard to see in the graph, being mostly hidden behind whichever component has the highest spectrum. Also this decreases with increasing frequency, becoming particularly low for frequencies above the inverse of 4 h.

Figure 12 shows the spectrum of the (full) baseline again, together with those of the raw measured B_x field and the field remaining after removal of the baseline (i.e. the disturbance magnetic field). This graph shows that the raw data have a flat spectrum up to about the inverse of half a month, after which it decreases with frequency. Because the baseline contains mainly low-frequency components, mainly those components are removed from the data. As a result, up to

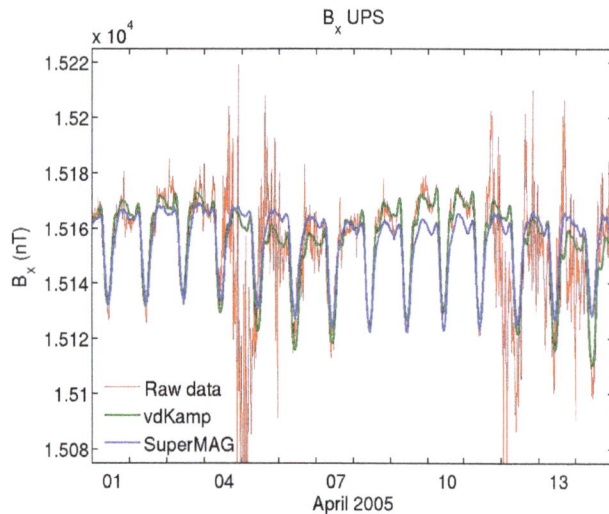

Fig. 13. B_x field at Uppsala from 1 to 15 April 2005 (red), the baseline derived as described in this paper (green), and the SuperMAG baseline (blue).

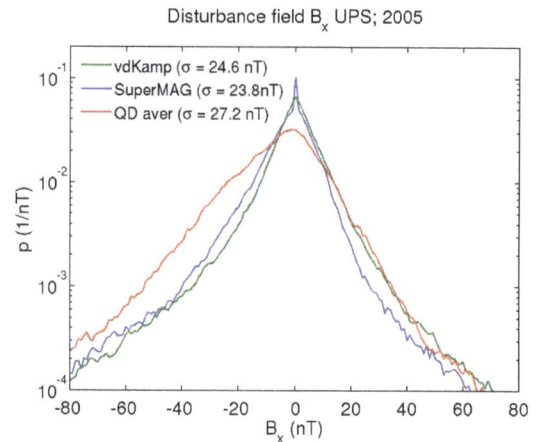

Fig. 14. Probability density function of the B_x disturbance field at Uppsala over the year 2005: using the baseline from this paper (green), using the SuperMAG baseline (blue), and using as a baseline the average values of quiet days (red).

$f = 1/(4\,h)$ the disturbance data have a fairly flat spectrum, resembling that of white noise, which indicates that in that frequency range all regular components have been removed from the data. The disturbance field gets closest to the original data for the highest frequencies.

3.2 Comparison to other methods

It is useful to compare the baseline method of this paper to other baseline methods, such as the ones described in Sect. 1.4: those by Janzhura and Troshichev (2008), Stauning (2011), and/or Gjerloev (2012). Unfortunately, the full procedure of those methods is not available to the author of this paper. However, the baseline method by Gjerloev (2012) was used for the data in the Super-MAG database; see http://supermag.jhuapl.edu/mag/ or http://supermag.uib.no/mag/. This database contains data from magnetometers distributed all over the globe. The baseline used there (hereafter referred to as "SuperMAG baseline") has been retrieved from data downloaded from one of these websites. A comparison with that baseline is presented in this section.

Firstly, a comparison is made of the B_x baseline at Uppsala over the entire year of 2005. This year was about midway between the solar maximum and minimum, and shows therefore moderate magnetic disturbances. The raw data in Uppsala are of good quality: no equipment effects can be noted in the data for the entire year 2005. Figure 13 shows the SuperMAG baseline in April 2005 (blue), compared to the the baseline from this paper (green, marked as "vdKamp"), and the raw data (red). (These data are partly the same as the ones shown in Fig. 9.)

This figure shows that the baseline from this paper follows the day-to-day variations more than the SuperMAG baseline (e.g. see on 9–11 April). Variations on this timescale are mostly tidal (i.e. the result of interference between S_q and L variations (see Sect. 2.2)). This suggests that some of these tidal variations are likely to remain in the field after removal of the SuperMAG baseline. On the other hand, as explained in Sect. 2.4, it might be that some quiet parts of solar storms are mistaken for tidal variations and included in the baseline of this paper; this could be the case in Fig. 13 on 6–7 April. However, since those quiet parts of solar storms are hard to tell apart from real tidal S_q variations, it is not evident from this observation which baseline is better in this case.

Some statistics of these two baselines over 2005 has been calculated: the standard deviation of the baseline of this paper is equal to 13.1 nT, while that of the SuperMAG baseline is 10.3 nT. This difference is a consequence of the fact that also over the entire year, the baseline according to this paper follows the variations of the raw data more than the SuperMAG baseline, as seen in Fig. 13. Still, the difference between the standard deviations is small. Furthermore, the standard deviation of the difference between the two baselines is only 7.5 nT. This demonstrates that the results of the two methods are actually very similar.

Figure 14 shows a probability density function (pdf) of the disturbance field B_x (i.e. the B_x field after subtraction of the baseline) according to both methods. For comparison, a third baseline method is also included in this graph: this is the "traditional" baseline method, using the average value of the quietest day of each month as a baseline for the entire month (marked as "QD aver").

Also this graph demonstrates the similarity in the results of the baseline method from this paper and the SuperMAG baseline: the shapes of the two distributions are very similar,

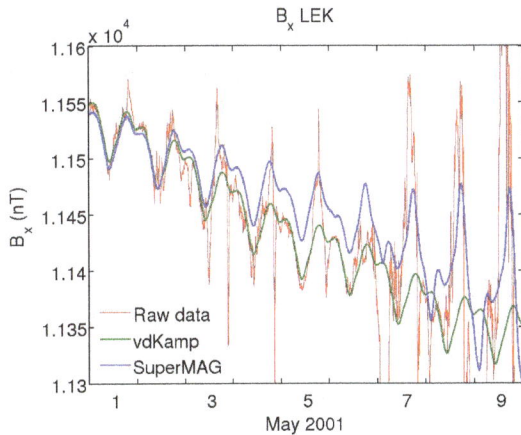

Fig. 15. B_x field measured in Leknes on 1–9 May 2001 (red), the baseline from this paper (green), and the SuperMAG baseline (blue).

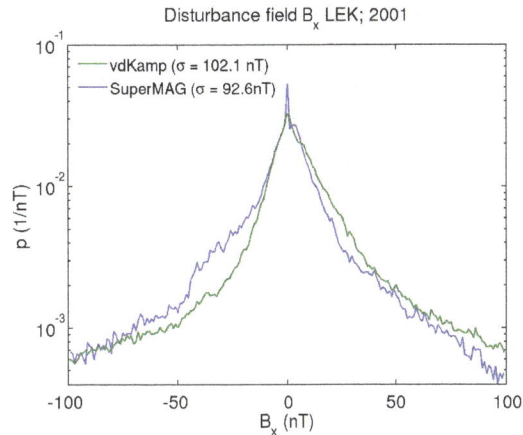

Fig. 16. Pdf of the B_x disturbance field at Leknes over February–May 2001: using the baseline from this paper (availability 2877.1 h), and the SuperMAG baseline (availability 2770.1 h).

as well as the standard deviations (marked in the graph). A notable difference is however that the disturbance field according to SuperMAG shows a peak in the centre: it is equal to 0 nT (within the 1 nT resolution of the SuperMAG data) for a larger portion of time than would be expected from the shape of the rest of the distribution. Apparently, during this portion of time (10.3 % of this period of data), the SuperMAG baseline is exactly equal to the raw measured data. The same was noted for other periods of data.

The disturbance field using the "QD aver" baseline clearly shows a wider spread than using the other two methods (this can also be seen from its standard deviation). This effect is due to the S_q variations, which occur every day, and using this baseline method are included in the resulting field. The effect of this on the pdf is that near-zero values of B_x become spread out over a larger range.

A second comparison was made for a different period of data: the B_x field at Leknes, during February–May 2001. This station exhibited during this time some strong equipment effects. In addition, the raw data contain a few gaps, together 2.9 h over these 4 months.

Figure 15 shows the raw data and the two baselines of B_x at this station on 1–9 May 2001. During this time, this magnetometer showed some strong equipment drift. It can be seen in this figure that the baseline of this paper follows the drift better than the SuperMAG baseline, especially on 4–6 May. The result of this can also be seen in the pdf of the disturbance field according to the two methods over February–May 2001, shown in Fig. 16. The increased probability for values between −15 and −44 nT using the SuperMAG baseline is due to the effect seen in Fig. 15.

It should however be noted that cases like this are rare in the databases. The more significant difference between the two methods is their availability. In some other parts of these data (not shown here), the magnetometer showed

strong equipment effects which caused the data to be qualified as "invalid" in the SuperMAG database, and as a result 110 h of the data of B_x in Leknes during February–May 2001 are unavailable there. However, the unavailability of the disturbance field using the baseline according to this paper is equal to that of the raw data in the IMAGE database over this period: 2.9 h. The data patches with equipment effects are therefore included in the pdf of the disturbance field using the baseline of this paper, shown in Fig. 16. Still, this pdf shows a curve without any significant irregularities (and similar in shape to the one in Fig. 14), which demonstrates that these equipment effects were adequately removed from the data.

In addition, it can be noted that from many other magnetometers which showed equipment effects, such as Nordkapp and Sørøya, no or very few data are available from the SuperMAG website. Presumably, this is because these data were there classified as "invalid" due to the poor quality. (Note that the SuperMAG project concentrates more on the global magnetic field than on high spatial resolution, making the loss of a few measuring points not vital.)

It can be concluded from this comparison that the baseline according to this paper is in general not very different from that used in SuperMAG. It follows the day-to-day variations of the field slightly more. The main difference between the two is that the method from this paper is better able to deal with equipment effects without having to remove much data. This allows a higher spatial resolution in critical areas, such as the auroral oval boundary north of Scandinavia.

3.3 Error estimate

A reliable evaluation of the error of the baseline procedure of this paper is not possible, since there is no "correct" baseline known in any magnetic field measurement. However, an estimate of the worst-case error can be made. This should be expected in the case of strong equipment effects, such as the equipment drift which happened in December 2009 in Sørøya, shown in Figs. 2 and 10. This is one of the strongest and most sudden variations due to equipment drift found in the database.

In Fig. 10, the B_x field is seen to decrease quite suddenly between 22 and 24 December, but because of the other irregularities, it is hard to say exactly when the change started and ended. The long-term baseline, because of the way it is set up, shows a drift from 22 December, 12:00 UT, to 24 December, 12:00 UT.

However, visual inspection of the graph seems to suggest that the equipment drift might also have started at 23 December, 00:00 UT, and reached its end value on 24 December, 00:00 UT. As a test case, this development was used for an alternative long-term baseline. A full baseline based on this alternative long-term baseline led to a difference with the baseline according to this paper with a maximum of 27 nT (which maximum occurred at 23 December, 00:00 UT), and being above 15 nT for 16 h. Hence, if the trace described above of the equipment drift is assumed correct, then that is the error in the baseline of this paper caused by this most extreme case of equipment drift. Note that other baseline procedures which do not take equipment drift into account would probably encounter due to this drift a much larger error, for a longer time: ignoring this equipment effect completely would lead to an error with a maximum of about 80 nT, and which would be above 15 nT for about 18 days. (However, in most other projects, this section of data would probably be labelled as "invalid" because of the equipment drift.)

Since naturally occurring S_q variations are much smaller than this equipment drift, and occur much less suddenly, the error caused by those should be expected to be much smaller than the values mentioned above. This is why the error estimate above can be seen as a worst-case error, with probably only a single occurrence in the entire database.

4   Other applications

Now that the harmonic coefficients X_h (e.g. in Eq. 1) and Y_h of the magnetometer baselines have been derived over 10 yr and for 32 magnetometer stations, these can be used to examine their long-term behaviour, which represent the solar quiet-time (S_q) diurnal variation of the magnetic field B_x and B_y. Like the disturbance field, also this field is mainly caused by currents in the ionosphere. However, unlike the disturbance field, this field varies with diurnal variations of electron density in the ionosphere, which in turn depends on solar radiation.

Several statistical studies of the harmonic coefficients of S_q diurnal variations, dependent on season, year, location and solar activity, have already been made throughout the 20th century, revealing information about the long-term dependencies of the geomagnetic field (e.g. see the overviews by Matsushita (1968) and Campbell (1989)). The results obtained in the study of the current paper can help to verify those studies, and help to improve prediction models of the S_q magnetic field. This will be the subject of a later paper.

5   Conclusions

A novel method of determining the baselines of magnetometer data has been presented, which makes magnetometer data suitable for the calculation of ionospheric equivalent currents. The full baseline is composed of three components:

- a jump baseline, which contains only the artificial jumps in the data and is otherwise constant.

- a long-term baseline, linearly interpolated between daily median values of the raw data. The daily median values are calculated for all days except the most disturbed ones.

- a diurnal baseline, linearly interpolated between "templates", derived from the quiet days. These templates consist of the first seven harmonics of the diurnal variation of the magnetic field on the quiet days. This method ensures a smooth baseline at all times, avoiding any discontinuities at transitions between days or months.

The baseline derived according to this method is able to follow medium- to long-term variations in the measured magnetic field, such as tidal and secular variations, as well as equipment drifts of individual instruments. The method avoids many problems associated with traditional baseline methods and some of the other recently published methods, and is simpler in procedure than most other recent ones.

The resulting disturbance field (i.e. field after removal of the baseline) has a fairly flat frequency spectrum up to $f = 1/(4\,\mathrm{h})$, and decreases after that.

The baseline follows the day-to-day variations of the magnetic field slightly more than the SuperMAG baseline; the resulting disturbance field distribution is quite similar between these two baselines.

The main advantage of the baseline method of this paper over others is that it removes equipment drift and other artefacts efficiently without discarding too much data, so that even low-quality data from remote unmanned magnetometers can be made suitable. This can give valuable contributions to the equivalent current database, especially in the area near the polar cap boundary.

Acknowledgements. The research leading to these results has received funding from the European Community's Seventh Framework Programme (FP7/2007-2013) within the call for "Exploitation of Space Science and Exploration Data", under grant agreement no. 263325 (ECLAT project).

For the IMAGE ground magnetometer data, the author would like to thank E. Tanskanen and the institutes that maintain the IMAGE magnetometer network.

For the SuperMAG baseline data, the author would like to thank J. Gjerloev, the Johns Hopkins Applied Physics Laboratory and the University of Bergen, Norway.

Edited by: M. Rose

References

Amm, O. and Viljanen, A.: Ionospheric disturbance magnetic field continuation from the ground to the ionosphere using spherical elementary current systems, Earth Planets Space, 51, 431–440, 1999.

Boteler, D. H., Pirjola, R. J., and Nevanlinna, H.: The effects of geomagnetic disturbances on electrical systems at the earth's surface, Adv. Space Res., 22, 17–27, 1998.

Campbell, W. H.: An Introduction to Quiet Daily Geomagnetic Fields, Pure Appl. Geophys., 131, 315–331, 1989.

Chapman, S. and Bartels, J.: Geomagnetism, Clarendon Press, Oxford, UK, 1940.

Davis, T. N. and Sugiura, M.: Auroral electrojet activity index AE and its universal time variations, J. Geophys. Res., 71, 785–801, 1966.

Gjerloev, J. W.: The SuperMAG data processing technique, J. Geophys. Res., 117, A09213, doi:10.1029/2012JA017683, 2012.

Janzhura, A. S. and Troshichev, O. A.: Determination of the running quiet daily geomagnetic variation, J. Atmos. Sol.-Terr. Phy., 70, 962–972, doi:10.1016/j.jastp.2007.11.004, 2008.

Kamide, Y. and Baumjohann, W.: Magnetosphere-Ionosphere Coupling, Physics and chemistry in space planetology, vol. 23, Springer-Verlag, New York, 178 pp., 1993.

Laakso, H., Perry, C., McCaffrey, S., Herment, D., Allen, A. J., Harvey, C. C., Escoubet, C. P., Gruenberger, C., Taylor, M. G. G. T., and Turner, R.: Cluster Active Archive: Overview, in: The Cluster Active Archive, edited by: Laakso, H., Taylor, M. G. T. T., and Escoubet, C. P., Astrophysics and Space Science Proceedings, Springer, Dordrecht, Heidelberg, London, New York, doi:10.1007/978-90-481-3499-1_1, 2009.

Matsushita, S.: Sq and L Current Systems in the Ionosphere, Geophys. J. Astr. Soc., 15, 109–125, 1968.

Pedatella, N. M., Forbes, J. M., and Richmond:, A. D.: Seasonal and longitudinal variations of the solar quiet (Sq) current system during solar minimum determined by CHAMP satellite magnetic field observations, J. Geophys. Res., 116, A04317, doi:10.1029/2010JA016289, 2011.

Price, A. T. and Wilkins, G. A.: New methods for the analysis of geomagnetic fields and their application to the Sq field of 1932–33, Philos. T. Roy. Soc. A, 156, 31–98, 1963.

Pulkkinen, A., Amm, O., Viljanen, A., and BEAR Working Group: Ionospheric equivalent current distributions determined with the method of spherical elementary current systems, J. Geophys. Res., 108, 1053, doi:10.1029/2001JA005085, 2003.

Stauning, P.: Determination of the quiet daily geomagnetic variations for polar regions, J. Atmos. Sol.-Terr. Phy., 73, 2314–2330, doi:10.1016/j.jastp.2011.07.004, 2011.

Sugiura, M.: Hourly values of equatorial Dst for the IGY, Annals of the International Geophysical Year, Vol. XXXV, Part I, International Council of Scientific Unions, Pergamon Press, Oxford, London, Edinburgh, New York, Paris, Frankfurt, 9–45, 1964.

Untiedt, J. and Baumjohann, W.: Studies of polar current systems using the IMS Scandinavian magnetometer array, Space Sci. Rev., 63, 245–390, 1993.

Viljanen, A. and Häkkinen, L.: IMAGE magnetometer network, in: Satellite-Ground Based Coordination Sourcebook, edited by: Lockwood, M., Wild, M. N., and Opgenoorth, H. J., ESA publications SP-1198, ESTEC, Noordwijk, the Netherlands, 111–117, 1997.

Viljanen, A., Nevanlinna, H., Pajunpää, K., and Pulkkinen, A.: Time derivative of the horizontal geomagnetic field as an activity indicator, Ann. Geophys., 19, 1107–1118, doi:10.5194/angeo-19-1107-2001, 2001.

9

In-flight calibration of double-probe electric field measurements on Cluster

Y. V. Khotyaintsev[1], P.-A. Lindqvist[2], C. M. Cully[1,*], A. I. Eriksson[1], and M. André[1]

[1] Swedish Institute of Space Physics, Uppsala, Sweden
[2] Royal Institute of Technology, Stockholm, Sweden
[*] now at: Department of Physics and Astronomy, University of Calgary, Calgary, Canada

Correspondence to: Y. V. Khotyaintsev (yuri@irfu.se)

Abstract. Double-probe electric field instrument with long wire booms is one of the most popular techniques for in situ measurement of electric fields in plasmas on spinning spacecraft platforms, which have been employed on a large number of space missions. Here we present an overview of the calibration procedure used for the Electric Field and Wave (EFW) instrument on Cluster, which involves spin fits of the data and correction of several offsets. We also describe the procedure for the offset determination and present results for the long-term evolution of the offsets.

1 Introduction

Double-probe electric field experiments have been flown on a number of spacecraft (see review by Pedersen et al., 1998) including Cluster (Gustafsson et al., 1997, 2001), and calibration of the direct-current (DC) electric field has always been a challenging and time-consuming task. Main reasons for this are strong influence of the ambient plasma and the spacecraft itself on the measurements. Other techniques to measure electric fields at low time resolution as electron drift instruments and ion spectrometers, such as Electron Drift Instrument (EDI) (Paschmann et al., 1997, 2001) and Cluster Ion Spectrometry (CIS) (Rème et al., 2001) on Cluster, are immune to some of the problems affecting the double-probe measurements, but have their own limitations. With the Cluster Active Archive a semi-automatic approach to in-flight calibration of the DC electric field data has been developed (Khotyaintsev et al., 2010), and the purpose of this paper is to describe the main elements of this calibration procedure under nominal operations.

An example of a clear deviation from nominal performance is the non-geophysical electric field detected by Electric Field and Wave experiment (EFW) due to the wake behind the spacecraft caused by cool (eV) outflowing ionospheric ions drifting at supersonic velocities (Eriksson et al., 2006). Such ions are common in the magnetospheric tail lobes. Careful investigation of this "problem" has resulted in a new method to detect positive low-energy ions, otherwise invisible to detectors on a sunlit spacecraft positively charged to several volts (Engwall et al., 2006, 2009a, b; André and Cully, 2012). However, it is not possible to recover the ambient geophysical electric field in such cases.

2 Short instrument description

The detector of the Cluster EFW instrument consists of four spherical sensors numbered 1 to 4 deployed orthogonally on 42.5 m long wire booms in the spin plane of the spacecraft (see Fig. 1). The spacecraft makes one rotation in approximately 4 s. The potential drop between two sensors, separated by 88 m (or 62 m in case of using non-opposing probes) tip to tip, is measured to provide an electric field measurement. The probe difference signals are normally routed through 10 Hz anti-aliasing low-pass filters when sampled at $25\,s^{-1}$, and through 180 Hz low-pass filters when sampled at $450\,s^{-1}$. The potential difference between each sensor and the spacecraft is measured separately with a sampling frequency of $5\,s^{-1}$ after routing through low-pass filters with

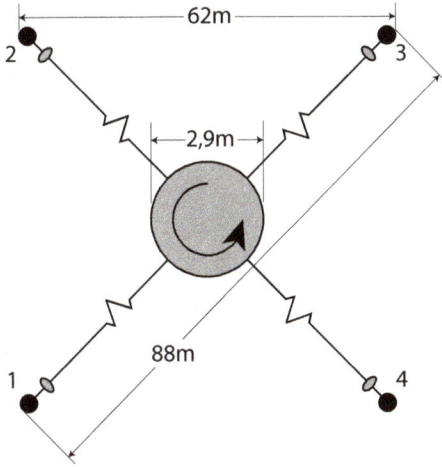

Figure 1. Cluster EFW double-probe electric field instrument.

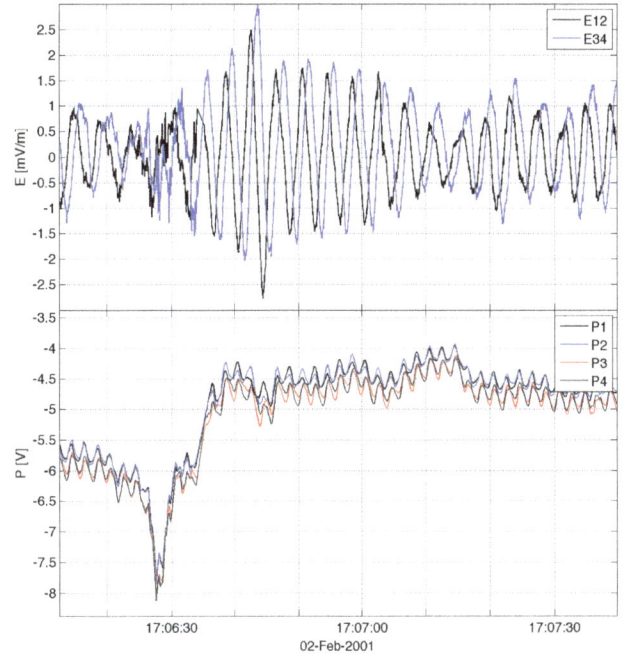

Figure 2. Raw data of the electric field, E_{12}, E_{34} (upper panel) and of the probe-to-spacecraft potential P_1, P_2, P_3, P_4 (bottom panel) measured by Cluster 1.

a cut-off frequency of 10 Hz. A detailed description of the EFW instrument can be found in Gustafsson et al. (1997, 2001).

3 Calibration procedure

The goal of the calibration procedure is to obtain geophysical DC electric field in the spacecraft spin plane in a despun reference frame. On Cluster we used the ISR2 (Inverted Spin Reference) system, also known as DSI (Despun System Inverted). The x and y axes are in the spin plane, with x axis pointing as near sunward as possible and y axis perpendicular to the sunward direction, positive towards dusk. The z axis is along the (negative) spacecraft spin axis, towards the north ecliptic. The coordinate system is called "inverted" because the actual spin axis of Cluster is pointing towards the south ecliptic. The ISR2 system thus is identical to Geocentric Solar Ecliptic system (GSE) if the satellite spin axis angle to ecliptic north is zero, and is a good approximation to GSE for the usual case of this angle being a few degrees.

3.1 Raw data

The raw data available from EFW under normal circumstances are the two orthogonal electric field components in the spinning frame (E_{12} and E_{34}) sampled at 25 or 450 Hz, as well as potentials of the individual probes (P_1, P_2, P_3 and P_4) sampled at 5 Hz. In case of probe 1 failure (for dates of permanent failures on C1, C2 and C3, see Lindqvist et al., 2013), instead of E12 we use E32. An example of raw data is shown in Fig. 2.

As the first stage of calibration, it is necessary to perform initial cleaning of the data at which we remove intervals with bad data due to issues with electronics, probe saturations due to low plasma density (often occurring in in the magnetospheric lobes), and saturations due to non-optimal

bias current settings occurring in dense plasmas such as magnetosheath and plasmasphere (Khotyaintsev et al., 2010). If the spacecraft is in the solar wind, we apply a correction for the wakes usually present in the raw data (Eriksson et al., 2007).

3.2 Spin fits

After initial cleaning of the data, a spin fitting procedure is performed; the output of this procedure provides basic parameters that are used later in the calibration procedure. In the presence of a constant ambient electric field, the raw data signal (probe potential difference) is a sine wave (see Fig. 2, upper panel) where the amplitude and phase give the electric field magnitude and direction. A least-squares fit to the raw data of the form

$$y = A + B\sin(\omega t) + C\cos(\omega t) + D\sin(2\omega t)$$
$$+ E\cos(2\omega t) + \ldots \tag{1}$$

is done once every 4 s ($2\pi/\omega \approx 4$ s is approximately the spacecraft spin period) and the fit is applied to 4 s long time intervals.

The standard deviation of the raw data from the fitted sine wave can be used as an indication of high-frequency variations in the data. Higher order terms, D, E, $\ldots$, may be used for diagnostics of data quality: normally the higher order terms are much smaller than B and C, and the opposite situation would indicate problems with the measurements.

3.3 Offsets

The sine and cosine terms, B and C, after correction for ISR2 offsets provide the 4 s (spin) resolution electric field in ISR2:

$$E_{x4s} = \alpha \left(B - \Delta E_x \right), \tag{2}$$

$$E_{y4s} = \alpha \left(C - \Delta E_y \right), \tag{3}$$

where α is the amplitude correction factor due to the ambient electric field being "short-circuited" by the presence of the spacecraft and wire booms (see Sect. 4.1). And ΔE_x (sunward offset) and ΔE_y (duskward offset) are the *ISR2 offsets*, which represent the difference between the measured and geophysical electric fields in the despun frame and are discussed in detail later.

As the spin fitting procedure would typically yield different values for the electric field from the two different probe pairs, it is useful to introduce an additional offset which describes the difference between the two measurements, Δ_{p12p34}, which we call the *delta offset*:

$$\Delta_{xp12p34} = E_{x4s}\left(E_{12}\right) - E_{x4s}\left(E_{34}\right), \tag{4}$$

$$\Delta_{yp12p34} = E_{y4s}\left(E_{12}\right) - E_{y4s}\left(E_{34}\right). \tag{5}$$

The despun full-resolution electric field is obtained as follows:

$$E_x = Re\left[\varepsilon_{12}\right] - \Delta_{xp12p34} + Re\left[\varepsilon_{34}\right], \tag{6}$$

$$E_y = Im\left[\varepsilon_{12}\right] - \Delta_{yp12p34} + Im\left[\varepsilon_{34}\right], \tag{7}$$

where $\varepsilon_{12} = (E_{12} - \Delta_{raw\ 12})\,e^{i\phi12}$, $\varepsilon_{34} = (E_{34} - \Delta_{raw\ 34})\,e^{i\phi34}$, and $\phi_{12} = \phi_{34} + \pi/2$ is the spin phase of probe 1 with respect to the sun; *raw data DC offset*, $\Delta_{raw} = \langle A \rangle$, is based on parameter A of the fit (Eq. 1). Ideally, the DC level of the raw data should be zero. However, small differences between the probe surfaces and in the electronics create a DC offset in the raw data. If not corrected, it shows up as a signal at the spin frequency in the despun electric field.

It must be noted that asymmetries due to the direction to the sun have the dominant contribution to the offsets, so that the following inequalities are typically satisfied:

$$\Delta E_x \gg \Delta E_y, \tag{8}$$

$$\Delta_{xp12p34} \gg \Delta_{yp12p34}. \tag{9}$$

4 Results

In this section we summarize the main results concerning the various offsets defined above. Raw data DC offset and delta offsets are obtained from spin fits, while the amplitude correction factor and ISR2 offsets are obtained based on inter-spacecraft calibration as well as cross-calibration with CIS (Rème et al., 2001) and EDI (Paschmann et al., 1997, 2001).

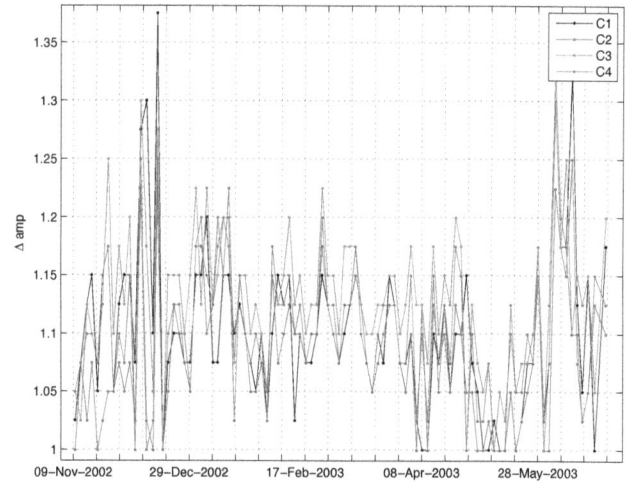

Figure 3. Amplitude correction factor for the electric field measured by EFW on Cluster 1–4 during a solar wind season from November 2002 to June 2003.

4.1 Amplitude factors

Amplitude factors are needed since the electric field is partially "short circuited" by the spacecraft potential, which is also the potential of the wire booms, extending out to a large distance from the spacecraft (Cully et al., 2007).

We have used the ISR2 y components of the electric field to determine the amplitude correction factor. This component of E is generally free from offsets, and thus by comparing E_y from EFW and CIS-HIA we are able to deduce the amplitude correction factor. Results of such computations for the spring season of 2002 are shown in Fig. 3. Every point in the plot corresponds to one orbit of data. One should mention that variations seen in the data are not caused by changes in the factor, but rather by "bad data" and insufficient data coverage.

On the basis of simulations and comparisons with other Cluster instruments, it has been determined that the measured electric field magnitude needs to be multiplied by a factor of $\alpha = 1.1$. We use this constant value through the entire mission. This value is consistent with valued obtained from simulations of the spacecraft–plasma interaction (Cully et al., 2007).

4.2 Raw data DC offset

The raw data DC offset, Δ_{raw}, from the both probe pairs is used to calculate the full-resolution E-field. It is applied to E_{12} and E_{34} prior to despinning. Variations in the electric field will result in small changes to A computed from spin fits for different 4 s intervals. So if Δ_{raw} depended only on the electronics, one could compute a long-term average of A and use it as Δ_{raw}. But we find that A also depends on the surrounding plasma environment as illustrated in Fig. 4, where A is plotted as a function of the spacecraft potential.

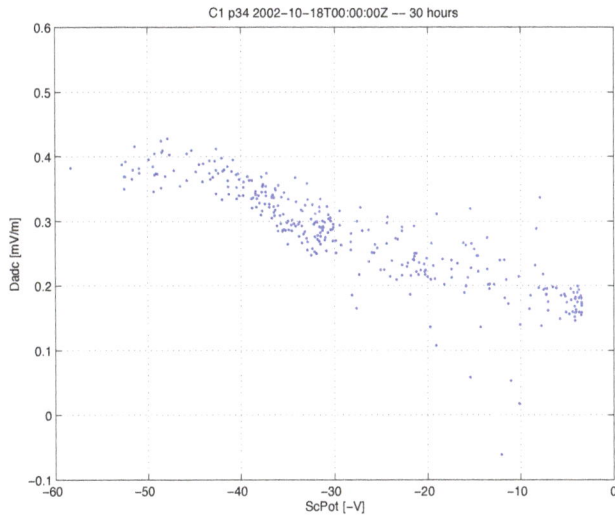

Figure 4. Dependence of the raw data DC offset, Δ_{raw}, computed from the spin fits on the spacecraft potential showing that the offset decreases when the potential is close to zero (characteristic of dense plasmas).

Therefore we want a smoothened value and at the same time to track changes in the plasma environment. That is why the DC offset is smoothed, $\Delta_{\text{raw}} = \langle A \rangle$, using a weighted average over seven spins using weights [0.07, 0.15, 0.18, 0.2, 0.18, 0.15, 0.07]. This approach was selected based on an empirical basis after testing several different lengths of the averaging interval.

4.3 ISR2 offsets

Our main assumption in the study of ISR2 offsets is that the offsets depend on the instrument configuration, spacecraft attitude and that the dependence on surrounding plasma parameters is weak, i.e., being in the same kind of plasma environment (for example plasma sheet) and having the same instrument settings and probe properties for two different time intervals, and the difference between ISR2 offsets for the two intervals must be within the uncertainty of the offset determination (fraction of mV m^{-1}). As the offsets still depend on the plasma environment, we decided to split the data set into two groups – "solar wind/magnetosheath" and "magnetosphere" – which correspond to two situations with "cold and dense" and "hot and rarefied" plasmas. To split every orbit into these two groups, we have used the Shue magnetopause model (Shue et al., 1997) with realistic solar wind parameters measured by the ACE spacecraft. For each of the groups we statistically determine offsets over a period of several weeks to several months in order to account for changes in the instrument setting, spacecraft attitude, solar UV flux, etc. Then, based on the position along the orbit, observed value of the spacecraft potential and manual inspection, we

determine which of the two offsets is to be applied at each point.

4.3.1 ISR2 offsets in the solar wind and magnetosheath

For the solar wind/magnetosheath intervals, we first perform the inter-spacecraft calibration under assumption that all the spacecraft observe the same large-scale electric field, which is the case in the solar wind. As a result for each interval (from outbound magnetopause crossing to the inbound, typically several hours long), we get relative offset between the spacecraft, which are the differences in E_x and E_y between the different spacecraft averaged over the entire interval. Figure 5 shows an example of such an interval, and the two upper panels show E_x and E_y from all four spacecraft.

Then by using CIS-HIA from C1 and/or C3 as reference data, we find the ISR2 offsets for EFW for each of the spacecraft. We get one value for offsets per orbit. The procedure can be controlled visually by using a type of plot presented in Fig. 5. The two upper panels show all the available EFW and CIS-HIA data (E_x and E_y in ISR2). Then we construct the reference E-field from CIS-HIA by averaging data from the spacecraft where CIS-HIA data are available. Such averaging is possible as the difference between the spacecraft in the solar wind/magnetosheath is typically small. Then we compute the difference between the EFW E_x on all spacecraft and the reference E-field; this difference is plotted in the third panel. Average of the difference over the entire interval gives the local IRS2 offset. This offset is then applied to the EFW on different spacecraft. The resulted corrected and reference E-fields are plotted in the two bottom panels.

Evolution of ISR2 offsets in the solar wind and magnetosheath during mission lifetime is shown in Fig. 6. One can see that the offsets are rather steady and slowly decreasing with approach of the solar minimum ($\sim$ 2009). The only striking feature is the sudden increase of the offset in C3 in 2005. This change is not yet understood.

4.3.2 ISR2 offsets in the magnetosphere

The problem of determining the offsets in the magnetosphere is significantly more complicated in comparison to the solar wind/magnetosheath. Data from the other instruments, which could have been used as a reference, are of very low quality in large areas of the magnetosphere due to low counts (the CIS instrument) or low magnetic fields (the EDI instrument). Also the EFW data are subject to frequent problems, such as electrostatic wakes, and the data affected by wakes need to be excluded from the data set used to determine the offsets.

In the ISR2 offset determination procedure, we decided not to use any reference data, but rather to use a condition of zero electric field $\langle E_x \rangle = 0$, as most of the time the electric fields are very weak in the magnetotail ($X\,\text{GSE} < 0$ and $R > 5\,\text{RE}$), and the average over a tail season must be very close to zero, and the difference from zero gives a rather good

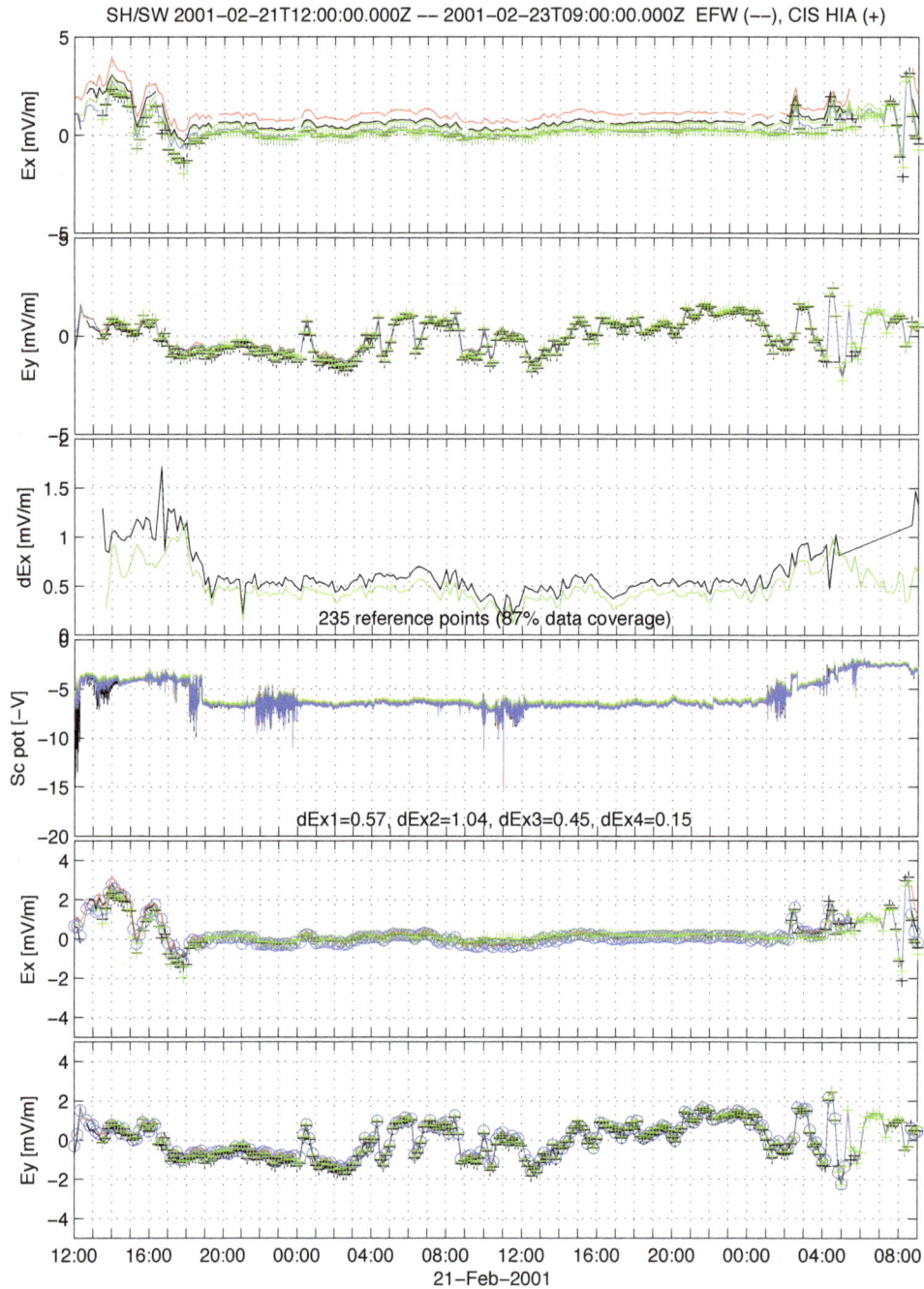

Figure 5. Inter-spacecraft calibration and cross-calibration with CIS in the solar wind/magnetosheath. Panels from top to bottom show E_x and E_y measured by EFW (solid lines) on the four spacecraft and by CIS-HIA (+) on C1 and C3, difference in E_x between CIS and EFW (median value from all spacecraft), the spacecraft potential, and the two bottom panels show the same data as on the top, but with the offsets applied to the EFW data. Data from the four Cluster spacecraft shown by black (C1), red (C2), green (C3) and blue (C4).

estimate for the ISR2 E_x offset. The resulting offsets were verified against the CIS data for a large number of cases, and in particular in the central plasma sheet the agreement is very good.

Results for Cluster 4 for years 2002–2005 are summarized in Fig. 7. One can see that there is a prominent peak around

$1.3 \, \mathrm{mV \, m^{-1}}$ for all years. However, there is also some group of points giving rise to a broadening towards lower offset values. Therefore we can conclude that the offset value is rather stable, and the broadening is due to actual geophysical electric fields present in the magnetosphere.

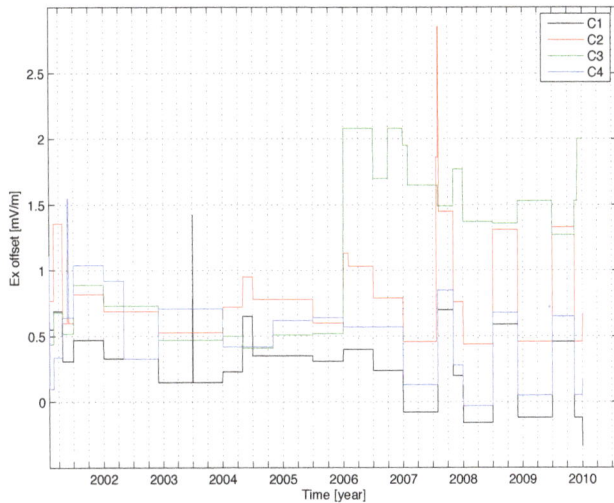

Figure 6. Long-term evolution of the ISR2 E_x (sunward) offset in the solar wind/magnetosheath from 2001 to 2009.

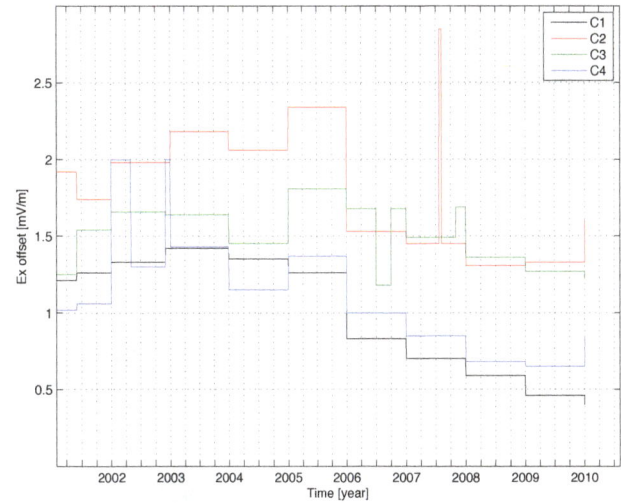

Figure 8. Long-term evolution of the ISR2 offsets in the magnetosphere from 2001 to 2009.

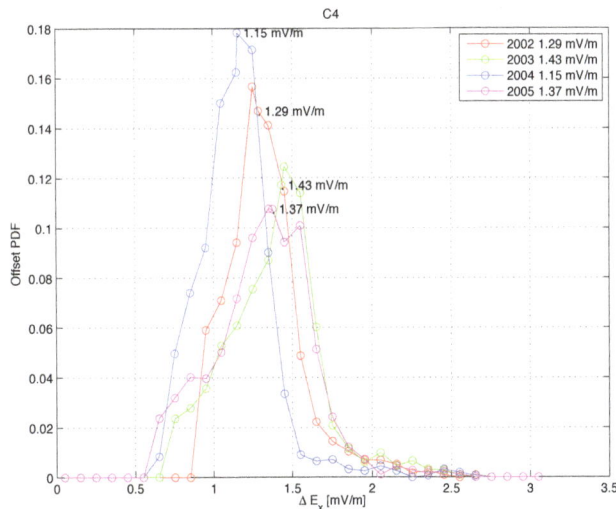

Figure 7. Probability distribution function of ISR2 E_x (sunward) offset for Cluster 4 in the magnetosphere for 2002–2005.

Evolution of ISR2 offsets in the magnetosphere during mission lifetime is shown in Fig. 8. The offsets are steady and slowly decreasing with approach of the solar minimum.

4.4 Delta offset

Given the two identical probe pairs, we are able to estimate the electric field at the timescale of the spacecraft spin from each of them, and in principle these estimates should be identical. In reality the probes are not identical, and the estimates of the electric fields differ. Such a difference is described by the delta offset. Figure 9 shows how the delta offsets change over time. The curves show the raw data, i.e., the difference between the electric fields computed from the two probe pairs averaged over 1.5 h long intervals of data. One can see

that the offset varies very slowly, at a typical timescale of several months, and with some sudden jumps typically related to spacecraft manoeuvres. Therefore for the calibration purposes we use a smoothened version of the offset, i.e., median over approximately two orbits. This approach allows us to get rid of the outliers, which can be caused by intervals with non-optimal instrument performance or strong geophysical electric fields.

Figure 10 shows the long-term evolution of the delta offsets for all four spacecraft. Variations in the offset are caused by a number of factors. First is the solar cycle. One can see that the offset is rather small and steady in the beginning of the mission and starts to grow with approach of the solar minimum, reaching its maximum in spring 2006. This behavior is caused by non-optimal bias current settings, and the situation became significantly improved by lowering the bias current in June 2006. The second cause is the probe failures, which forced usage of P32 (shorter base and asymmetric with respect to the spacecraft) instead of P12 (see Fig. 1).

5 Discussion and conclusions

Here we presented an overview of the calibration procedure used for the DC electric field measurements by EFW on Cluster, which is applied for production of the data for the Cluster Active Archive (CAA). EFW measures potential difference between probes mounted on long wire booms, which, after some corrections, can be used to construct the electric field in the spacecraft spin plane. We show that the calibration procedure leading to an estimate of the geophysical electric field can be described by a set of offsets, which are determined from the symmetry considerations enabled by rapid spacecraft spin as well as from statistical comparisons with other measurements of the electric field on Cluster. We show

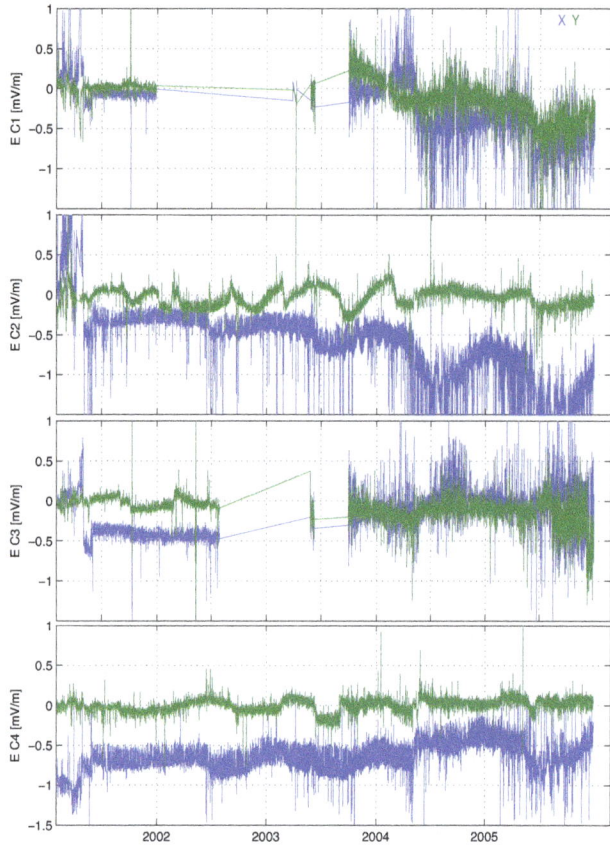

Figure 9. Evolution of the "raw" delta offsets on Cluster 1–4 from 2001 to 2005. The blue and green lines show X and Y components of the difference between the electric fields computed from the different probe pairs, i.e., the "raw" delta offsets. In order to get rid of the outliers (spikes), we compute the median of this difference, which is then used as the delta offset applied to data during the calibration process.

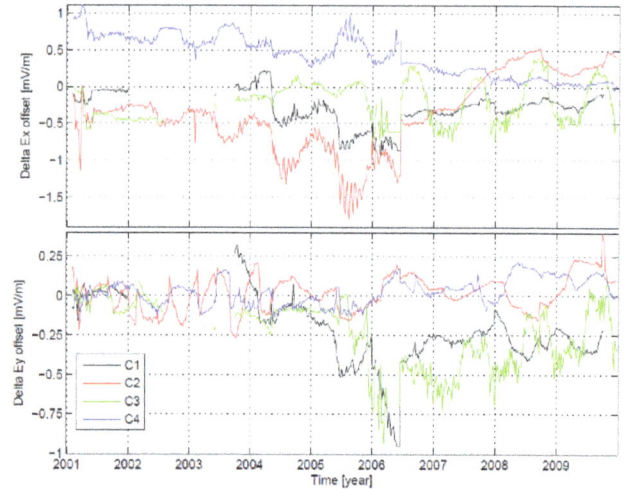

Figure 10. Long-term evolution of delta offsets from 2001 to 2009 for C1 (black), C2 (red), C3 (green) and C4 (blue).

that most of the offsets have a rather slow variation with time on time–space timescales of weeks and month.

Both the amplitude factor (boom shortening) and the sunward offset depend on the Debye length and the spacecraft potential. For the Debye length shorter than the probe-to-puck distance (1.5 m for EFW), there is no boom shortening (i.e., amplitude factor = 1); for the longer Debye length the effective boom length is shorter than the physical length, and such a shortening in some way depends on the spacecraft potential. We were unable to establish an empirical relation of the shortening factor with the spacecraft potential for EFW and used a constant amplitude factor for the CAA calibration as the measurement is performed in the long Debye length regime most of the time. We note that even determination of the Debye length on a routine basis is a challenging task for Cluster, due to high uncertainties and insufficient time resolution of the electron data.

Factors determining the sunward offset are much less understood. Simulations by Cully et al. (2007) suggested that

the offset appears due to an asymmetric photoelectron cloud around the spacecraft. However, analyzing changes in the sunward offset for EFW we found that the offset may be strongly driven by other factors, not only for the photoelectron cloud asymmetry. For example the offset in the solar wind depends on the solar wind speed in such a way that it decreases with an increase of the wind speed, and for sufficiently fast solar wind the offset turns into anti-sunward. Qualitative and quantitative understanding of the dependence of the amplitude factor and sunward offset on the surrounding plasma remains an open question, which will be addressed in the future by advanced numerical simulations and possibly by empirical comparison of the electric field measurements by the double probes to EDI and particle instruments on Magnetospheric Multiscale (MMS) Mission.

Using particle measurements to determine time-dependent offsets and to correct the EFW data on a routine basis showed to be practically impossible for Cluster, as such a correction would typically introduce a large random error into the electric field measurement. However, we do no exclude a possibility for such a correction for case studies where one has full control of the quality of the reference data used to determine the offsets. Our choice of the calibration procedure for the CAA which is outlined in this paper is to rely solely on the EFW data for high-time-resolution calibrations, and then to use statistical offsets determined by averaging large amounts of data, and among others by comparison with the particle measurements.

We should note that the presented approach is not capable of correcting for fast changes of the offsets due to rapid crossing of plasma boundaries such as the bow shock and magnetopause. At present, calibration of such events usually requires manual calibration in order to achieve precision on the DC electric field below $1 \, \mathrm{mV \, m^{-1}}$, and reliable calibration is only possible when all probes respond to changes in the plasma environment in a very similar way, and the effect of these changes on offsets is not too drastic. Developing a procedure that would produce reliable results also for such cases during routine data production remains a challenging task.

As a concluding remark, we note that the described calibration procedure applies to data acquired when the instrument operates close its optimal regime, so that one can reconstruct the ambient electric field present in the plasma by applying relatively small corrections. However, a major effort during the CAA production goes into detection of strong deviations from the nominal operations (Khotyaintsev et al., 2010), which can be caused by both changes of the plasma environment surrounding the spacecraft and non-optimal instrument settings.

Acknowledgements. The authors thank Andris Vaivads and other colleagues for continuing support and discussion around the coffee breaks.

Edited by: H. Laakso

References

André, M. and Cully, C. M.: Low-energy ions: a previously hidden solar system particle population, Geophys. Res. Lett., 39, L03101, doi:10.1029/2011GL050242, 2012.

Cully, C. M., Ergun, R. E., and Eriksson, A. I.: Electrostatic structure around spacecraft in tenuous plasmas, J. Geophys. Res., 112, A09211, doi:10.1029/2007JA012269, 2007.

Engwall, E., Eriksson, A. I., André, M., Dandouras, I., Paschmann, G., Quinn, J., and Torkar, K.: Low-energy (order 10 eV) ion flow in the magnetotail lobes inferred from spacecraft wake observations, Geophys. Res. Lett., 33, L06110, doi:10.1029/2005GL025179, 2006.

Engwall, E., Eriksson, A. I., Cully, C. M., André, M., Torbert, R., and Vaith, H.: Earth's ionospheric outflow dominated by hidden cold plasma, Nat. Geosci., 2, 24–27, 2009a.

Engwall, E., Eriksson, A. I., Cully, C. M., André, M., Puhl-Quinn, P. A., Vaith, H., and Torbert, R.: Survey of cold ionospheric outflows in the magnetotail, Ann. Geophys., 27, 3185–3201, doi:10.5194/angeo-27-3185-2009, 2009b.

Eriksson, A. I., André, M., Klecker, B., Laakso, H., Lindqvist, P.-A., Mozer, F., Paschmann, G., Pedersen, A., Quinn, J., Torbert, R., Torkar, K., and Vaith, H.: Electric field measurements on Cluster: comparing the double-probe and electron drift techniques, Ann. Geophys., 24, 275–289, doi:10.5194/angeo-24-275-2006, 2006.

Eriksson, A. I., Khotyaintsev, Y., and Lindqvist, P.-A.: Spacecraft wakes in the solar wind, in: Proceedings of the 10th Spacecraft Charging Technology Conference (SCTC-10), available at: http://www.space.irfu.se/aie/publ/Eriksson2007b.pdf (last access: 30 January 2014), 2007.

Gustafsson, G., Boström, R., Holback, B., Holmgren, G., Lundgren, A., Stasiewicz, K., Åhlén, L., Mozer, F. S., Pankow, D., Harvey, P., Berg, P., Ulrich, R., Pedersen, A., Schmidt, R., Butler, A., Fransen, A. W. C., Klinge, D., Thomsen, M., Fälthammar, C.-G., Lindqvist, P.-A., Christenson, S., Holtet, J.,

Lybekk, B., Sten, T. A., Tanskanen, P., Lappalainen, K., and Wygant, J.: The Electric Field and Wave Experiment for the Cluster Mission, Space Sci. Rev., 79, 137–156, 1997.

Gustafsson, G., André, M., Carozzi, T., Eriksson, A. I., Fälthammar, C.-G., Grard, R., Holmgren, G., Holtet, J. A., Ivchenko, N., Karlsson, T., Khotyaintsev, Y., Klimov, S., Laakso, H., Lindqvist, P.-A., Lybekk, B., Marklund, G., Mozer, F., Mursula, K., Pedersen, A., Popielawska, B., Savin, S., Stasiewicz, K., Tanskanen, P., Vaivads, A., and Wahlund, J.-E.: First results of electric field and density observations by Cluster EFW based on initial months of operation, Ann. Geophys., 19, 1219–1240, doi:10.5194/angeo-19-1219-2001, 2001.

Khotyaintsev, Y., Lindqvist, P.-A., Eriksson, A. I., and André, M.: The EFW Data in the CAA, the Cluster Active Archive, Studying the Earth's Space Plasma Environment, in: Astrophysics and Space Science Proceedings, edited by: Laakso, H., Taylor, M. G. T. T., and Escoubet, C. P., Springer, Berlin, 97–108, 2010.

Lindqvist, P.-A., Cully, C. M., and Khotyaintsev, Y.: User Guide to the EFW measurements in the Cluster Active Archive (CAA), available at: http://caa.estec.esa.int/caa/ug_cr_icd.xml (last access: 30 January 2014), 2013.

Paschmann, G., Melzner, F., Frenzel, R., Vaith, H., Parigger, P., Pagel, U., Bauer, O., Haerendel, G., Baumjohann, W., Sckopke, N., Torbert, R., Briggs, B., Chan, J., Lynch, K., Morey, K., Quinn, J., Simpson, D., Young, C., McIlwain, C., Fillius, W., Kerr, S., Mahieu, R., and Whipple, E.: The electron drift instrument for cluster, Space Sci. Rev., 79, 233–269, 1997.

Paschmann, G., Quinn, J. M., Torbert, R. B., Vaith, H., McIlwain, C. E., Haerendel, G., Bauer, O. H., Bauer, T., Baumjohann, W., Fillius, W., Förster, M., Frey, S., Georgescu, E., Kerr, S. S., Kletzing, C. A., Matsui, H., Puhl-Quinn, P., and Whipple, E. C.: The Electron Drift Instrument on Cluster: overview of first results, Ann. Geophys., 19, 1273–1288, doi:10.5194/angeo-19-1273-2001, 2001.

Pedersen, A., Mozer, F., and Gustafsson, G.: Electric field measurements in a tenuous plasma with spherical double probes, in: Measurement Techniques in Space Plasmas – Fields: Geophysical Monograph 103, edited by: Pfaff, R. F., Borovsky, J. E., and Young, D. T., published by the American Geophysical Union, Washington, D.C., USA, 1–12, 1998.

Rème, H., Aoustin, C., Bosqued, J. M., Dandouras, I., Lavraud, B., Sauvaud, J. A., Barthe, A., Bouyssou, J., Camus, Th., Coeur-Joly, O., Cros, A., Cuvilo, J., Ducay, F., Garbarowitz, Y., Medale, J. L., Penou, E., Perrier, H., Romefort, D., Rouzaud, J., Vallat, C., Alcaydé, D., Jacquey, C., Mazelle, C., d'Uston, C., Möbius, E., Kistler, L. M., Crocker, K., Granoff, M., Mouikis, C., Popecki, M., Vosbury, M., Klecker, B., Hovestadt, D., Kucharek, H., Kuenneth, E., Paschmann, G., Scholer, M., Sckopke, N., Seidenschwang, E., Carlson, C. W., Curtis, D. W., Ingraham, C., Lin, R. P., McFadden, J. P., Parks, G. K., Phan, T., Formisano, V., Amata, E., Bavassano-Cattaneo, M. B., Baldetti, P., Bruno, R., Chionchio, G., Di Lellis, A., Marcucci, M. F., Pallocchia, G., Korth, A., Daly, P. W., Graeve, B., Rosenbauer, H., Vasyliunas, V., McCarthy, M., Wilber, M., Eliasson, L., Lundin, R., Olsen, S., Shelley, E. G., Fuselier, S., Ghielmetti, A. G., Lennartsson, W., Escoubet, C. P., Balsiger, H., Friedel, R., Cao, J.-B., Kovrazhkin, R. A., Papamastorakis, I., Pellat, R., Scudder, J., and Sonnerup, B.: First multispacecraft ion measurements in and near the Earth's magnetosphere with the identical Cluster ion spectrometry (CIS) experiment, Ann. Geophys., 19, 1303–1354, doi:10.5194/angeo-19-1303-2001, 2001.

Shue, J.-H., Chao, J. K., Fu, H. C., Russell, C. T., Song, P., Khurana, K. K., and Singer, H. J.: A new functional form to study the solar wind control of the magnetopause size and shape, J. Geophys. Res., 102, 9497–9511, doi:10.1029/97JA00196, 1997.

10

In-flight calibration of the Cluster PEACE sensors

N. Doss[1], **A. N. Fazakerley**[1], **B. Mihaljčić**[1], **A. D. Lahiff**[1,*], **R. J. Wilson**[1,**], **D. Kataria**[1], **I. Rozum**[1,***], **G. Watson**[1], and **Y. Bogdanova**[1,*]

[1]Mullard Space Science Laboratory, University College London, Dorking, UK
[*]now at: Rutherford Appleton Laboratory, Oxford, UK
[**]now at: University of Colorado Boulder, Colorado, USA
[***]now at: European Centre for Medium-Range Weather Forecasts, Reading, UK

Correspondence to: N. Doss (n.doss@ucl.ac.uk)

Abstract. The Plasma Electron and Current Experiment (PEACE) instruments operate on all four of the Cluster spacecraft and measure the 3-D velocity distribution of electrons in the energy range from 0.59 eV to 26.4 keV during each spacecraft spin. Pitch angle distributions and moments of the velocity distribution are also produced. As the mission has progressed, the efficiency of the detectors has declined. Several factors may play a role in this decline such as exposure to radiation, high electron fluxes and spacecraft thruster firings. To account for these variations, continuous in-flight calibration work is essential. The purpose of this paper is to describe the PEACE calibration parameters, focussing in particular on those that vary over time, and to describe the methods which are used to determine their evolution.

1 Introduction

A detailed description of the Plasma Electron and Current Experiment (PEACE) instrument is not provided here but can be found in Johnstone et al. (1997) and Fazakerley et al. (2010a).

Each of the Cluster spacecraft carries an identical PEACE instrument which consists of two sensors and a data processing unit. Both sensors are capable of covering the full energy range of the instrument, but each sensor usually covers about 70 % of the instrument energy range in any given spin. The LEEA (Low Energy Electron Analyser) sensor has a smaller geometric factor appropriate for the higher fluxes that are normally found at the lower energies such as in the solar wind and magnetosheath. The HEEA (High Energy Electron Anal-

yser) sensor has a larger geometric factor better suited for the weaker fluxes seen in the magnetosphere. Used together the sensors can cover the full energy range every spin.

Each of the sensors is a "top hat" electrostatic analyser (Carlson et al., 1982), whose operational principle is illustrated in Fig. 1. A voltage is applied across the hemispheres of the analyser which diverts electrons of a specific energy and acceptance angle (shown in blue) through the analyser to the semi-annular micro-channel plate (MCP) detector. Sunlight passes through the aperture and out again; measures are taken to minimise the amount of light reflected within the analyser and reaching the MCP. Electrons which do not have the selected energy (shown in red) strike the analyser hemispheres and are not counted. When an electron reaches the MCP, the signal is amplified and the resulting charge cloud is detected in one of the 12 segments of the anode beneath, providing information about the direction in which the electrons were travelling. The number of electrons that result for each incident electron is defined as the gain of the MCP. A voltage is applied across the MCP in order to produce charge amplification.

The two PEACE sensors are mounted on opposite sides of the spacecraft with their field-of-view fans lying perpendicular to the spacecraft surface as illustrated in Fig. 2. The field of view of each PEACE sensor perpendicular to the spacecraft frame is 3.8° (HEEA) and 2.9° (LEEA). The azimuthal angle is measured in the spacecraft spin plane, while the polar angle is measured in the plane orthogonal to the spin plane. Each individual sensor has an 180° field of view and covers a 4π field of view in one spacecraft spin. The combined field of view of the two sensors covers the complete

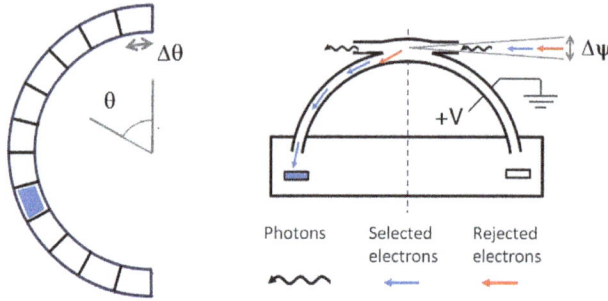

Fig. 1. Illustration of the principle of the PEACE electrostatic analyser.

4π solid angle range during half a spacecraft spin, in the energy range overlapped by the two sensors. The sensor numbering in Fig. 2 shows the direction from which the arriving electrons are counted on the 12 anodes (0 to 11); e.g. zone 0 looks toward $-X_b$ and sees electrons travelling with velocities along the spin axis direction $+X_b$.

2   The calibration parameters

During a spin each PEACE sensor can sample the velocity distribution of the plasma electrons by making a series of individual measurements in a set of different look directions and energies. Such a measurement gives the velocity space density of the electrons, f_{ijk}, in the small region of velocity space defined by polar angle i, azimuthal angle j and energy (speed) interval k. The velocity space density is related to measured quantities and calibration factors as follows:

$$f_{ijk} = \frac{P_{ijk}}{t_{acc} v_k^4 G_i \varepsilon_{ik}}, \qquad (1)$$

where P_{ijk} is the number of electrons counted after dead time correction (related to instrument electronics not the MCP); t_{acc} is the data accumulation time, a fixed fraction of the spin period; v_k is the mean value of the measured electron speed during time t_{acc}; G_i is the geometric factor for the ith polar angle sector, which in a perfectly concentric analyser reduces to a single value G for all sectors (the geometric factor is different for HEEA and LEEA sensors due to different mechanical designs for the electrostatic analyser entrance aperture and collimator); and $\varepsilon_{ik} = \varepsilon_0 \varepsilon(v_k^2)_i$ is the detector efficiency, which varies with time, position on the detector and electron energy. It is defined as the probability that a particle reaching the detector is actually registered. ε_0 is an energy- and position-independent efficiency term, and $\varepsilon(v_k^2)_i$ is the relative sensitivity of the detector as a function of position (i.e. anode segment) and electron energy.

2.1   Ground calibration

Ground calibration work established values for the non-time-varying parameters for each individual sensor before launch.

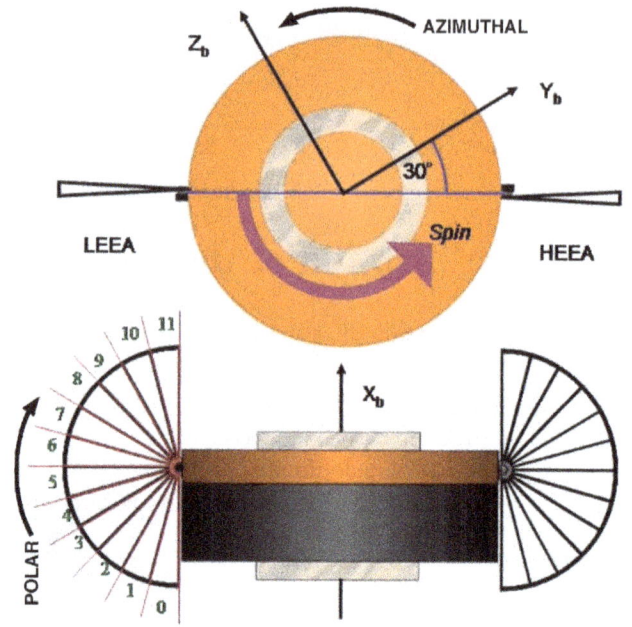

Fig. 2. The physical deployment of the PEACE LEEA and HEEA sensors on the spacecraft. The spacecraft body co-ordinate system is shown (X_b, Y_b, Z_b). In orbit the spacecraft spin axes are maintained roughly anti-parallel to the GSE z axis.

The electrostatic analysers of the four HEEA sensors were made as mutually identical as possible, similarly for the four LEEA sensors. All eight sensors use the same equipment to control the electron energy selection and to count detected electrons. The least controllable aspect of the design is the efficiency of the individual MCP detectors in each sensor. Values were obtained for the geometric factor G_i, the energies measured during energy sweeps v_k, and the relative sensitivity of the detector $\varepsilon(v_k^2)_i$, under conditions of optimum detector performance. Due to the difficulty in measuring the current in an electron beam with sufficient accuracy it is challenging to establish good values for ε_0 in a calibration facility. Therefore values for ε_0 were determined in flight through cross-calibration of PEACE and WHISPER (Waves of High Frequency and Sounder for Probing of Density by Relaxation) density measurements (Fazakerley et al., 2010b).

These parameters were measured in a test chamber to ensure the sensors were identical to within specified tolerances before being accepted for flight. The values obtained are used as the baseline from which any in-flight calibration corrections are made.

2.2   Calibration correction factors

It was expected that the performance of the detector would vary over time due to degradation of the MCP and thus require correction through in-flight calibration. There are two aspects of the instrument calibration which may vary during flight operations. The energy/angle-dependent detector efficiency, ε_{ik}, can be described by correction factors (α, β) that account for these variations over time as follows:

$$\varepsilon_{ik} = \alpha(t)\varepsilon_0 \beta_{ik}\varepsilon(v_k^2)_i, \qquad (2)$$

where $\alpha(t)$ is the time-dependent correction factor for the energy/angle-independent part of ε_{ik}. It describes the effect of the sub-optimal detector sensitivity as it declines over time. By definition this correction factor applies equally for all anodes.

β_{ik} represents corrections for each anode to $\varepsilon(v_k^2)_i$, the relative sensitivity of the 12 individual anodes of each detector. Even small errors in the inter-anode calibration can result in large errors in the spin axis component of the bulk velocity which are determined by integration of the velocity space density collected during a spin. This correction factor varies with gain of the MCP.

The parameters α and β were each equal to 1 in ground test conditions, when the MCP gain was well above 2×10^6 electrons.

It is possible to increase the detector efficiency by increasing the voltage applied across the MCP, and this has been performed in small steps periodically throughout the mission.

In this paper we describe two methods which have been used for determining how the α factor term changes throughout the mission. The first method, detailed in Sect. 3, uses data from weekly in-orbit tests of the PEACE sensor performance. However over time, some of the PEACE MCPs have degraded to the point that we cannot collect the input information needed to apply this method, and an alternative technique was developed. This second method uses comparisons of electron densities measured by PEACE LEEA and by other instruments to determine α for the LEEA sensor, and comparisons of PEACE HEEA and PEACE LEEA to determine α for the HEEA sensor. The latter method is currently in use and is described in Sect. 4. We also present the results of these studies of the detector sensitivity evolution and comparisons of alpha determined by both methods.

The method for determination of the β correction factors has been described elsewhere (see Fazakerley et al., 2010b) and not provided here.

2.3 Relationship between calibration factors and moments

It can be shown that the electron density measured by PEACE is inversely proportional to the calibration factors $G\alpha\varepsilon_0$. Thus these calibration factors need to be well characterised to achieve good densities from PEACE. It also has a more complex dependency on $\beta_{ik}\varepsilon(v_k^2)_i$. In our experience so far the required correction to α is usually much greater than β_{ik}. The effect of the β_{ik} correction on the density has been checked after β_{ik} is applied by repeating the α work. The subsequent correction to α was found to be very small.

The electron bulk velocity is independent of $G\alpha\varepsilon_0$ but has a strong dependency on $\beta_{ik}\varepsilon(v_k^2)_i$; therefore to achieve good

plasma bulk flow velocity vectors requires accurate determination of the relative sensitivity of the 12 detector polar zones as a function of energy and gain. The independence of velocity from $G\alpha\varepsilon_0$ means that $\beta_{ik}\varepsilon(v_k^2)_i$ can be determined independently from $G\alpha\varepsilon_0$ using the method described in Fazakerley et al. (2010b). The validity of β_{ik} corrections can be confirmed by comparing PEACE velocity measurements with CIS (Cluster Ion Spectroscopy experiment) (Rème et al., 2001) velocity measurements.

3 Determination of α using in-orbit MCP tests

3.1 MCP gain–voltage characteristic before launch

The gain for each individual MCP as a function of the voltage applied was characterised in ground calibration tests prior to launch. Using a radioactive tritium source which has a well-known emission rate, the flux of beta particles (i.e. electrons which have a maximum energy of 18.6 keV) was measured while operating across a range of MCP voltage levels. At a given MCP voltage the number of electrons emerging from the MCP in response to an incident electron is not always the same. The spread of values of measured charge produced by the MCP for a given voltage is characterised by a pulse height distribution (PHD) as shown by the sketch in Fig. 3. As the voltage increases, larger signals are generated and the PHD peak moves to higher gains. From the ground test results, the peak of each pulse height distribution was taken to be the modal gain for that voltage. Using these points a characteristic gain-versus-voltage curve, as illustrated by the black curve in Fig. 4, was produced for each MCP. The PEACE instruments have a threshold level, $\sim 0.45 \times 10^6$ electrons, below which the counter electronics ignore the signal. The voltage at which the PHD modal gain is equal to the threshold level is referred to as the threshold voltage, shown by V_{ref} in Fig. 3.

3.2 MCP gain evolution monitoring: dual-sensor technique

We choose to operate the MCPs with a specific operational voltage level in order to achieve a desired gain. The MCP efficiency at a given voltage decreases with increased operating time. We compensate for the performance decline of the MCP over time by periodically raising the operational voltage level applied across the analyser to maintain the desired gain. In between these periodic voltage level raises we need to correct for the sub-optimal detector sensitivity, which requires us to know how the MCP is performing at a specific time. It is not possible to perform in-orbit tests in which we measure PHDs directly, so an alternative method was required. A unique technique was developed to estimate the PHD modal gain though not the full PHD shape. The technique makes use of the fact that we have two sensors on each spacecraft and is based on two assumptions: firstly,

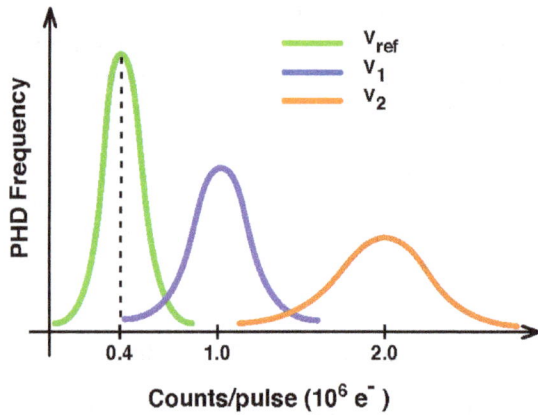

Fig. 3. Sketch of the pulse height distributions (PHDs) of measured values of charge from tritium tests for different voltages applied across the MCP during ground calibration tests. The peak of the PHD gives the modal gain for that voltage. As the voltage across the MCP is raised, e.g. from V_1 to V_2, the modal gain also increases. The increase in PHD spread with voltage is characteristic MCP behaviour. The ratio of the PHD FWHM to the PHD peak is typically observed to be roughly constant; thus as the voltage increases, so does the spread.

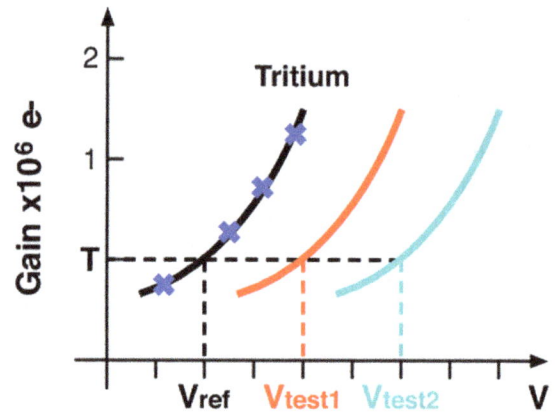

Fig. 4. Sketch of the characteristic gain–voltage curve. The black curve is the trend produced from the tritium tests prior to launch, where the blue crosses are the modal gains for particular voltages and V_{ref} is the threshold voltage that has a corresponding gain equal to the electronic threshold of the instrument, T. The red and cyan curves show the gain–voltage curves shifted to higher threshold voltages, V_{test1} and V_{test2}, obtained from in-orbit MCP tests.

that the PHD for any given voltage is symmetric about the modal gain, and, secondly, that the gain–voltage curve does not change shape significantly as the MCP ages. The first assumption is based on measurements of the PHDs for the sensors during ground tests. The PHDs were typically neither perfectly symmetric nor far from being symmetric. They also varied slightly from sensor to sensor, and as a function of gain. Nonetheless, in a normal operating regime we consider that the assumption of symmetry is a good first approximation. The second assumption is based in part on literature that shows examples of similar MCPs that have been tested before and after "scrubbing", e.g. Eberhardt (1979). Our assumptions also seems to us to be well justified by the agreement between alpha factors determined using our technique and those inferred via density comparisons with WHISPER. At any given time the peak of the PHD moves to higher gains with increasing voltage level (Fig. 5a). As the modal gain rises, the fraction of the PHD which lies above the electronic threshold of the instrument, and so the fraction registered by the counter electronics, increases (Fig. 5b). The fraction of the PHD above the electronic threshold gives the efficiency of the MCP. In the case of perfectly symmetric PHDs, the position of the point of inflection on the "cumulative distribution function above threshold" curve is where the MCP is counting 50 % of the electrons entering its pores, and so the gain can be determined and is equal to the electronic threshold.

An in-orbit MCP test procedure was created where we can apply the principle illustrated in Fig. 5. During a test both sensors, HEEA and LEEA, are set to observe the same energy range and thus the same plasma electrons. The MCP on

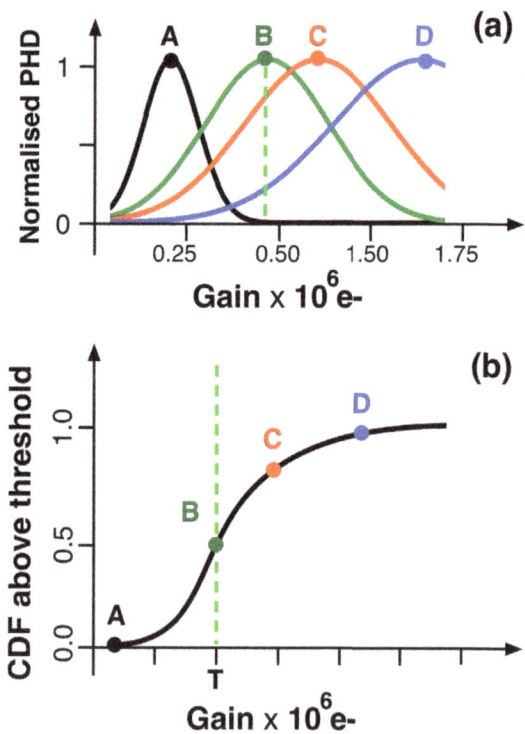

Fig. 5. Sketch of the normalised PHD and cumulative distribution function above threshold. Panel **(a)** shows how the normalised PHD changes for increasing modal gain, asymptotically approaching 1. Panel **(b)** shows the fraction of the PHD that lies above the electronic threshold of the instrument, shown by the vertical green dashed line, assuming a PHD symmetric about the peak.

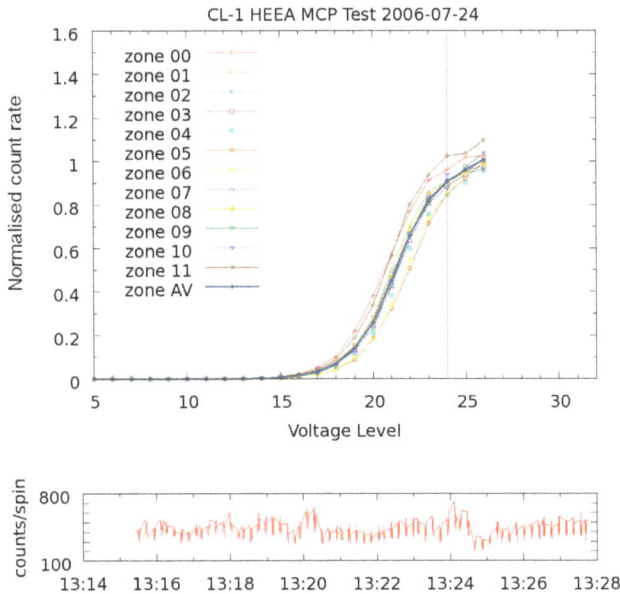

Fig. 6. Typical MCP test result for the Cluster-1 HEEA sensor. The upper panel shows the count rate from the HEEA (test) sensor collected over a range of MCP voltages. The count rates have been normalised using the count rates from the LEEA (monitoring) sensor shown in the lower panel. Curves are provided for each anode together with an average (thick dark blue line).

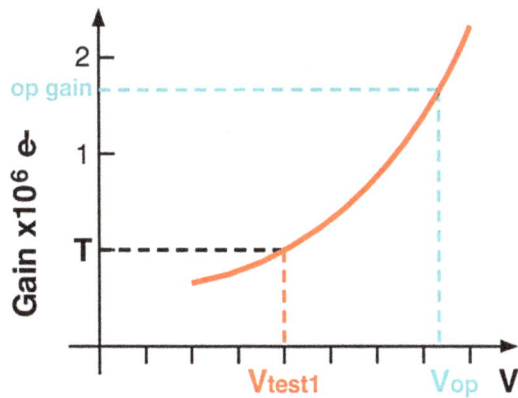

Fig. 7. Sketch of obtaining the MCP gain for the normal operational voltage level, V_{op}, after shifting the characteristic gain–voltage curve to apply at the time of the weekly MCP test by requiring that the gain $= T$ when the MCP voltage equals the threshold value determined from the test.

one sensor is swept through a range of voltage levels, sitting at each level for around 40 s (10 spacecraft spins), and the counts per spin at each level are measured for each anode. Simultaneously the other sensor is kept at a suitable voltage level in order to monitor variations of the plasma environment. The key advantage of this two-sensor technique is that the response of the measured count rate to increasing MCP voltage in the test sensor can be separated from the measured count rate changes due to flux variations in the ambient plasma, using data from the monitoring sensor. The test is then repeated for the other sensor by exchanging their roles. A typical weekly MCP test result is shown in Fig. 6 (for C1 HEEA). The upper panel shows the count rate from the HEEA (test) sensor collected over a range of MCP voltages. The count rates have been normalised using the count rates from the LEEA (monitoring) sensor shown in the lower panel. In this case the ambient plasma fluxes seen by the LEEA sensor are varying (bottom panel), showing the importance for normalisation. Curves are provided for each anode together with an average (thick dark blue line).

We assume a symmetrical PHD distribution, in which case the point of inflection on the averaged test curve gives the voltage V_{ref} at which the gain is equal to the electronic threshold, the threshold voltage, at the time of the test. We now need to determine the gain at the normal operating MCP voltage, which is not usually the same as the threshold voltage. The vertical red line in Fig. 6 shows the normal operating voltage level of the HEEA sensor at the time of that test.

To determine what the gain is for the operational voltage level, we apply our second assumption that the characteristic voltage-versus-gain curves obtained from the ground calibration tests do not change in shape over time but can simply be shifted to higher voltages, as shown by the red and blue curves in Fig. 4. We know that higher voltages are required to produce the same gain as the MCP ages, so our assumption is the simplest way to address this. We shift the curve so that the voltage V_{ref} lines up with the threshold voltage obtained from the test, and we can then infer the gain at the time of the test for any other voltage level as shown in Fig. 7.

Once we know the gain of the MCP corresponding to the normal operational voltage level, we can use the cumulative distribution function (CDF) curve to determine what proportion of the PHD distribution lies above the electronic threshold for that value of gain (Figs. 8 and 5). This fraction is represented in our calibrations as the α factor, which is a measure of the number of electrons entering the MCP which are above threshold and thus are being counted.

In early calibration releases we used the mathematically derived Gaussian model of the cumulative distribution function shown by the red curve in Fig. 9. This worked well for high gains, which we verified by comparing PEACE densities with WHISPER densities; however it did not work so well at low gains (<0.6). A new alpha factor model curve was created by using results from in-orbit MCP tests. An empirical curve was fitted to data of the normalised counts ratios from a large number of MCP tests, both HEEA / LEEA and LEEA / HEEA as shown in Fig. 9. This empirical alpha factor model deviates from the Gaussian model at lower gains, and it was found to give better results and is now used for calibrations instead of the Gaussian model. It was also used to recalibrate the earlier data sets. A single empirical curve was made for all spacecraft sensors; however future work is planned to check if different sensors can be better

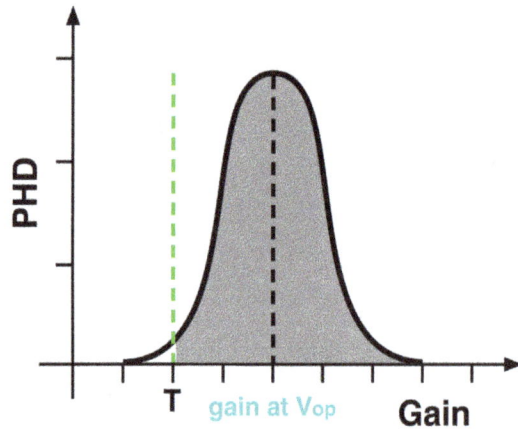

Fig. 8. Example sketch of the fraction of the PHD lying above the electronic threshold of the instrument, T, shown by the shaded area. The position of the peak of the PHD is at the gain corresponding to the normal operational voltage level, V_{op}.

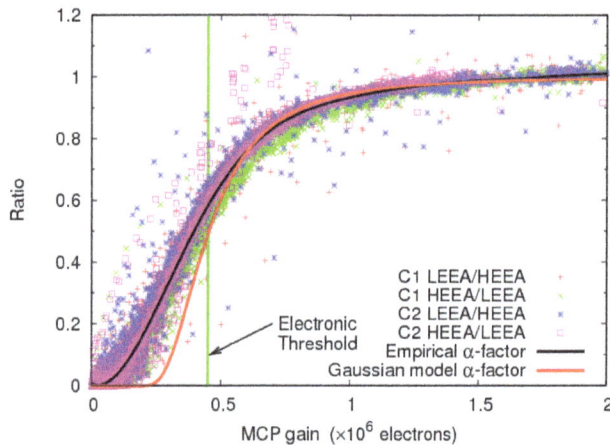

Fig. 10. Time history of the threshold voltage obtained from weekly MCP tests for all eight sensors up to 2012.

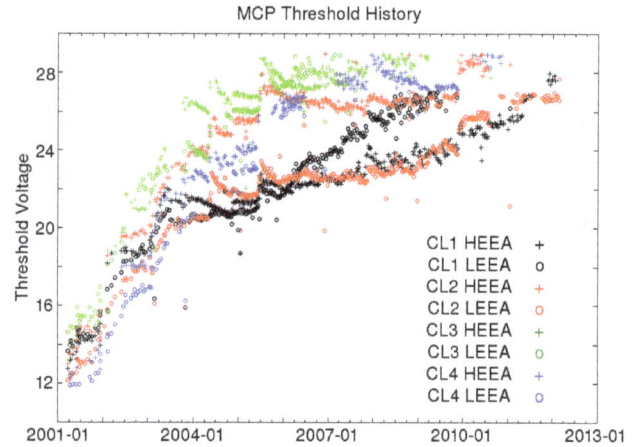

Fig. 9. Alpha-versus-gain curve. This plot shows the difference between the Gaussian model (red) and the empirical model (black) obtained from fitting to counts ratios from a large number of MCP tests, normalised to 1 at high gains.

characterised using individual curves. It may also be the case that the actual PHDs at low gains are not only not Gaussian but also not symmetrical, in which case our method may be less reliable as its assumptions are no longer completely applicable.

3.3 Limitations of the dual-sensor in-orbit MCP test method

In-orbit MCP tests are routinely carried out every week. An analysis of these tests gives a detailed time history of the detector sensitivity variations for each sensor. A time history of the threshold voltage determined from MCP test curves for the eight sensors is shown in Fig. 10.

At the time of writing this method has ceased to be effective for the majority of the sensors. As noted above, we compensate for the efficiency decline of an MCP over time by periodically raising the operational voltage level applied across the MCP. Similarly, we have to use a set of voltage levels during the MCP tests that have higher voltage values than earlier in the mission, to produce the range of gain values that we wish to cover for the test, as illustrated in Fig. 5. The decline in efficiency has developed more rapidly across some sensors than others, and so the number of required voltage level increases has varied from sensor to sensor. On some sensors the highest available voltage level is now in use during normal operations, so it is no longer possible to raise them further. Also the threshold voltage is now close to the maximum level, so it is no longer possible to obtain a complete MCP test curve and identify the point of inflexion. This evolution is shown for two sensors in Fig. 11. These plots demonstrate the more rapid efficiency decline of the Cluster-3 LEEA sensor in comparison to the Cluster-1 HEEA sensor. By 2006 C3 LEEA is already operating at one of the highest levels and the full curve is unobtainable. By 2012 this is also the case for C1 HEEA. In 2012 C3 LEEA is operating at its highest MCP voltage level, so compensating for efficiency decline in the future is no longer possible.

Although we no longer use this method for calibration purposes, we still routinely perform the weekly tests as we expect that they will allow us to improve the low-gain statistics for the α-versus-gain curve (Fig. 9), and they may provide clues to the evolution of the PHD.

4 Determination of α using PEACE–WHISPER density comparisons

Since it is no longer possible to calibrate the MCP sensitivity in flight using the method described in Sect. 3, we have adopted an alternative procedure in which we adjust the α

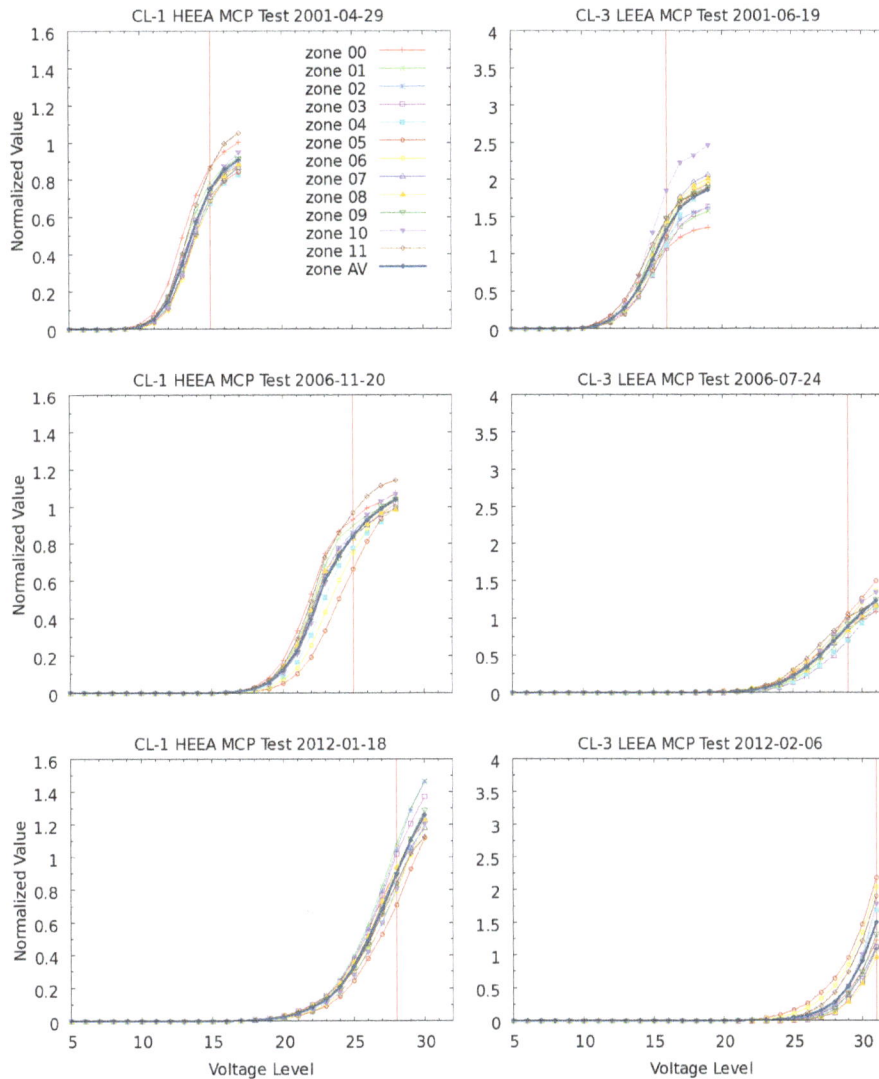

Fig. 11. Failure of the MCP test method. These plots show how the weekly MCP test results have evolved for Cluster-1 HEEA and Cluster-3 LEEA over the mission. The red vertical line shows the operational voltage level of the test sensor used around the time of the test. The normalisation becomes unreliable if the operating voltage of the monitoring sensor is not sufficient to give $\alpha \sim 1$ for that sensor. This problem is clearly apparent in some cases, for example C3 LEEA in 2012.

factor in order to achieve agreement in electron density values from PEACE LEEA sensors with results from the WHIS-PER experiment (Décréau et al., 2001). This is only possible because of the high quality of the WHISPER total density data, optimised in active sounding mode, available in some of the plasma environments visited by Cluster.

The calibrations are extended to the HEEA sensors by comparing the densities from the HEEA and LEEA sensors in the energy overlap region. This method has so far been applied for the period November 2004 through to January 2012. This includes an overlap period with the results from in-orbit MCP tests, providing a check on the accuracy of our PEACE-only technique, which would be relevant in future missions with no sounder.

4.1 LEEA sensors: LEEA–WHISPER density comparisons

For the WHISPER densities we use WHISPER active mode electron data from the magnetosheath, available from the Cluster Active Archive (CAA). Corresponding PEACE LEEA electron densities are produced using the ground calibration geometric factor. To calculate electron densities from PEACE data also requires knowledge of the spacecraft potential. In this analysis the EFW (Electric Field and Wave experiment) (Gustafsson et al., 2001) spin resolution probe-spacecraft potential from the CAA is used. A correction of +1 eV, which is suitable in the dense magnetosheath plasma environment, is applied to the EFW probe potential to give

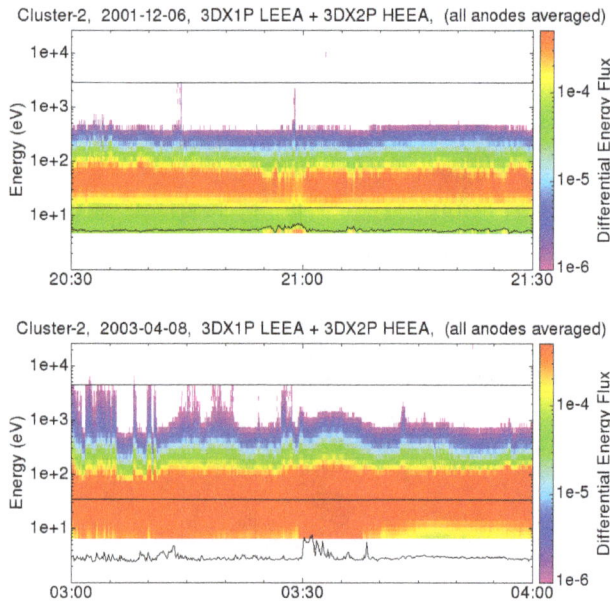

Fig. 12. Example of full coverage (top panel) and partial coverage (bottom panel) of the plasma energy range. The two black horizontal lines show the energy overlap region of the two PEACE sensors. Energies above the top black line are being measured by the HEEA sensor only. Energies below the bottom black line are being measured by the LEEA sensor only. Energies in between the two lines are measured by both sensors. If LEEA does not measure energies as low as the spacecraft potential we have "partial coverage".

the true value of the spacecraft electric potential. When calculating PEACE moments we try to eliminate photoelectrons in the plasma distribution by increasing the lower cut-off in the energy integration. We reject the energy bin containing the EFW probe-spacecraft potential and the one above as the 1 eV correction mentioned above may put the true spacecraft potential in this bin. Sometimes "spikes" can be seen in the PEACE moments time series data. These "spikes" are observations of photoelectrons that are briefly energised by a few eV during WHISPER soundings, which occur periodically, at intervals of 52 s or 104 s. PEACE moments data were not filtered for WHISPER soundings in this study because the contribution from "spikes" is not significant in the magnetosheath regions, compared to the very high plasma electron fluxes observed there at the same energies. In contrast, the additional flux associated with spikes in regions with lower plasma electron fluxes, such as the magnetotail plasmasheet, does significantly add to the plasma electron fluxes at low energy and hence to the phase space density, leading to clear variation in the moments. However, data from such regions was not used in this study.

4.1.1 Event selection and partial coverage

For this study carefully selected magnetosheath intervals are used. These are only available between November and June

each year. Thus MCP degradation during the magnetotail crossing region (July to October) is not determined, however the decline is less severe in the magnetotail. It is important to determine the total electron density, thus we require that the energy range selected for the PEACE LEEA sensors is such that the sensor measures all of the plasma. If this is not the case then we would naturally expect the PEACE partial densities to be smaller than WHISPER densities. An example of this is shown in Fig. 12. In the upper spectrogram the LEEA sensor is measuring the energy range 4.7–2880.0 eV. It can be seen from the spectrogram that the LEEA sensor is measuring all energies above the spacecraft potential thus seeing all of the plasma, so this event would be selected for use in the study. In the lower spectrogram the LEEA sensor does not measure below 9.5 eV however the spacecraft potential is ~ 3 eV, so LEEA does not measure some of the plasma fluxes above the spacecraft potential which is expected to result in underestimated densities. Events of this kind are not used in our study. Useful events are selected manually by looking through spectrograms similar to those in Fig. 12.

4.1.2 Filtering out possible errors in WHISPER & EFW data

Compromised points in both EFW and WHISPER data which could cause errors in the analysis are filtered out by using comparisons of these data. It is assumed that there should be a characteristic curve relating the EFW probe-spacecraft potential data and the WHISPER density data as shown by Pedersen et al. (2008). Figure 13 shows a plot of this type using selected events of Cluster-4 magnetosheath data between November 2002 and June 2003. Any points that are far from the trend are not used in the study.

4.1.3 Determination of alpha

A time history of the α factor with weekly time steps is obtained by extracting the peak (modal) value of weekly averaged LEEA/WHISPER density ratios. In order to produce daily values we interpolate between weekly points to give an α factor for each day. Special attention is paid to times when there is expected to be a sharp rise or fall in the α factor, such as when there is an MCP level raise or a thruster firing, by looking in detail at the density ratios of individual events instead of using the weekly average and interpolation technique. Also careful analysis is applied to the calibration intervals used for any dubious events which do not follow the α factor trend and which are deemed not to be real.

For the magnetotail seasons where WHISPER density data is not routinely available, the α factor is obtained by linear interpolation between the two values on either side of the gap, accounting for any MCP level raises and thrusters firings during these intervals where possible. Cross-calibrations of PEACE densities with WBD (Wide Band Data receiver) (Gurnett et al., 2001) densities in the plasmasheet have been

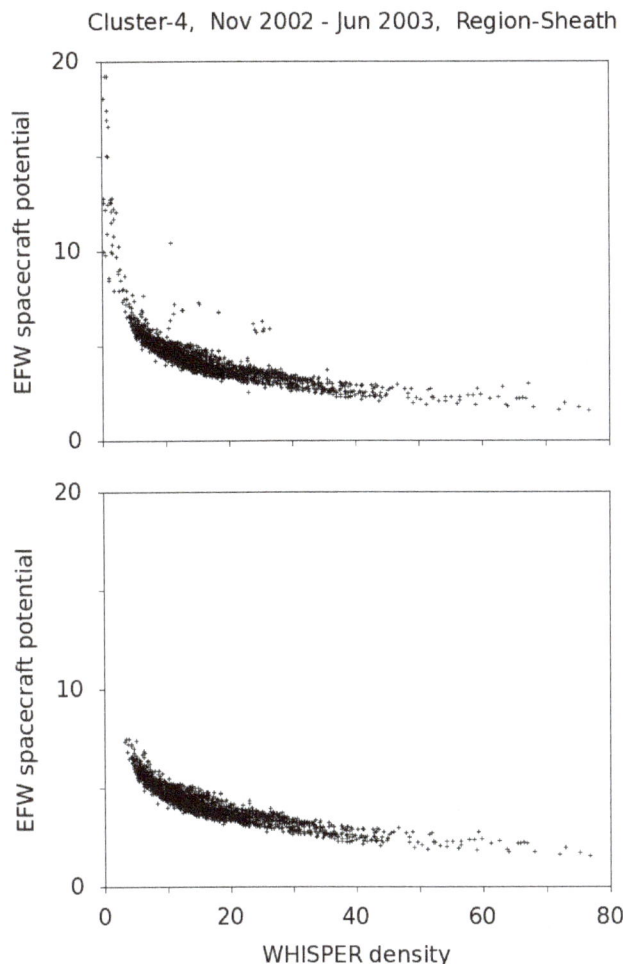

Cluster-4, Nov 2002 - Jun 2003, Region-Sheath

Fig. 13. Correlation between EFW spacecraft potential and WHISPER density for Cluster-4 using magnetosheath data between November 2002 and June 2003. Points which do not fit the trend (top plot) are removed and not included in the study (bottom plot).

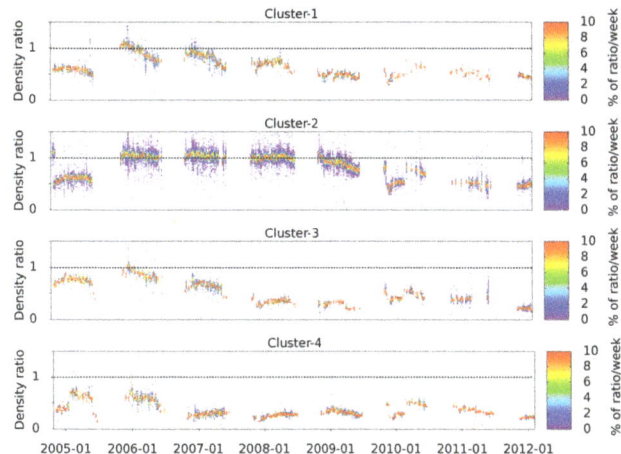

Fig. 14. Degradation of the four LEEA sensors between November 2004 and January 2012. These plots show the PEACE LEEA / WHISPER density ratio frequency. Each vertical strip of data is the frequency histogram for one week's worth of data.

used, where available, to validate these α factors. However there are very few intervals available for these studies. For the most part agreement was found. Disagreement for Cluster-4 in the 2007 tail season allowed us to fine-tune alpha factors which were inaccurate due to MCP gain degradation following thruster firings at the end of the tail season.

The α factors extracted from these density ratios apply for the MCP voltage level used at the time for the magnetosheath intervals (often the nominal $V_{sheath} = V_{op} - 1$ level). The α factors for the normal operational and other voltage levels are then calculated by using the gain-versus-voltage and alpha-versus-gain curves described in Sect. 3.

4.2 HEEA sensors: HEEA–LEEA comparisons

To calibrate the HEEA sensors PEACE HEEA densities are compared with PEACE LEEA densities calculated using data only from the energy overlap region of the two sensors. It

is necessary to filter the events used in the study and reject those cases where the energy overlap of the sensors is small and/or the energy overlap covers tenuous plasmas which result in poor counting statistics. Poor count rates are becoming more common in later years as the MCP gains decrease. The PEACE HEEA densities are produced with only the ground calibration geometric factor applied. The PEACE LEEA densities are produced after applying the α correction factor determined from PEACE–WHISPER density comparisons.

4.3 Results

4.3.1 LEEA sensitivity degradation

The PEACE LEEA / WHISPER density ratios from November 2004 to January 2012 for each LEEA sensor are shown in Fig. 14. Each vertical strip in the plots is the frequency histogram of the density ratio for one week of data. These plots effectively show the sensitivity degradation of the LEEA sensors over this time period. The regular gaps in the plots are for the magnetotail months (July–October) where we do not see the magnetosheath. There is also a gap in the data for Cluster-3 between March and May 2011 where the Wave Experiment Consortium (WEC) instrument suite, which includes WHISPER and EFW, was non-operational.

Figure 15 shows the α factor history for the four LEEA sensors for November 2004 to January 2012. The α factors determined from both the in-orbit MCP test method (black) and PEACE–WHISPER density comparison method (red) are shown. By design, the alpha factors determined from PEACE–WHISPER comparisons include the correction required to refine the value of the time-independent ε_0 calibration parameter determined in ground tests (see Sect. 2). The correction to ε_0 is not included in the alpha obtained

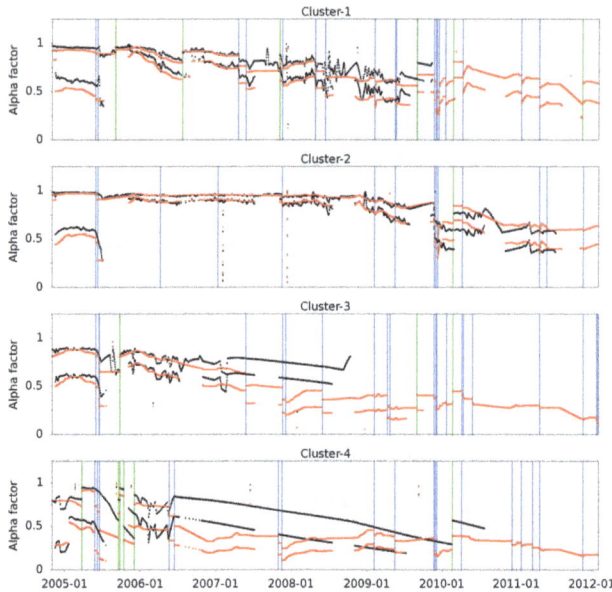

Fig. 15. The relative MCP sensitivity time history for the four LEEA sensors. The red points are the alpha factors inferred from the PEACE–WHISPER density comparison method including estimates in the magnetotail months. The black points are the alpha factors produced using the weekly MCP test method. For each data set the upper line is for operational MCP voltage level and the lower line for lowered MCP voltage level (used in high-flux environments). Commanded MCP voltage level changes are shown by green vertical lines. Thruster firings are shown by blue vertical lines.

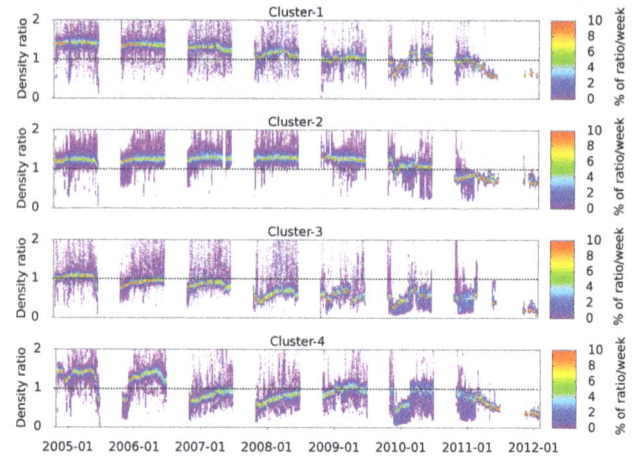

Fig. 16. Degradation of the four HEEA sensors between November 2004 and January 2012. These plots show the PEACE HEEA / PEACE LEEA density ratio frequency. Each vertical strip of data is the frequency histogram for one week's worth of data.

from in-orbit MCP tests and is applied separately. Therefore in order to compare like with like alpha factors from the two methods, the correction for ε_0 has been factored out of the alpha factor from PEACE–WHISPER comparisons.

For each set of α factors the upper line is the α factor for the normal operational level and the lower line is the α factor for the lowered MCP voltage levels used in the magnetosheath and solar wind. MCP level raises (green vertical lines) and the thruster firings (blue vertical lines) which have had an observable impact on the α factor are also indicated. Very good agreement between the two methods is seen for the earlier years at normal operational voltage level. The weekly MCP test method is available in all months, unlike the LEEA–WHISPER comparison method which relies on observations in the magnetosheath. It can be clearly seen where the in-orbit MCP test method begins to fail for each sensor.

4.3.2 HEEA sensitivity degradation

The PEACE HEEA / PEACE LEEA density ratios from November 2004 to January 2012 for each HEEA sensor are shown in Fig. 16. The data shown in these plots are from intervals where both sensors are operating at their normal operational level (i.e. not lowered). These plots show the degrada-

tion of the HEEA sensor efficiencies over this time period. As with the corresponding plot for the LEEA sensors, changes in the density ratio due to MCP voltage level raises and thrusters firings can be seen in the plots.

The α factor history for the four HEEA sensors extracted from the density ratio results are shown in Fig. 17 together with the α factors determined using the in-orbit method.

4.3.3 Discussion

There are several features that can be seen in these plots which require further discussion:

1. MCP operational voltage level raises: due to decline in MCP efficiency over time, the operational MCP voltage levels on all sensors have been raised at various times throughout the mission in order to recover a desired sensitivity level. These MCP voltage level changes have occurred at different times for different sensors, with some sensors requiring more level raises than others. We tend to aim for $\alpha \sim 0.5$ to 1 since the degradation rate has been observed to be greater when $\alpha > 1$. In Figs. 14 and 15 an example of this level raise can be seen on 20 February 2010 where the voltage level was raised on all four LEEA sensors. The increase in gain/sensitivity is clearly evident.

2. Thruster firings: spacecraft thruster firings of a variety of durations have been performed throughout the mission. There is clear evidence that PEACE MCP sensitivity declines following some thruster firings. For example, firings which have had severe effects on the sensitivity can be seen in Figs. 15 and 17 in November 2007 and November/December 2009. However it has been observed that after some firings the MCP

sensitivity does recover somewhat and in some cases to pre-firing levels. A separate study into the effect of thruster firings on the MCP efficiency is ongoing.

3. MCP voltage level lowering: from November 2003 the MCP voltage levels were routinely lowered by 1 or 2 levels when operating in the magnetosheath or solar wind as it was believed that the performance decline would be more gradual if the MCPs were operated at a lower gain when in high-flux environments. As of November 2009 the MCP voltage levels on Cluster-3 and Cluster-4 are no longer lowered as their performance at the operational level from this time was considered as "low gain", so there was no need to lower them further. The increase in the density ratio in Fig. 14 for Cluster-3 and Cluster-4 at the start of the 2009/2010 dayside is due to this change in commanding of the voltage level and is not a real sensitivity increase.

4. MCP voltage lowering to estimated half gain: during the 2004/2005 dayside season the MCP voltage levels were lowered in order to achieve an estimated gain $= 0.5 \times 10^6$ electrons on the sensors. This was performed to protect the MCPs as described in point 3 above. However the gains went lower than intended, so the attempted targeting of a preferred gain was not repeated.

4.4 Limitations of the PEACE–WHISPER density comparison method

There are several issues which cause problems for this method:

1. Availability of good magnetosheath calibration intervals for LEEA/WHISPER work: this is sometimes a major problem. Between November 2003 and June 2008 the PEACE sensors on only one spacecraft would operate throughout the magnetosheath and solar wind with the other spacecraft sensors powered off except during predicted times of bow shock and magnetopause crossings. The "observational spacecraft" role rotated between Cluster-1, 2 and 4 (only Cluster-1 and 2 between March 2005 and June 2008). From November 2008 Cluster-2 was always, and still is, used as the observational spacecraft with Cluster-1, 3 and 4 routinely turned off in the magnetosheath except around the bow shock and magnetopause crossings. For a short period between January and June 2012 Cluster-1 was used as the observational spacecraft instead of Cluster-2. Thus for the non-observational spacecraft only magnetosheath data collected during bow shock and magnetopause crossings are available for the study.

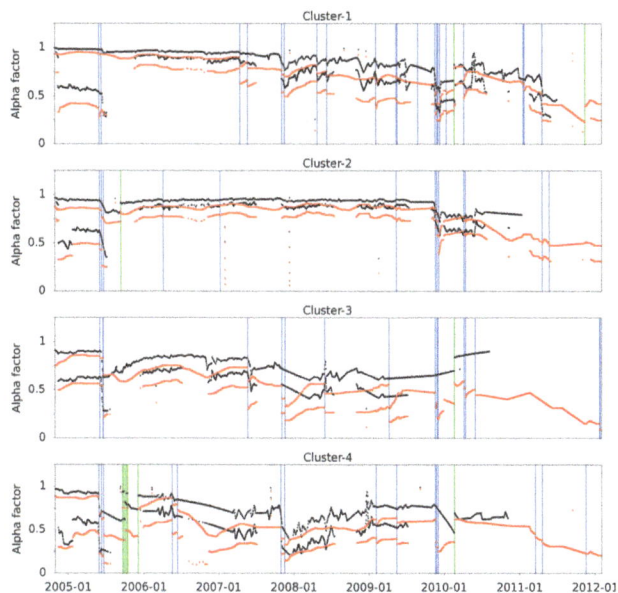

Fig. 17. The relative MCP sensitivity time history for the four HEEA sensors. The red points are the alpha factors inferred from HEEA–LEEA density comparisons. The black points are the alpha factors produced using the weekly MCP test method. For each data set the upper line is for operational MCP voltage level and the lower line for lowered MCP voltage level (used in high-flux environments). Commanded MCP voltage level changes are shown by green vertical lines. Thruster firings are shown by blue vertical lines.

2. Partial coverage of the plasma velocity distribution: even when the sensors are on in the magnetosheath not all intervals can be used. Depending on the energy range measured by the LEEA sensor, which is chosen to try and avoid the high fluxes of the photoelectrons, partial coverage of the plasma velocity distribution (see Sect. 4.1.1) can be a major problem too. This was particularly an issue in 2009/2010.

3. HEEA/LEEA energy overlap: a small energy overlap region can cause large spreads in the ratios. As a result many intervals need to be eliminated from the study.

4. Low counting statistics: this can also cause large spreads in the ratios.

5. Low gains: at very low gains the gain-versus-voltage and alpha-versus-gain curves (Sect. 3) are not as accurate as for higher gains, so determining alpha for different MCP voltage levels carries some error, which is compounded when comparing HEEA and LEEA.

To help with points 1, 2 and 3, special calibration intervals have been introduced into the routine commanding since 2011, to ensure full plasma coverage in the magnetosheath for the LEEA sensor and large energy overlap region for the two sensors.

5 Conclusions

It has been shown in this paper that it has been essential to perform continuous in-flight calibrations to monitor the health and correct for the evolution of the MCP detector performance on each of the eight PEACE sensors. This builds on careful ground calibration work to define many parameters that cannot be determined in flight. We have described two independent methods to determine the detector sensitivity variations. The LEEA–WHISPER comparison method relies on accurate density determination by WHISPER and PEACE. The dual-sensor MCP test method uses only PEACE data and is based on some assumptions which have been discussed. The MCP test method works in all months, unlike the LEEA–WHISPER comparison method which cannot be used in the magnetotail season. Although the MCP test method can no longer be applied, we continue to have good knowledge of α factor evolution thanks to PEACE–WHISPER cross-calibrations. The good agreement of the two methods in earlier years validates the PEACE-only method, which may be of interest for future missions that do not carry a sounder.

Acknowledgements. The authors thank ESA and UKSA/STFC for funding the team, members of other PI teams for providing data required to carry out the work reported here and also the CAA team at ESTEC for supporting cross-calibration activities. We would also like to thank Dave Walton and Barry Hancock for their valuable technical assistance.

Edited by: A. Masson

References

Carlson, C. W., Curtis, D. W., Paschmann, G., and Michael, W.: An Instrument for Rapidly Measuring Plasma Distribution Functions with High Resolution, Adv. Space Res., 2, 67–70, doi:10.1016/0273-1177(82)90151-X, 1982.

Décréau, P. M. E., Fergeau, P., Krasnoselskikh, V., Le Guirriec, E., Lévêque, M., Martin, Ph., Randriamboarison, O., Rauch, J. L., Sené, F. X., Séran, H. C., Trotignon, J. G., Canu, P., Cornilleau, N., de Féraudy, H., Alleyne, H., Yearby, K., Mögensen, P. B., Gustafsson, G., André, M., Gurnett, D. C., Darrouzet, F., Lemaire, J., Harvey, C. C., Travnicek, P., and Whisper experimenters (Table 1): Early results from the Whisper instrument on Cluster: an overview, Ann. Geophys., 19, 1241–1258, doi:10.5194/angeo-19-1241-2001, 2001.

Eberhardt, E. H.: Gain model for microchannel plates, Appl. Optics, 18, 1418–1423, 1979.

Fazakerley, A. N., Lahiff, A. D., Wilson, R. J., Rozum, I., Anekallu, C., West, M., and Bacai, H.: PEACE Data in the Cluster Active Archive, The Cluster Active Archive: Studying the Earth's Space Plasma Environment, 129–144, doi:10.1007/978-90-481-3499-1_8, 2010a.

Fazakerley, A. N., Lahiff, A. D., Rozum, I., Kataria, D., Bacai, H., Anekallu, C., West, M., and Asnes, A.: Cluster-PEACE In-flight Calibration Status, The Cluster Active Archive: Studying the Earth's Space Plasma Environment, 281–299, doi:10.1007/978-90-481-3499-1_19, 2010b.

Gurnett, D. A., Huff, R. L., Pickett, J. S., Persoon, A. M., Mutel, R. L., Christopher, I. W., Kletzing, C. A., Inan, U. S., Martin, W. L., Bougeret, J.-L., Alleyne, H. St. C., and Yearby, K. H.: First results from the Cluster wideband plasma wave investigation, Ann. Geophys., 19, 1259–1272, doi:10.5194/angeo-19-1259-2001, 2001.

Gustafsson, G., André, M., Carozzi, T., Eriksson, A. I., Fälthammar, C.-G., Grard, R., Holmgren, G., Holtet, J. A., Ivchenko, N., Karlsson, T., Khotyaintsev, Y., Klimov, S., Laakso, H., Lindqvist, P.-A., Lybekk, B., Marklund, G., Mozer, F., Mursula, K., Pedersen, A., Popielawska, B., Savin, S., Stasiewicz, K., Tanskanen, P., Vaivads, A., and Wahlund, J.-E.: First results of electric field and density observations by Cluster EFW based on initial months of operation, Ann. Geophys., 19, 1219–1240, doi:10.5194/angeo-19-1219-2001, 2001.

Johnstone, A. D., Alsop, C., Burge, S., Carter, P. J., Coates, A. J., Coker, A. J., Fazakerley, A. N., Grande, M., Gowan, R. A., Gurgiolo, C., Hancock, B. K., Narheim, B., Preece, A., Sheather, P. H., Winningham, J. D., and Woodliffe, R. D.: PEACE: A Plasma Electron and Current Experiment, Space Sci. Rev., 79, 351–398, doi:10.1023/A:1004938001388, 1997.

Pedersen, A., Lybekk, B., André, M., Eriksson, A., Masson, A., Mozer, F. S., Lindqvist, P.-A., Décréau, P. M. E., Dandouras, I., Sauvaud, J.-A., Fazakerley, A., Taylor, M., Paschmann, G., Svenes, K. R., Torkar, K., and Whipple, E.: Electron density estimations derived from spacecraft potential measurements on Cluster in tenuous plasma regions, J. Geophys. Res., 113, A07S33, doi:10.1029/2007JA012636, 2008.

Rème, H., Aoustin, C., Bosqued, J. M., Dandouras, I., Lavraud, B., Sauvaud, J. A., Barthe, A., Bouyssou, J., Camus, Th., Coeur-Joly, O., Cros, A., Cuvilo, J., Ducay, F., Garbarowitz, Y., Medale, J. L., Penou, E., Perrier, H., Romefort, D., Rouzaud, J., Vallat, C., Alcaydé, D., Jacquey, C., Mazelle, C., d'Uston, C., Möbius, E., Kistler, L. M., Crocker, K., Granoff, M., Mouikis, C., Popecki, M., Vosbury, M., Klecker, B., Hovestadt, D., Kucharek, H., Kuenneth, E., Paschmann, G., Scholer, M., Sckopke, N., Seidenschwang, E., Carlson, C. W., Curtis, D. W., Ingraham, C., Lin, R. P., McFadden, J. P., Parks, G. K., Phan, T., Formisano, V., Amata, E., Bavassano-Cattaneo, M. B., Baldetti, P., Bruno, R., Chionchio, G., Di Lellis, A., Marcucci, M. F., Pallocchia, G., Korth, A., Daly, P. W., Graeve, B., Rosenbauer, H., Vasyliunas, V., McCarthy, M., Wilber, M., Eliasson, L., Lundin, R., Olsen, S., Shelley, E. G., Fuselier, S., Ghielmetti, A. G., Lennartsson, W., Escoubet, C. P., Balsiger, H., Friedel, R., Cao, J.-B., Kovrazhkin, R. A., Papamastorakis, I., Pellat, R., Scudder, J., and Sonnerup, B.: First multispacecraft ion measurements in and near the Earth's magnetosphere with the identical Cluster ion spectrometry (CIS) experiment, Ann. Geophys., 19, 1303–1354, doi:10.5194/angeo-19-1303-2001, 2001.

11

The AmeriFlux data activity and data system: an evolving collection of data management techniques, tools, products and services

T. A. Boden, M. Krassovski, and B. Yang

Oak Ridge National Laboratory, Carbon Dioxide Information Analysis Center, Oak Ridge, TN 37831-6290, USA

Correspondence to: T. A. Boden (bodenta@ornl.gov)

Abstract. The Carbon Dioxide Information Analysis Center (CDIAC) at Oak Ridge National Laboratory (ORNL), USA has provided scientific data management support for the US Department of Energy and international climate change science since 1982. Among the many data archived and available from CDIAC are collections from long-term measurement projects. One current example is the AmeriFlux measurement network. AmeriFlux provides continuous measurements from forests, grasslands, wetlands, and croplands in North, Central, and South America and offers important insight about carbon cycling in terrestrial ecosystems. To successfully manage AmeriFlux data and support climate change research, CDIAC has designed flexible data systems using proven technologies and standards blended with new, evolving technologies and standards. The AmeriFlux data system, comprised primarily of a relational database, a PHP-based data interface and a FTP server, offers a broad suite of AmeriFlux data. The data interface allows users to query the AmeriFlux collection in a variety of ways and then subset, visualize and download the data. From the perspective of data stewardship, on the other hand, this system is designed for CDIAC to easily control database content, automate data movement, track data provenance, manage metadata content, and handle frequent additions and corrections. CDIAC and researchers in the flux community developed data submission guidelines to enhance the AmeriFlux data collection, enable automated data processing, and promote standardization across regional networks. Both continuous flux and meteorological data and irregular biological data collected at AmeriFlux sites are carefully scrutinized by CDIAC using established quality-control algorithms before the data are ingested into the AmeriFlux data system. Other tasks at CDIAC include reformatting and standardizing the diverse and heterogeneous datasets received from individual sites into a uniform and consistent network database, generating high-level derived products to meet the current demands from a broad user group, and developing new products in anticipation of future needs. In this paper, we share our approaches to meet the challenges of standardizing, archiving and delivering quality, well-documented AmeriFlux data worldwide to benefit others with similar challenges of handling diverse climate change data, to further heighten awareness and use of an outstanding ecological data resource, and to highlight expanded software engineering applications being used for climate change measurement data.

1 Introduction and background

1.1 Brief overview of the AmeriFlux network

The AmeriFlux network is a collection of more than 150 past and present flux towers (Fig. 1) located mostly in the US, but with limited sites in Canada, Central America, and South America, making continuous measurements of water vapor, carbon dioxide (CO_2), energy fluxes, and related environmental variables using eddy covariance techniques (Baldocchi, 2003). The network covers a large variety of ecosystem types including forests, grasslands, croplands, shrublands, wetlands, savannas, and others (e.g., urban) (Fig. 2). The Carbon Dioxide Information Analysis Center (CDIAC) at Oak Ridge National Laboratory (ORNL), USA serves as the AmeriFlux permanent data archive and focal point for dissemination of AmeriFlux data (http://public. ornl.gov/ameriflux). As scientists and the public try to better understand climate change, AmeriFlux data have detailed

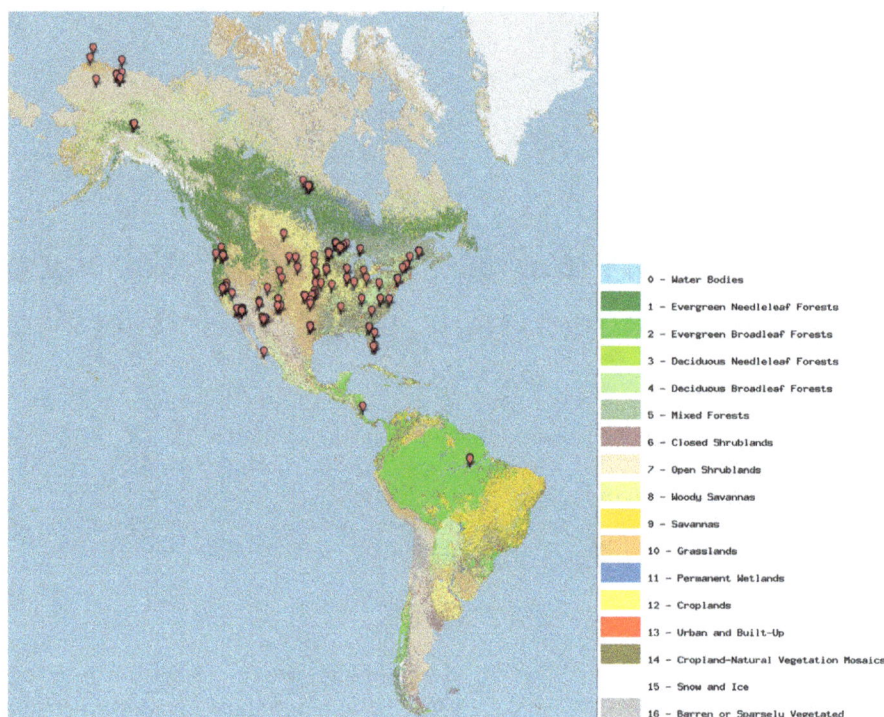

Fig. 1. Geographic distribution of registered flux sites in the AmeriFlux network as of January 2013 superimposed on a map depicting ecosystem representations from the International Geosphere-Biosphere Programme (IGBP).

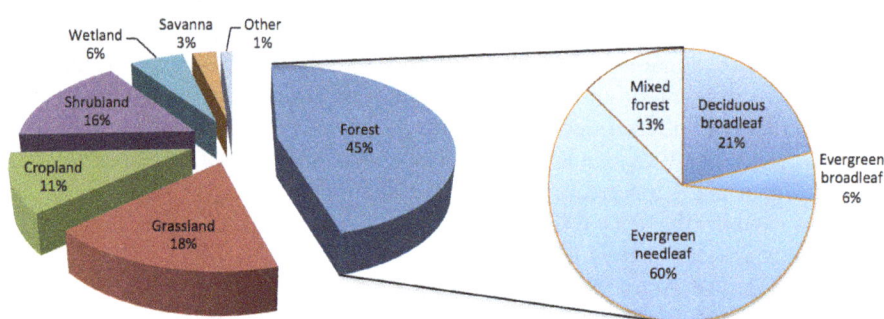

Fig. 2. Ecosystem representation, according to IGBP ecosystem classifications, of the AmeriFlux network based on registered sites as of January 2013.

how diverse ecosystems respond to changes in their physical environment and how ecosystems, in turn, affect their environments. AmeriFlux data have improved understanding of changes in net carbon uptake with interannual variation in climate and the influence of disturbance on carbon storage and fluxes. Models are fundamental to understanding and predicting biogeochemical cycles and future climates. AmeriFlux data are proving invaluable in improving and calibrating models used in climate science (Randerson et al., 2009; Schwalm et al., 2010; Williams et al., 2012 among many others).

1.2   Challenges, design criteria, and requirements

Numerous challenges face CDIAC in organizing, managing, and distributing the AmeriFlux data collection from network operation challenges to software engineering challenges. The breadth and diversity of measurements made at individual sites, the variability of data processing performed by individual AmeriFlux sites, the absence of standardized nomenclature and metadata standards for handling ecological and micrometeorological data, the need to track data provenance, and the difficulties of securing detailed metadata necessary to fully understand and evaluate site measurements serve as examples of challenges addressed by CDIAC in constructing the network-wide AmeriFlux data system.

AmeriFlux sites are funded by multiple sources, typically US federal agencies. Recommendations are made by an AmeriFlux Scientific Steering Committee (SSC) regarding standard operating procedures, but there are no mandatory measurement and instrumentation protocols. Science objectives and corresponding measurement strategies are often site-specific, resulting in great variance, and richness, in the measurements made across the collective AmeriFlux network. Other regional flux networks exist worldwide, and coordination with these networks is important to optimize the utility of the AmeriFlux data collection and to better serve the larger climate change scientific community.

The AmeriFlux SSC and CDIAC offer guidance on data submission requirements to site investigator teams for continuous measurements and infrequent, complementary biological measurements. Raw data captured by the site data loggers for continuous measurements are maintained by the site teams and are not submitted to CDIAC. Site investigators process high-frequency samples (usually at 10 Hz or higher) to produce half-hourly or hourly estimates of measured variables. Independent biological and ecological measurements (e.g., soil carbon content, foliage nitrogen concentration, tree allometry) are important for interpreting and validating the continuous AmeriFlux eddy covariance measurements and for parameterizing models. These companion biological measurements pose a challenge in assembling a consistent, network-wide resource due to the range of measurement methodologies and frequencies, differences in spatial representativeness, and the detailed metadata needed to fully characterize the measurement.

The breadth and diversity of measurements and data processing at individual sites, along with site differences in resources and staff dedicated to data handling, makes uniform data submission difficult. Data quality and consistency vary across AmeriFlux sites and measurement groups. As a result, CDIAC is tasked with assembling a network-wide database based on varying submissions from individual sites.

Adequate description of AmeriFlux measurements is crucial to proper use and understanding of reported AmeriFlux data. Metadata must be coupled with the primary environmental measurements to understand the reported data and expand use. Many micrometeorological and ecological terms commonly reported by AmeriFlux investigators are not covered by ongoing attempts to unify and standardize environmental data and metadata collections. For example, soil temperature and moisture measurements at various depths below the surface, which are commonly reported by AmeriFlux sites and fundamental to the AmeriFlux data collection, are not covered in the present NetCDF Climate and Forecast (CF) metadata convention.

AmeriFlux data records can be updated frequently as improvements (e.g., new screening algorithm) and adjustments (e.g., apply a correction to a degrading photosynthetically active radiation sensor) are made to site data records. These changes must be documented, making data provenance an essential ingredient of the AmeriFlux data system.

No single commercial software package or software engineering technique satisfies all the requirements for the AmeriFlux data system, although the prototype Microsoft ($MS^{®}$) DataCube was considered. Below we provide an overview of a comprehensive, data management approach used by CDIAC to handle Earth system data from the AmeriFlux network. We detail the software engineering techniques and tools used in the AmeriFlux data effort in the hopes other scientific data efforts will benefit from our experiences and these tools.

2 Data management methods and approaches

2.1 Data submission, quality control and standardization

As mentioned previously, CDIAC is charged with assembling a network-wide AmeriFlux database from data submissions by individual site teams. CDIAC developed data submission guidelines (http://public.ornl.gov/ameriflux/data-guidelines.shtml) for the continuous half-hourly or hourly estimates of flux and meteorological variables to improve efforts in automating data processing and to invoke standardization of reported variables, units, and conventions (e.g., terrestrial carbon source and sink sign conventions).

To encourage data submissions, CDIAC accepts data in the manner most convenient for the submitting site team. Hourly or half-hourly summaries are received and obtained in numerous ways, including mirroring scripts (wget and cron jobs), commercial transfer packages (e.g., Dropbox), dedicated areas on secure File Transfer Protocol (FTP) servers, or as e-mail attachments.

Due to differences in measurement strategies, data processing rigor, and data quality, CDIAC invests considerable effort to first evaluate submissions by individual site teams and then reformat, convert, or derive variables according to the prescribed submission guidelines (Table 1). A key initial task is to fully understand the content and context of the submitted data before producing a consistent network resource and this elementary step requires micrometeorological expertise. Fundamental groundwork is laid during this initial evaluation towards standardizing variable names, units, and reporting intervals and towards capturing and enhancing metadata.

Given the importance and diversity of AmeriFlux data and recent scrutiny towards climate change data and model results, it is imperative that scientific data archives address and document data quality. Permanent scientific data archives, like CDIAC, cannot accept data at face value without evaluating data quality or trusting quality assessments by data providers or outside assessment groups. CDIAC scrutinizes the continuous meteorological and flux data submitted by

Table 1. Initial checks, evaluations, and clarifications performed by CDIAC on incoming AmeriFlux meteorological and flux data. These checks are necessary to determine the content and quality of the submitted data, as well as to identify corrections and derivations that must be made in subsequent processing.

Categories used to evaluate incoming AmeriFlux data	Considerations and examples
Time representation	Are the data time-stamped using local time or coordinated universal time (UTC)? Are values reported at the beginning or ending of a sampling period?
Variable nomenclature, definition, and symbolic convention	What measured and derived variables are provided and how are they named or represented symbolically? For example, water vapor flux vs. evapotranspiration or different acronyms (PAR vs. PPFD) for the same measurement of photosynthetically active radiation. How is each variable defined? For example, is soil heat flux defined as heat flux through the soil heat plate or as the sum of the heat flux through the soil heat plate and the heat stored in the soil layer above the plate?
Unit	What unit is used for each reported variable and is it consistent with the prescribed data submission guidelines? For example, reporting CO_2 concentration in $\mu mol\ mol^{-1}$ vs. $mg\,m^{-3}$.
Sign convention	What sign conventions are used by the reporting team? For example, carbon uptake is often reported as negative values for downward CO_2 fluxes but as positive values for upward releases to the atmosphere from photosynthetic production.
Multiple and redundant measurements	Are multiple and redundant measurements reported and, if so, how are they represented? For example, are they reported as independent measurements or an average from multiple sensors?
Measurement height/depth	Is each measurement properly associated with a measurement height or depth? Are values reported at a single level or reflecting an integral measurement of multiple levels? For example, soil temperature measured at 10 cm vs. within a vertical column (0–30 cm).
Data handling and processing	What adjustments have been applied to the reported variables? Were the data filtered, screened, or gap-filled? Were corrections applied to the flux terms and, if so, which ones (e.g., planar vs. two-dimensional coordinate rotation)?
Metadata	Are the submitted data accompanied by proper metadata? Different measurement teams report different levels of details in their metadata, from simple file headers (symbolic variable names and units only) to comprehensive documentation including measurement techniques, variable definitions, sign conventions, etc.
Basic data quality assessment	Are the data submitted of "publication" quality or just one step after being retrieved from the data-logger?

AmeriFlux investigators for quality and completeness (Table 2). Resident micrometeorological expertise benefits our programming efforts to develop robust and credible quality checks. Issues identified during the data quality evaluations are documented and resolved interactively between CDIAC and the site investigator teams. The checks identified in Table 2 are currently implemented in C, FORTRAN, and SAS® (Statistical Analysis Software, Inc.) computer programs and applied to all incoming data. These checks are subject to further development, improvement, and expansion as needed. After these quality checks are complete, CDIAC applies standard naming and unit conventions to the incoming data to facilitate the generation of a uniform and consistent network dataset.

Table 2. Quality evaluations and checks performed by CDIAC on incoming meteorological and flux data submissions.

Quality evaluations/checks	Detail and action
Missing or repeated entry	Determine if the total number of yearly data records matches the expected total (e.g., 17 520 half-hourly records for a non-leap year), if the time increment between consecutive entries matches the sampling intervals (30 or 60 min); and if data entries are in chronological order.
Time-stamp	Check for consistency between time-stamp entries, for example, between a set of month, day, hour, minute and Julian day entries.
Threshold	Conservative thresholds are determined for each variable and site (e.g., the likely maximum air temperature at an Alaskan site), and out-of-bound values are reset to a prescribed missing value (e.g., −9999).
Nighttime radiation	Determine the daily nighttime period using a sunrise and sunset calculator and the percentage of nighttime radiation values beyond a tolerance level; these invalid values (for shortwave and photosynthetically active radiation) are reset to zero.
Biological and meteorological inter-relationship	Some variables or redundant measurements should be inherently correlated, for example, photosynthetically active radiation and global radiation, air temperature and soil temperature, etc. Poor correlations typically indicate problems and are investigated on a case-by-case basis.
Spike detection	Generally speaking, a data value is considered a "spike" if the value deviates from the mean by ±3 standard deviations. For variables with little expected variation (e.g., pressure), we check for spikes over the entire time series. For variables showing distinct seasonal and diurnal patterns (e.g., air temperature), we check for spikes against the diurnal means within a moving time window (e.g., 20 days).
Stationarity	Check if time series lack reasonable variation during a day or over a short period (10–20 days).
Diurnal and seasonal cycles	Check the monthly mean diurnal cycles and seasonal cycles where these cycles are known and expected; diurnal cycle for a single day is checked by evaluating the correlation and consistency between this day and the mean diurnal cycles in the current and neighboring months.
Discontinuity and inter-annual variation	Detect the discontinuities and trends in time series across multiple years. For example, a declining trend in radiation measurement over years may indicate an instrument calibration drift.

Former AmeriFlux Science Chair, Dr. Beverly Law (Oregon State University), and others developed guidelines for measurement teams to submit summaries of biological measurements collected at flux sites along with important site ancillary, disturbance, and management information (Law et al., 2008). These guidelines continue to evolve and have been sorely needed by ecological measurement networks attempting to amass ecological data reportings. Templates, referred to as BADM (Biological-Ancillary-Disturbance-Management) templates, are available to potential submitters as multi-worksheet Microsoft Excel® spreadsheets (http://public.ornl.gov/ameriflux/AmeriFlux_BiologicalDataTemplates_2009.xls). The guidance serves an important first step in defining and standardizing highly variable, irregularly measured ecological data essential for model parameterization.

Due to the irregular and sometimes non-numeric nature of biological data, an automated system for quality evaluation is not practical and requires considerable manual effort. CDIAC ecological experts evaluate the biological data based on their own field measurement experience, by cross-examining the submitted records over multiple years at the same site, or by making comparisons with published data from other sites of the same ecosystem type. Suspicious entries and other issues are usually reported back to data providers and resolved in an interactive and collaborative manner. CDIAC also invests effort to integrate the biological data from individual sites into a uniform network dataset. Specifically, CDIAC edits data entries and documents the changes if data providers fail to follow the submission guidelines in their original submissions. Typical examples are unit conversions (from g dry biomass m^{-2} to $g\,C\,m^{-2}$), adjustments to conform to prescribed definitions (from double-sided and all-sided leaf area indices to single-sided), and data code translations (replacing the scientific and common names of tree species by their standard codes defined by the US Department of Agriculture, Natural Resources Conservation Service).

2.2 Data products

CDIAC offers four different product levels for the continuous AmeriFlux meteorological and flux measurements. These products differ in origin, content, and level of data processing. The four product levels are:

– Level 1: processed data provided by the site investigators. Level 1 data files originally provided by the site measurement teams are evaluated and corrected by CDIAC as described in Sect. 2.1 and further processed by CDIAC to produce Level 2 data products. The Level 1 data files are posted in their original form on the publically-available CDIAC anonymous FTP server (ftp://cdiac.ornl.gov/pub/ameriflux/data/) and include regular and continuous half-hourly or hourly measurements of flux and meteorological variables and irregular measurements of biological variables in the BADM template or site specific format.

– Level 2: data checked and formatted by CDIAC. Data received from individual sites are reviewed, quality-controlled, reformatted, and incorporated into a network-wide AmeriFlux database. The review or evaluation process includes checks for consistent units, naming conventions, reporting intervals and others (see Sect. 2.1) and reformatting is often necessary to maintain consistency within the larger network-wide database. Level 2 data include both regular and continuous measurements of flux and meteorological variable and irregular measurements of biological variables.

Considering the breadth and diversity of measurements at AmeriFlux sites, it is necessary for CDIAC

to generate Level 2 standardized files to accommodate network-wide synthesis studies. Presently, Level 2 standardized files are generated for each year and individual site for flux and meteorological measurements only, and contain a core suite of approximately 40 variables. Each standardized file contains headers providing vital site, citation, and file content information. There may be two sets of Level 2 standardized files for each site (i.e., a gap-filled set and a set with gaps in records) depending on the site submissions. The gap-filled Level 2 standardized files contain variables gap-filled by the site investigators or teams using site-specific methods.

– Level 3: processed data with quality flags assigned. Level 3 files are created in conjunction with ICOS (Integrated Carbon Observation System) in order to provide uniform flux tower records and data files across all regional flux networks, including the AmeriFlux network. These Level 3 files contain the same values as CDIAC's Level 2 data files but with quality flags assigned. Quality flags are only applied to select flux and meteorological variables.

– Level 4: gap-filled and adjusted data files with estimates of gross primary production (GPP) and total ecosystem respiration (Re). Two sets of Level 4 data products exist. The first set contains ICOS filtered and gap-filled flux records (CO_2, sensible heat and latent heat fluxes), as well as derived GPP and Re terms. Data filtering, gap-filling and flux-partitioning are accomplished using widely accepted techniques and algorithms in the flux community (Papale and Valentini, 2003; Reichstein et al., 2005). Level 4 data differ from gap-filled Level 2 standardized files in that the selected variables are filled using the same techniques for all sites while Level 2 gap-filled records are generated by site-specific methods. After being gap-filled, hourly or half-hourly Level 4 data are aggregated into longer time intervals including daily, weekly, and monthly reporting intervals.

The second set of Level 4 data products are produced by CDIAC and contain gap-filled meteorological records for AmeriFlux sites only. To satisfy the needs of driving ecosystem models with gap-free meteorological data, CDIAC recently developed a method (Table 3) to gap-fill commonly reported meteorological variables (e.g., relative humidity, air temperature, radiation) by adopting and enhancing the algorithms used in support of a model-data synthesis activity (Schwalm et al., 2010). This new Level 4 data product is now available for public use and evaluation, and CDIAC plans to further improve the methodology based on user feedback from this initial release.

According to the AmeriFlux site registry (http://ameriflux.ornl.gov/sitelocations.php) in January 2013, 98 AmeriFlux sites are active while 58 sites are inactive. One hundred fifty

Table 3. Flow diagram for the methodology used by CDIAC to gap-fill meteorological records for the AmeriFlux network.

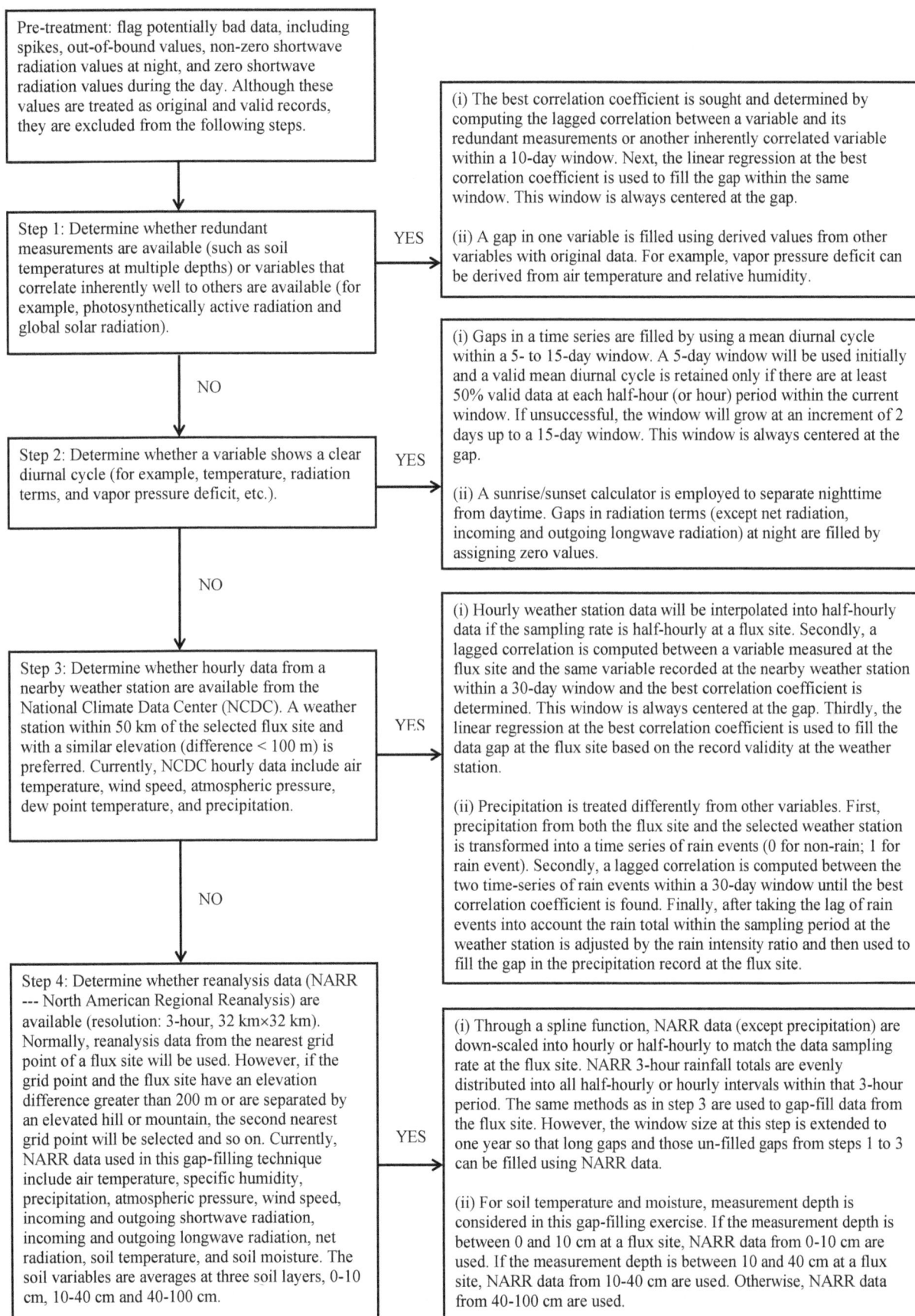

Pre-treatment: flag potentially bad data, including spikes, out-of-bound values, non-zero shortwave radiation values at night, and zero shortwave radiation values during the day. Although these values are treated as original and valid records, they are excluded from the following steps.

↓

Step 1: Determine whether redundant measurements are available (such as soil temperatures at multiple depths) or variables that correlate inherently well to others are available (for example, photosynthetically active radiation and global solar radiation).

→ YES →

(i) The best correlation coefficient is sought and determined by computing the lagged correlation between a variable and its redundant measurements or another inherently correlated variable within a 10-day window. Next, the linear regression at the best correlation coefficient is used to fill the gap within the same window. This window is always centered at the gap.

(ii) A gap in one variable is filled using derived values from other variables with original data. For example, vapor pressure deficit can be derived from air temperature and relative humidity.

NO ↓

Step 2: Determine whether a variable shows a clear diurnal cycle (for example, temperature, radiation terms, and vapor pressure deficit, etc.).

→ YES →

(i) Gaps in a time series are filled by using a mean diurnal cycle within a 5- to 15-day window. A 5-day window will be used initially and a valid mean diurnal cycle is retained only if there are at least 50% valid data at each half-hour (or hour) period within the current window. If unsuccessful, the window will grow at an increment of 2 days up to a 15-day window. This window is always centered at the gap.

(ii) A sunrise/sunset calculator is employed to separate nighttime from daytime. Gaps in radiation terms (except net radiation, incoming and outgoing longwave radiation) at night are filled by assigning zero values.

NO ↓

Step 3: Determine whether hourly data from a nearby weather station are available from the National Climate Data Center (NCDC). A weather station within 50 km of the selected flux site and with a similar elevation (difference < 100 m) is preferred. Currently, NCDC hourly data include air temperature, wind speed, atmospheric pressure, dew point temperature, and precipitation.

→ YES →

(i) Hourly weather station data will be interpolated into half-hourly data if the sampling rate is half-hourly at a flux site. Secondly, a lagged correlation is computed between a variable measured at the flux site and the same variable recorded at the nearby weather station within a 30-day window and the best correlation coefficient is determined. This window is always centered at the gap. Thirdly, the linear regression at the best correlation coefficient is used to fill the data gap at the flux site based on the record validity at the weather station.

(ii) Precipitation is treated differently from other variables. First, precipitation from both the flux site and the selected weather station is transformed into a time series of rain events (0 for non-rain; 1 for rain event). Secondly, a lagged correlation is computed between the two time-series of rain events within a 30-day window until the best correlation coefficient is found. Finally, after taking the lag of rain events into account the rain total within the sampling period at the weather station is adjusted by the rain intensity ratio and then used to fill the gap in the precipitation record at the flux site.

NO ↓

Step 4: Determine whether reanalysis data (NARR --- North American Regional Reanalysis) are available (resolution: 3-hour, 32 km×32 km). Normally, reanalysis data from the nearest grid point of a flux site will be used. However, if the grid point and the flux site have an elevation difference greater than 200 m or are separated by an elevated hill or mountain, the second nearest grid point will be selected and so on. Currently, NARR data used in this gap-filling technique include air temperature, specific humidity, precipitation, atmospheric pressure, wind speed, incoming and outgoing shortwave radiation, incoming and outgoing longwave radiation, net radiation, soil temperature, and soil moisture. The soil variables are averages at three soil layers, 0-10 cm, 10-40 cm and 40-100 cm.

→ YES →

(i) Through a spline function, NARR data (except precipitation) are down-scaled into hourly or half-hourly to match the data sampling rate at the flux site. NARR 3-hour rainfall totals are evenly distributed into all half-hourly or hourly intervals within that 3-hour period. The same methods as in step 3 are used to gap-fill data from the flux site. However, the window size at this step is extended to one year so that long gaps and those un-filled gaps from steps 1 to 3 can be filled using NARR data.

(ii) For soil temperature and moisture, measurement depth is considered in this gap-filling exercise. If the measurement depth is between 0 and 10 cm at a flux site, NARR data from 0-10 cm are used. If the measurement depth is between 10 and 40 cm at a flux site, NARR data from 10-40 cm are used. Otherwise, NARR data from 40-100 cm are used.

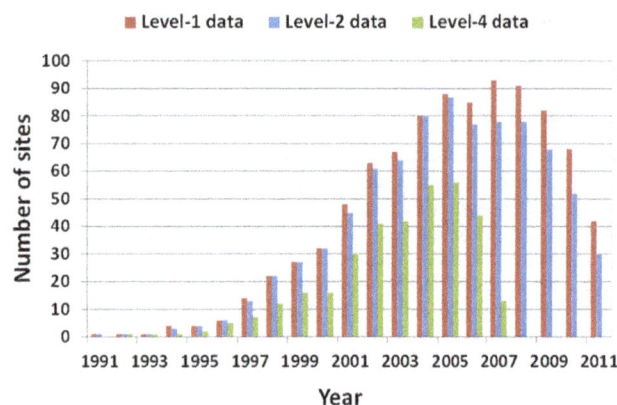

Fig. 3. Number of AmeriFlux sites with data for each year from 1991 to 2011 based on the CDIAC AmeriFlux data collection in January 2013.

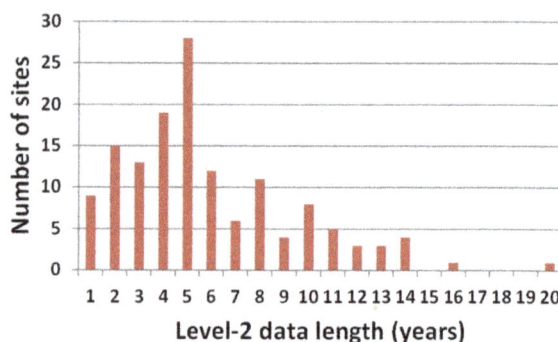

Fig. 4. Distribution of data record lengths in years for the 142 AmeriFlux sites with available Level 2 data products based on the CDIAC AmeriFlux data collection in January 2013.

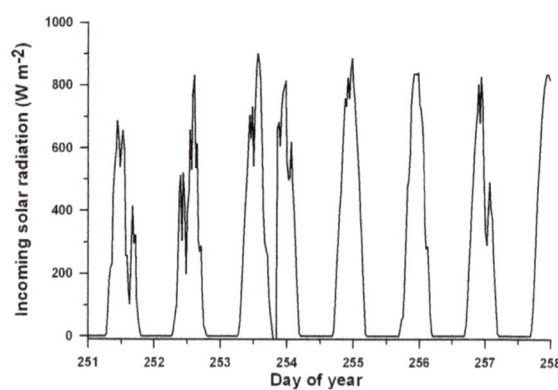

Fig. 5. CDIAC uses graphics and visualization tools to review time series for all measured AmeriFlux variables, assess data quality, and identify possible errors. This plot shows an erroneous time shift beginning on day 254 for incoming solar radiation.

sites have submitted Level 1 flux and meteorological data in various lengths. Of these sites, 142 site submissions have been processed into Level 2 products by CDIAC while the rest either have lingering quality issues or are being processed and evaluated. Seventy-seven sites have contributed biological data. Level 3 and Level 4 gap-filled flux data are available for 78 sites, while Level 4 gap-filled meteorological records are available for 102 sites. Data availability for each year from 1991 to 2011 and the distribution of site record lengths are illustrated in Figs. 3 and 4. The average Level 2 data record length is 5.9 yr with 25 sites contributing records of 10 yr or longer. Among all registered sites, the Harvard Forest site has the longest data length and is the only AmeriFlux site presently with a record exceeding twenty years.

2.3 Database, archive, and distribution

Most computer programs used by CDIAC to evaluate the integrity of submitted AmeriFlux data are written in SAS®. SAS® is a proven, well-documented statistical analysis software package with powerful sort and merge capabilities and flexibilities in handling file formats. SAS/IntrNet® enables web browsers to run a SAS® report application and make the results available to the browser thus providing the ability to create a wide range of web-ready, dynamic data reports. CDIAC has used this product and feature to produce an interactive data interface to help in evaluating incoming AmeriFlux data. One of the most powerful data evaluation tools is to review plots of individual measurement variables over time or plots of two correlated variables. The old adage, "A picture is worth a thousand words", is very true when evaluating scientific data. SAS® graphics capabilities complement the routine CDIAC quality checks summarized in Table 2 and often highlight obvious quality issues in the incoming data (Fig. 5). Another nice feature of the SAS/IntrNet® product is the ability to produce automated metadata reports in

HTML format documenting data provenance including variable mappings, calculations, and unit conversions, etc.

CDIAC archives and distributes AmeriFlux data across two systems: (1) a PHP-based web interface that draws on a MySQL relational database; and (2) an anonymous FTP server. The two systems are independent but connected by links in the PHP web interface pointing to the FTP server.

Once CDIAC data quality checks are complete, the AmeriFlux Level 2 data in the form of a SAS® database are ingested directly into an AmeriFlux MySQL relational database using a PHP script. PHP is an advanced web-programming language that enables programmers to construct complicated web sites without compromising server-side processing abilities, an important consideration for the AmeriFlux system given the anticipated multi-terabyte-level data volumes. MySQL at the same time is a powerful, robust, and flexible open-source product capable of handling large volumes of complex Earth system data. MySQL design architecture enables environmental programmers to partition large data volumes and thus increase query performance. For AmeriFlux data, it was logical to implement site-based

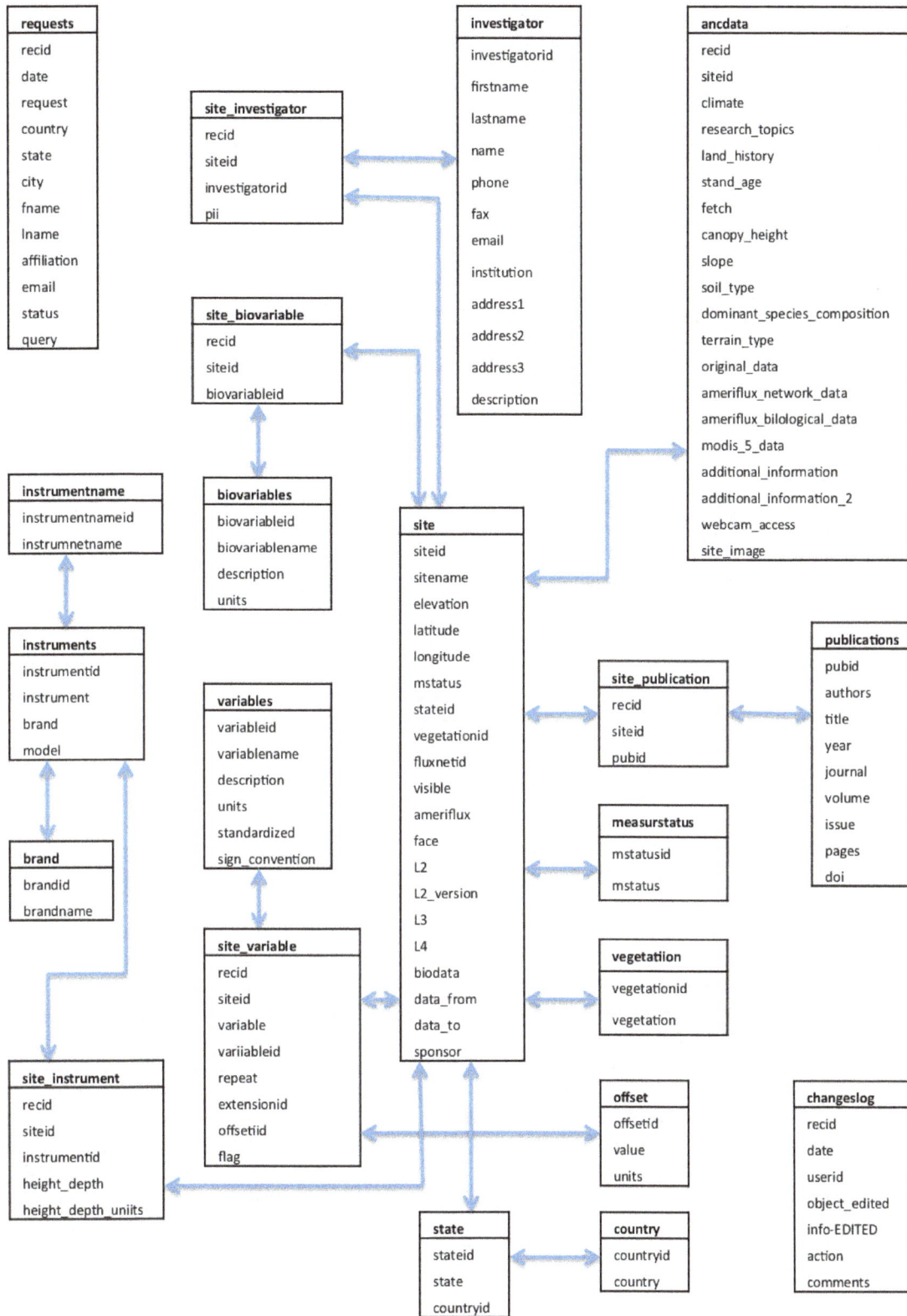

Fig. 6. Diagram identifying the tables comprising the CDIAC AmeriFlux MySQL relational database. Tables connected by arrows share two identical variables, the site identifier and site investigator, to enable queries across multiple tables. Tables not connected by arrows are stand-alone tables within the database. Table names are shown at the top of each table box in bold letters. Column names within each table are listed below the table names.

partitioning since the data collection is built on data streams provided by individual site teams. Site investigators and identifiers serve as common threads linking MySQL tables comprising the AmeriFlux relational database (Fig. 6). Another important decision was made early in the AmeriFlux data architecture design to handle all data additions and corrections via a PHP-developed, web-based editing interface. This strategy allows CDIAC to control the content of the production and archive databases, automate data movement, track data provenance (i.e., track changes, assign version numbers, etc.), and manage metadata content and standardization. The strategy also removed the need for client-side software distributions. Presently, over 500 million observations reside in the production-side AmeriFlux MySQL database available to the public.

The primary tool to deliver AmeriFlux data is a web-based interface referred to as the AmeriFlux Site and Data Exploration System (http://ameriflux.ornl.gov). The PHP interface allows users to query the AmeriFlux MySQL database in a variety of ways. Anticipated popular data and metadata search criteria were identified by CDIAC staff and incorporated into the interface to assist users to navigate through the AmeriFlux database. Examples of search criterion are product levels (see Sect. 2.2), dates of available data, ecosystem types, site operating status (i.e., active or inactive), site coordinates, and instruments. New scientific-based search criteria are being added (e.g., distinction of measurements between nighttime and daytime, growing and dormant seasons) based on emerging needs from climate change researchers. Once users have identified data of interest, the interface permits users to further subset, download, or visualize their selections. The same interface can be used to obtain fundamental site information for all registered AmeriFlux sites. Independent web pages are dedicated to every single site and users can browse a broad set of metadata for the site of interest. These metadata include, for example, site identification, coordinates, names of principal investigator(s), site instruments and publications, descriptions of site climate, vegetation, soil, terrain and land use history, and many others. Interactive tables list all measurements reported by a site including variable names, definitions, units, and sign conventions. Visualization capabilities are also available at this web interface. Data can be downloaded directly from the Ameri-Flux MySQL database through this web interface or from the links pointing to the CDIAC FTP server.

As described in Sect. 2.2, CDIAC generates a set of Level 2 standardized ASCII data files for flux and meteorological variables. These Level 2 files are posted on the CDIAC anonymous FTP server (ftp://cdiac.ornl.gov/ameriflux/data/Level2) after being created using SAS® codes. These files are posted in separate site folders identified by the site name or a unique site ID (e.g., US-UMB) assigned to all flux sites worldwide. To satisfy users interested in data from multiple AmeriFlux sites rather than an individual site,

"Unix tar balls" are created and posted on the same CDIAC anonymous FTP server. New "tar balls" are created whenever new AmeriFlux standardized files are posted to the FTP server using a Perl script, which executes based on new file names and creation dates. The "tar ball" file name conveys the creation date to promote easy mirroring and citation by users. These standardized files serve another critical function. They enable the AmeriFlux regional network to participate in a virtual global flux network (FLUXNET) and its cross-continent synthesis studies. To further aid modeling and synthesis activities, identical NetCDF files are produced from the Level 2 standardized files using a Perl script to reformat the standardized files into the file format required by the NetCDF Compiler. All header information provided in the standardized ASCII files is written as global attributes into the corresponding NetCDF files. These NetCDF files are posted to the CDIAC server and included in the above-mentioned "tar balls".

Also available at this FTP server are the higher level data products (i.e., Level 3 and 4 data) and biological data. After scrutinizing the biological data (see Sect. 2.1), individual BADM files are integrated into a single MS Excel® file with multiple spreadsheets dedicated to different categories of biological measurements, including leaf area index (LAI), soil respiration, stand properties, biomass production, vegetation C and N content, soil properties, and phenology. This integrated file, also referred to as Level 2 biological data, is entered into the AmeriFlux MySQL database by CDIAC staff through the PHP editing interface and also posted to the AmeriFlux FTP server. The complete AmeriFlux data life cycle, from site collection to distribution by CDIAC, is shown in Fig. 7.

Two additional expanding systems are used to further promote and broadcast CDIAC's AmeriFlux data products and services. First, CDIAC deploys a Mercury instance. Mercury is a web-based, distributed metadata management, data discovery, and data access system (http://mercury.ornl.gov) implemented using Internet standards, including XML, and supports international metadata standards including FGDC, Dublin-Core, EML, and ISO-19115. CDIAC produces metadata summaries for AmeriFlux data products for inclusion in the CDIAC Mercury instance (http://mercury.ornl.gov/cdiac/) using a customized online metadata editor (OME). OME produces XML files for ingest, cataloging, and indexing by the CDIAC Mercury instance and other disparate climate-change related Mercury instances. Secondly, CDIAC AmeriFlux data products are being published into the Earth System Grid Federation (ESGF). ESGF is a data distribution portal used primarily to distribute large-scale, modeling results (Williams et al., 2009). ESGF integrates supercomputers with petabyte-scale data and analysis servers located at national laboratories and research centers to create a powerful environment for next generation climate research. ESGF served as the focal point for disseminating

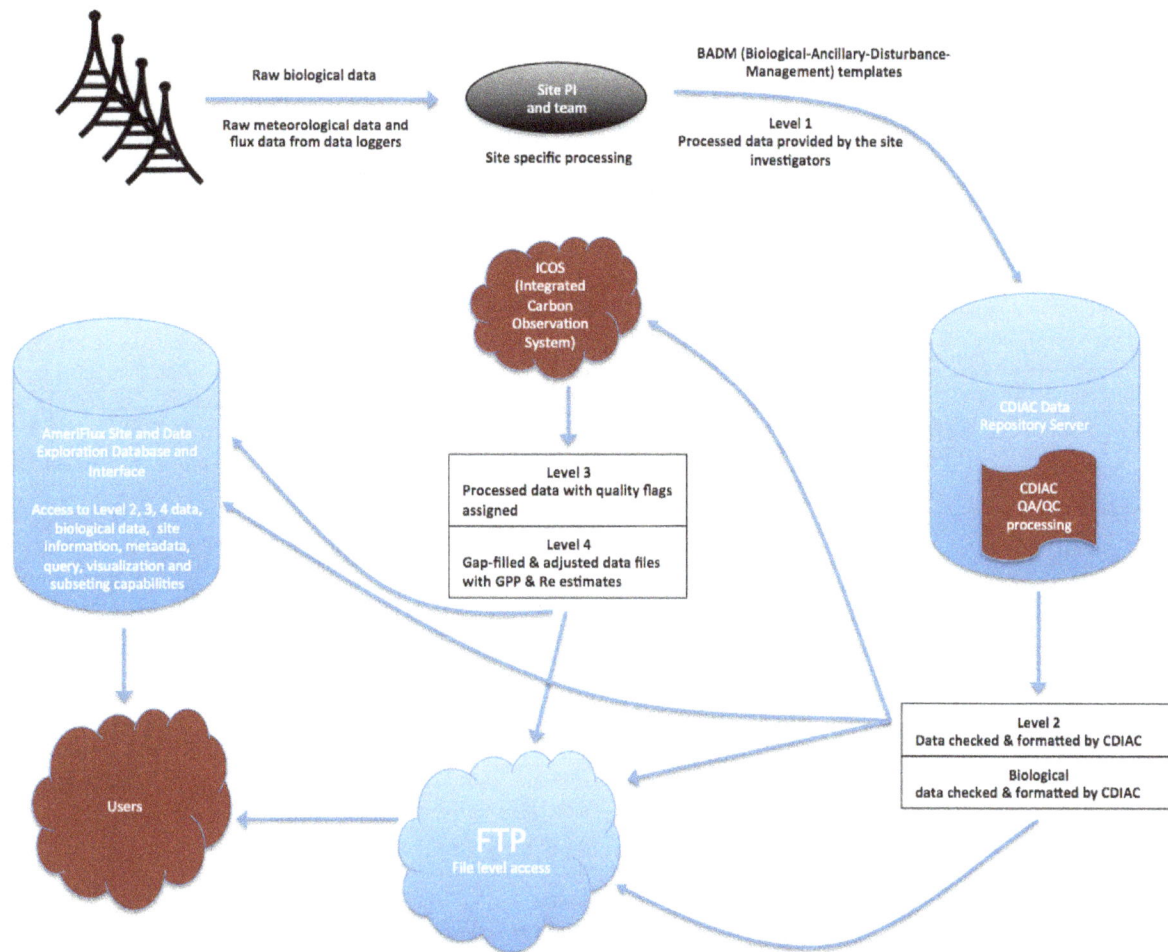

Fig. 7. The AmeriFlux data life cycle. The life cycle moves from site collection (upper left) to processing by the site teams (top center) to CDIAC processing and data product generation (right side) to Integrated Carbon Observing System processing and product generation to dissemination by CDIAC.

results from the Intergovernmental Panel on Climate Change (IPCC) Fourth Assessment Report (AR4) and will serve in a similar capacity for the IPCC Fifth Assessment Report (AR5). ESGF has historically focused on model output, but is now including observational data important for model testing and validation, including AmeriFlux data provided by CDIAC (http://esg2-gw.ccs.ornl.gov).

3   Conclusions

Like most scientific computing endeavors, there is no ready, commercial product that satisfies the full suite of data analysis, data management, metadata management, and data delivery requirements for the AmeriFlux data activity. The breadth, complexity, and variability of data and metadata are too great to be handled by a single tool or product. Instead, CDIAC has deployed a variety of data management tools, software, and software engineering techniques to support the AmeriFlux data activity. Where needed, tools and

applications have been adapted and modified to specific requirements driven by the data needs of the international climate change research community. Scientific data management approaches must be flexible, portable, and expandable in order to accommodate future users, increased data volumes, and new data products. It is virtually impossible to predict all future applications of scientific data making, it imperative that data system designs be flexible for unanticipated growth and, more importantly, that detailed metadata be captured to promote unexpected use and application.

Immediate needs for the AmeriFlux data system demonstrate the need for flexibility. There is a need for synchronization of regional flux databases, including common variable nomenclatures and units, common data products and metadata, uniform data processing, and node access to a "global" flux database. As publication tools for observational databases are developed within the ESGF, there will be a need to automate the ingest of AmeriFlux data to ESGF and, further, to launch new model simulations as revised and updated AmeriFlux data become available. Demand from the scientific community and general public has grown for inclusion of uncertainty estimates or error bounds on flux and meteorological data. CDIAC is now undertaking an effort to provide uncertainty estimates in the AmeriFlux data system and developing other products such as derivation of phenology stages from the existing measurements at AmeriFlux sites.

Acknowledgements. This research and CDIAC were supported by the US Department of Energy, Office of Science, Biological and Environmental Research (BER), and conducted at Oak Ridge National Laboratory (ORNL), managed by UT-Battelle, LLC, for the US Department of Energy under contract DE-AC05-00OR22725.

Edited by: W. Schmidt

References

Baldocchi, D. D.: Assessing the eddy covariance technique for evaluating carbon dioxide exchange rates of ecosystems: past, present and future, Global Change Biol., 9, 479–492, 2003.

Law, B. E., Arkebauer, T., Campbell, J. L., Chen, J., Sun, O., Schwartz, M., van Ingen, C., and Verma, S.: Terrestrial Carbon Observations: Protocols for Vegetation Sampling and Data Submission, Global Terrestrial Observing System, 55, Rome, Italy, 2008.

Papale, D. and Valentini, R.: A new assessment of European forests carbon exchanges by eddy fluxes and artificial neural network spatialization, Global Change Biol., 9, 525–535, 2003.

Randerson, J. T., Hoffman, F. M., Thornton, P. E., Mahowald, N. M., Lindsay, K., Lee, Y. H., Nevison, C. D., Doney, S. C., Bonan, G., Stockli, R., Covey, C., Running, S. W., and Fung, I. Y.: Systematic assessment of terrestrial biogeochemistry in coupled climate-carbon models, Global Change Biol., 15, 2462–2484, 2009.

Reichstein, M., Falge, E., Baldocchi, D., Papale, D., Aubinet, M., Berbigier, P., Bernhofer, C., Buchmann, N., Gilmanov, T., Granier, A., Grunwald, T., Havrankova, K., Ilvesniemi, H., Janous, D., Knohl, A., Laurila, T., Lohila, A., Loustau, D., Matteucci, G., Meyers, T., Miglietta, F., Ourcival, J. M., Pumpanen, J., Rambal, S., Rotenberg, E., Sanz, M., Tenhunen, J., Seufert, G., Vaccari, F., Vesala, T., Yakir, D., and Valentini, R.: On the separation of net ecosystem exchange into assimilation and ecosystem respiration: review and improved algorithm, Global Change Biol., 11, 1424–1439, 2005.

Schwalm, C. R., Williams, C. A., Schaefer, K., Anderson, R., Arain, M. A., Baker, I., Barr, A., Black, T. A., Chen, G. S., Chen, J. M., Ciais, P., Davis, K. J., Desai, A., Dietze, M., Dragoni, D., Fischer, M. L., Flanagan, L. B., Grant, R., Gu, L. H., Hollinger, D., Izaurralde, R. C., Kucharik, C., Lafleur, P., Law, B. E., Li, L. H., Li, Z. P., Liu, S. G., Lokupitiya, E., Luo, Y. Q., Ma, S. Y., Margolis, H., Matamala, R., McCaughey, H., Monson, R. K., Oechel, W. C., Peng, C. H., Poulter, B., Price, D. T., Riciutto, D. M., Riley, W., Sahoo, A. K., Sprintsin, M., Sun, J. F., Tian, H. Q., Tonitto, C., Verbeeck, H., and Verma, S. B.: A model-data intercomparison of CO_2 exchange across North America: Results from the North American Carbon Program site synthesis, J. Geophys. Res., 115, G00H05, doi:10.1029/2009JG001229, 2010.

Williams, C. A., Reichstein, M., Buchmann, N., Baldocchi, D., Beer, C., Schwalm, C., Wohlfahrt, G., Hasler, N., Bernhofer, C., Foken, T., Papale, D., Schymanski, S., and Schaefer, K.: Climate and vegetation controls on the surface water balance: Synthesis of evapotranspiration measured across a global network of flux towers, Water Resour. Res., 48, W06523, doi:10.1029/2011WR011586, 2012.

Williams, D. N., Ananthakrishnan, R., Bernholdt, D. E., Bharathi, S., Brown, D., Chen, M., Chervenak, A., Cinquini, L., Drach, R., Foster, I. T., Fox, P., Fraser, D., Garcia, J., Hankin, S., Jones, P., Middleton, D. E., Schwidder, J., Schweitzer, R., Schuler, R., Shoshani, A., Siebenlist, F., Sim, A., Strand, W. G., Su, M., and Wilhelmi, N.: The Earth System Grid: Enabling Access to Multimodel Climate Simulation Data, B. Am. Meteorol. Soc., 90, 195–205, 2009.

12

Auroral all-sky camera calibration

F. Sigernes[1,*], S. E. Holmen[1,*], D. Biles[2], H. Bjørklund[1], X. Chen[1,*], M. Dyrland[1,*], D. A. Lorentzen[1,*], L. Baddeley[1,*], T. Trondsen[3], U. Brändström[4], E. Trondsen[5], B. Lybekk[5], J. Moen[5], S. Chernouss[6], and C. S. Deehr[7]

[1]University Centre in Svalbard, Longyearbyen, Norway
[2]Magnetosphere Ionosphere Research Lab, University of New Hampshire, USA
[3]Keo Scientific Ltd, Calgary, Alberta, Canada
[4]Swedish Institute of Space Physics, Kiruna, Sweden
[5]Department of Physics, University of Oslo, Oslo, Norway
[6]Polar Geophysical Institute, Murmansk Region, Apatity, Russia
[7]Geophysical Institute, University of Alaska, Fairbanks, USA
[*]also at: Birkeland Centre for Space Science, University of Bergen, Bergen, Norway

Correspondence to: F. Sigernes (freds@unis.no)

Abstract. A two-step procedure to calibrate the spectral sensitivity to visible light of auroral all-sky cameras is outlined. Center pixel response is obtained by the use of a Lambertian surface and a standard 45 W tungsten lamp. Screen brightness is regulated by the distance between the lamp and the screen. All-sky flat-field correction is carried out with a 1 m diameter integrating sphere. A transparent Lexan dome at the exit port of the sphere is used to simulate observing conditions at the Kjell Henriksen Observatory (KHO). A certified portable low brightness source from Keo Scientific Ltd was used to test the procedure. Transfer lamp certificates in units of Rayleigh per Ångstrøm (R/Å) are found to be within a relative error of 2 %. An all-sky camera flat-field correction method is presented with only 6 required coefficients per channel.

1 Introduction

During the last decades, numerous ground-based all-sky cameras have been installed in both hemispheres to monitor aurora and airglow. In the northern hemisphere, the fields of view of these cameras overlap to cover large sections of the aurora oval (cf. Akasofu, 1964). The desire to, for example, estimate and compare auroral hemispherical power as measured by satellites (cf. Zhang and Paxton, 2008) requires unified and accurate calibration routines (Brändström et al.,

2012) to quantify the radiance in photometric units (Hunten et al., 1956). This paper presents a two-step procedure to calibrate to sensitivity the all-sky cameras at the Kjell Henriksen Observatory (KHO).

2 Experimental setup

The calibration tools are shown in Fig. 1. The fixed imaging compact spectrograph (FICS) is mounted on a height-adjustable table. The table can be moved on rails towards the Lambertian screen. The entrance optics of FICS is a 22° field-of-view-fused silica fibre bundle. The spectrograph is made by ORIEL (model 77443). It uses a concave holographic grating (230 grooves/mm). The nominal spectral range is 4000–11 000 Å, and the bandpass is approximately 80 Å with the 100 μm wide entrance slit. The detector is a 16 bit dynamic range thermoelectric cooled CCD camera from the company Andor (model DU 420A-OE). Our main calibration source, the 45 W tungsten lamp from Oriel (s/n 7-1867), is also mounted on the table. The lamp is a traceable National Institute of Standards (NIST) source. The lamp certificate is listed in Table 1. Both the Lambertian screen (SRT-99-180) and the 1 m diameter integrating sphere (CSTM-LR-40) are made by the company Labsphere. Note that in Fig. 1, the spectrograph is set up to measure the output of the integrating sphere.

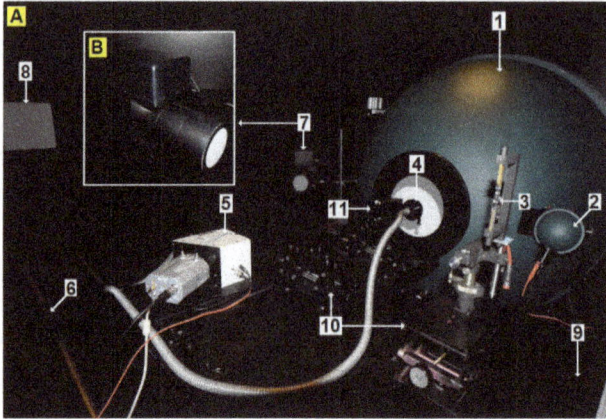

Figure 1. Experimental setup at the UNIS calibration lab. (a) (1) Labsphere 1m diameter low light level integrating sphere, (2) source sphere with tungsten lamp, (3) 45 W tungsten lamp, (4) FICS fiber bundle probe, (5) FICS, (6) rail road, (7) Keo Alcor-RC low light source, (8) Lambertian screen, (9) height adjustable table on rails, (10) table jacks and (11) rotary probe mount. (b) Keo Alcor-RC low light source.

Table 1. Oriel 45 W tungsten lamp certificate (s/n 7-1867). The spectral irradiance values are measured at a distance of $z_0 = 0.5$ m.

Wavelength λ [Å]	Irradiance $M_0(\lambda)$ [$mW\,m^{-2}\,nm^{-1}$]	Irradiance $M_0(\lambda)$ [$\#photons\,cm^{-2}\,s^{-1}\,Å^{-1}$]
4000	0.79670	1.60221×10^{10}
4500	1.71388	3.87755×10^{10}
5000	2.99143	7.51994×10^{10}
5550	4.65315	1.29839×10^{11}
6000	6.04915	1.82478×10^{11}
6546	7.64049	2.51456×10^{11}
7000	8.76666	3.08530×10^{11}
8000	11.1985	4.50416×10^{11}

The integrating sphere is modified by including a transparent dome at the exit port and a baffle to block light from the source sphere. A sketch of the modifications is shown in Fig. 2. The dome is made of the same material (Lexan) and thickness (5 mm) as the domes at the KHO. The all-sky cameras should be inserted into the dome in order to fill the total field of view of 180°. The baffle acts as a moon blocker. The net result is that observational and calibration conditions are the same for all optical instruments housed at the KHO.

The FICS is sensitivity-calibrated by the use of the Lambertian screen and the 45 W tungsten lamp. The distance and angle between the screen and the lamp regulate the brightness of the screen. This well-known method of calibrating narrow field-of-view instruments is described in detail by Sigernes et al. (2007).

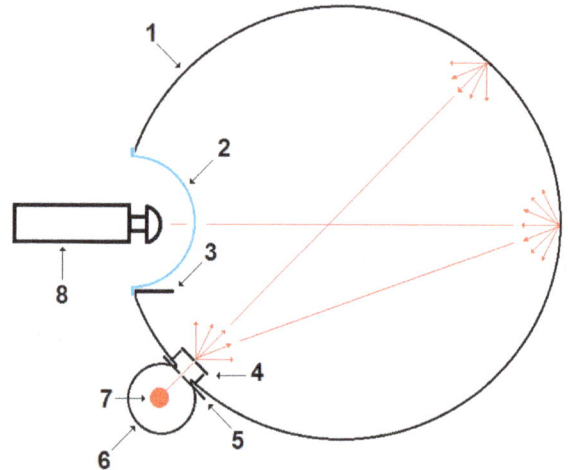

Figure 2. Sketch of modification to the 1m diameter integrating sphere: (1) Labsphere CSTM-LR-40, (2) transparent Lexan dome, (3) baffle, (4) transparent diffusor, (5) adjustable aperture, (6) source sphere, (7) tungsten lamp and (8) instrument with all-sky lens. Red arrows and lines indicate the effect of multiple scattering inside the sphere.

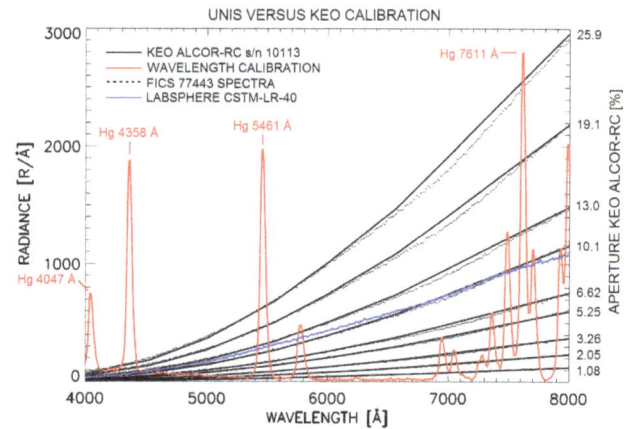

Figure 3. Absolute and wavelength calibration of the FICS. The line spectrum in red is from a mercury vapor lamp supplied by Edmund Optics Ltd (SN K60-908). The solid black spectra are from the Keo Alcor-RC certificate. The corresponding black dotted lines are spectra measured by the FICS. The spectrum of the integrating sphere is plotted in blue.

3 Test and tuning of calibration tools

A mobile low light source made by the company Keo Scientific is used to test the new calibration method. The head unit of the Keo Alcor-RC low brightness source (s/n 10113) is visible in Fig. 1. It contains a 100 W tungsten lamp, aperture wheels and diffusors to attenuate the brightness of the opal output surface. The source is certified by the National Research Council of Canada (NRC).

The FICS fiber probe is mounted head-on to the center of the output surface of the Keo-Alcor-RC source. The field of

Table 2. The difference in spectral calibration between UNIS – FICS and NRC of the Keo Alcor-RC low brightness source as a function of aperture. All numbers are in units of %.

Keo Alcor – RC aperture	25.9	19.1	13.0	10.1	6.62	5.24	3.26	2.05	1.08
Difference	2.00	1.56	1.41	0.40	1.42	0.04	0.90	0.26	1.96

Table 3. Fish-eye mapping functions. F is the effective focal length of the lens, and θ is the angle to the optical axis.

Type	Fish-eye mapping function
Linear scaled	$R = f \times \theta$
Orthographic	$R = f \times \sin(\theta)$
Equal area	$R = 2 \times f \times \sin(\theta/2)$
Stereographic	$R = 2 \times f \times \tan(\theta/2)$

view of the probe is then within a 1 cm diameter spot size of the diffuse surface. The setup is compatible to the spectral irradiance measured by Gaertner (2013). Figure 3 shows the measured FICS spectra and the corresponding certified spectra from NRC as a function of aperture setting. The Keo Alcor-RC aperture is given in units of percentage. 100 % is maximum brightness, while 10 % means that the brightness is one tenth of maximum.

The percent of error between the measured integrated FICS spectra and the NRC certificate is used to quantify the relative difference in the calibrations. The results are listed in Table 2 as a function of Keo aperture. Note that the relative errors are less than or equal to 2 %.

The next step in the test is to move the FICS fiber probe to the center of the Labsphere integrating sphere output port and tune the aperture of the source sphere down to a level that corresponds to the one tenth aperture brightness of the Alcor-RC source. The spectral radiance is then below the threshold of ~ 1 kR/Å, which is close to the intensity range of magnitude for auroras and airglow as indicated by the blue-colored spectrum in Fig. 3.

4 Basic camera equations

This section derives the equations needed for calibration of multiple wavelength filtered all-sky cameras. The experimental setup is as sketched in Fig. 2 using the integrating sphere as source. The type of filters used depends on center wavelength λ_c, bandpass BP and transmission T. The most traditional one is the Fabry–Perot filter design, where parallel transparent glass plates create interference due to multiple reflections between the plates.

The raw data counts of the camera at pixel position (x, y) are given as an integral over wavelength:

$$u = \int B(\lambda) \cdot S(\lambda) d\lambda, \, [\text{cts}], \tag{1}$$

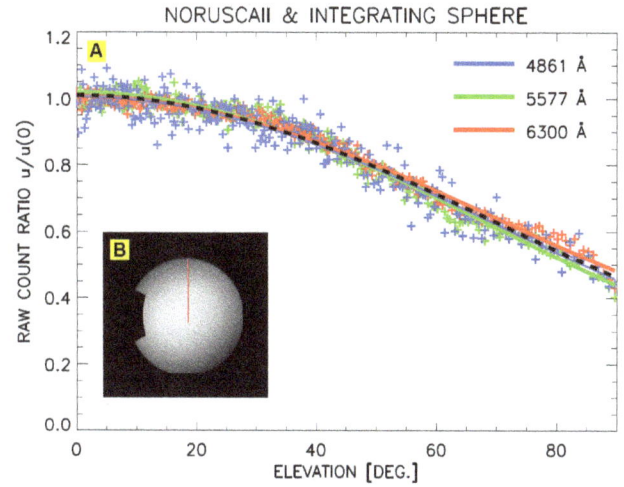

Figure 4. NORUSCA II all-sky camera data of integrating sphere by Labsphere (CSTM-LR-40). **(a)** Blue, red and green crosses are raw count ratios as function of elevation θ for wavelengths 4861, 5577 and 6300 Å, respectively. Exposure time is 1 sec and detector gain is 100. Center image count is defined as $u(\theta = 0)$. Solid colored lines are corresponding functional fits to the data. The dotted black line is the average fit over the wavelength. **(b)** Grayscale image of integrating the sphere at the center wavelength 5577 Å. The red-colored line marks center-to-edge coordinates used to obtain the count ratios.

Table 4. NORUSCA II all-sky camera center value calibration factors.

Channel #	Wavelength λ_c [Å]	Emission species	Bandpass BP [Å]	Calibration factor $B(\lambda_c) \cdot \text{BP}/u(0)$ [R cts^{-1}]
1	4278	N_2^+	54.4	110.2
2	4500	Background	57.3	88.7
3	4709	N_2^+	59.9	60.1
4	4861	H_β	61.9	44.4
5	5002	NII	63.7	38.1
6	5577	[OI]	71.0	25.1
7	5680	NII	72.3	23.7
8	5890	NaI	75.0	22.3
9	6300	[OI]	80.2	18.8
10	6364	[OI]	81.0	17.7
11	6563	H_α	83.5	16.6
12	6624	$N_2$1P(6-3)	84.3	15.7
13	6705	$N_2$1P(5-2)	85.3	14.4
14	6764	N_2	86.1	14.1
15	7000	Background	89.1	12.4

where S is defined as the spectral responsivity and B is the source spectrum in absolute units. The wavelength dependency of the spectral responsivity is assumed to be proportional to the transmission of the filter:

$$S(\lambda) \approx \varepsilon \cdot T(\lambda).[\text{cts R}^{-1}]. \qquad (2)$$

If the source B, lens transmissions and detector sensitivity vary slowly in the wavelength interval $\Delta\lambda$, then Eq. (1) becomes

$$u = B(\lambda_c) \cdot \varepsilon \cdot \int T(\lambda)\mathrm{d}\lambda = B(\lambda_c) \cdot \varepsilon \cdot A, \qquad (3)$$

where A is the area of the filter transmission curve. It is assumed that T is narrow and triangular in shape. Then

$$A = \int T(\lambda)\mathrm{d}\lambda \approx T_m \cdot \text{BP}, \qquad (4)$$

where T_m is the peak transmission of the filter at $\lambda = \lambda_c$. Furthermore, the spectral radiance of a discrete auroral emission line at wavelength λ_c is defined as

$$J_a(\lambda) \equiv J \cdot \delta(\lambda - \lambda_c)[\text{R/Å}], \qquad (5)$$

where δ is the Kronecker delta. The number of auroral raw data counts is then

$$\begin{aligned} u_a &= \int J_a(\lambda) \cdot S(\lambda)\mathrm{d}\lambda = \int J \cdot \delta(\lambda - \lambda_c) \cdot \varepsilon \cdot T(\lambda)d\lambda \\ &= J \cdot \varepsilon \cdot \int T(\lambda) \cdot \delta(\lambda - \lambda_c)\mathrm{d}\lambda = J \cdot \varepsilon \cdot T_m. \end{aligned} \qquad (6)$$

Finally, from Eqs. (3), (4) and (6),

$$J = u_a \times \left[\frac{B(\lambda_c) \cdot \text{BP}}{u}\right].[\text{R}]. \qquad (7)$$

Note that Eq. (7) is only valid when the transmission profile of the filter is narrow and triangular.

The (B/u) factor for each pixel in the all-sky image must be examined further. Let us introduce the radial center pixel distance, R, of a point in the image plane defined as

$$R \equiv \sqrt{(x - x_c)^2 + (y - y_c)^2}, \qquad (8)$$

where (x_c, y_c) are the center pixel coordinates. The relation between R and the zenith angle θ is known as the radial mapping function. Table 3 lists typical fish-eye mapping functions. In a more general form, Kumler and Bauer (2000) suggested that circular image fish-eye lenses have mapping functions equal to

$$R = k_1 \cdot f \cdot \sin(k_2 \cdot \theta).[\text{mm}]. \qquad (9)$$

For the hyperspectral all-sky camera at KHO named NORUSCA II (Sigernes et al., 2012), the coefficients are $f = 3.5$ mm, $k_1 = 1.2$ and $k_2 = 0.83$, using known star positions as input data.

The radiance B of the integrating sphere is per definition uniform in all directions of θ, and (x, y) points with equal R should, due to symmetry, have the same raw data count rate of u. As a consequence, it is useful to transform our (x, y) coordinates to (R, θ) coordinates with Eqs. (8) and (9). It must be emphasized that the above assumption is for an ideal conditioned sphere with no deterioration of the inside coating (Barium sulfate) over time. A functional fit to the data for each wavelength channel may then be found as

$$u = u(\theta) \approx u(0) \cdot [a_0 \cos(a_1 \cdot \theta) + a_2]. \qquad (10)$$

A 3rd degree polynomial fit may also be used. The final form of Eq. (7) becomes

$$J = u_a \times \left[\frac{B(\lambda_c) \cdot \text{BP}}{u(0)}\right] \times \left[\frac{1}{a_0 \cos(a_1 \cdot \theta) + a_2}\right]. \qquad (11)$$

Note that the left bracket in Eq. (11) only requires center ($\theta = 0$) values u, while the right bracket describes the off-axis ($\theta > 0$) behavior of the camera. It is known as flat-field correction of image u_a.

Based on the above equations, we propose the calibration to be undertaken in two steps. The first step is to find the coefficients a_0, a_1 and a_2 by using the integrating sphere. The second step is to measure the center pixel area counts of $u(0)$ by using the Lambertian screen setup instead of the integrating sphere. The screen to lamp distance z is very useful to regulate screen brightness to the same order of magnitude as that expected during sampling of the aurora. The exposure time and gain settings are identical for both the calibration and normal dark-sky operation of the cameras. The latter also cancels out any effect due to nonlinear behavior of count levels versus exposure time. The spectral radiance of the screen is given as

$$B(\lambda) = \left(\frac{4\rho}{10^6}\right) \times M_0(\lambda) \times \left(\frac{z_0}{z}\right)^2 \times \cos\alpha.[\text{R/Å}]. \qquad (12)$$

$M_0(\lambda)$ is the known irradiance (certificate) of the lamp in units of [#photons cm^{-2} s^{-1} Å^{-1}], initially obtained at a distance of $z_o = 0.5$ m (see Table 1). The diffuse reflectance factor ρ of the screen is nearly constant ($\rho = 0.98$) throughout the visible and near infrared regions of the spectrum. The angle α is between the screen and the tungsten lamp ($\alpha = 0$).

5   Results

Three 1 sec exposures at center wavelengths 4861, 5577 and 6300 Å of the integrating sphere were used to obtain the raw count ratio $u(\theta)/u(0)$ for the NORUSCA II camera. The results are shown in Fig. 4. A functional LMFIT in IDL (Interactive Data Language) based on the Levenberg-Marquardt algorithm results in coefficients $a_0 = 0.38$, $a_1 = 1.29$ and $a_2 = 0.63$ (see Eq. 10). Note that in our case there is no significant difference in shape and level of the raw count ratio between the three wavelength channels. This leads to the conclusion that the off-axis effect of the NORUSCA II camera is the same for all wavelengths across the visible spectrum. Or in other words, flat-field correction is independent of wavelength. All parameters in the right bracket of Eq. (11) are now found.

The center pixel calibration factors of the NORUSCA II camera $[B(\lambda_c) \cdot BP/u(0)]$ were found by the method described by Sigernes et al. (2007). 15 wavelength bands were selected to cover the most prominent auroral emissions within the spectral range of the camera. Table 4 lists the origin of the emissions and the corresponding calibration factors.

The net result is that the camera is sensitivity-calibrated and flat-field corrected with only six parameters per channel. For each pixel we obtain the radial center distance R using Eq. (8). Two parameters, k_1 and k_2, are used in Eq. (9) to calculate the elevation θ. Three more parameters, a_0, a_1 and a_2 in Eq. (10), are then used to correct for off-axis effects. The sixth parameter is the center pixel calibration factor $[B(\lambda_c) \cdot BP/u(0)]$ of Eq. (11).

The above procedure is fast to compute and ideal for real-time display of all-sky calibrated data.

6 Conclusions

A two-step method to calibrate and flat-field correct an all-sky camera is outlined. The center pixel spectral sensitivity is obtained and tested by a traditional method including a flat Lambertian screen and a 45 W tungsten lamp. Flat-field correction or off-axis response is conducted by the use of a modified 1 m diameter integrating sphere. The net result is that it is sufficient with only six parameters per channel to calibrate an all-sky camera.

Acknowledgements. We wish to thank Arnold A. Gaertner at NRC for assistance and checking our calibration equations. This work is financially supported by the Research Council of Norway through the project named: Norwegian and Russian Upper Atmosphere Co-operation on Svalbard part 2 # 196173/S30 (NORUSCA2).

Edited by: S. Szalai

References

Akasofu, S. I.: The latitudinal shift of the auroral belt, J. Atmosph. Terr. Phys., 26, 1167–1174, 1964.

Brändström, B. U. E., Enell, C.-F., Widell, O., Hanson, T., Whiter, D., Mäkinen, S., Mikhaylova, D., Axelsson, K., Sigernes, F., Gulbrandsen, N., Schlatter, N. M., Gjendem, A. G., Cai, L., Reistad, J. P., Daae, M., Demissie, T. D., Andalsvik, Y. L., Roberts, O., Poluyanov, S., and Chernouss, S.: Results from the intercalibration of optical low light calibration sources 2011, Geosci. Instrum. Method. Data Syst., 1, 43–51, 2012.

Gaertner, A.: Spectral radiance of low brightness source serial number 10113, Calibration report no. PAR-2012-3049, Institute for National Measurement Standards, National Research Council Canada, 1–11, 2013

Hunten, D. M., Roach, F. E., and Chamberlain, J. W.: A photometric unit for the airglow and aurora, J. Atmosph. Terr. Phys., 8, 345–346, 1956.

Kumler, J. J. and Bauer, M. L.: Fish-eye lens design and their relative performance, Proc. SPIE 4093, Current Developments in Lens Design and Optical Systems Engineering, 360 (24 October 2000), doi:10.1117/12.405226, 2000.

Sigernes, F., Holmes, J. M., Dyrland, M., Lorentzen, D.A., Chernouss, S. A., Svenøe, T., Moen, J., and Deehr, C. S.: Absolute calibration of optical devices with a small field of view, J. Opt. Technol., 74, 669–674, 2007.

Sigernes, F., Ivanov, Y., Chernouss, S., Trondsen, T., Roldugin, A., Fedorenko, Y., Kozelov, B., Kirillov, A., Kornilov, I., Safargaleev, V., Holmen, S., Dyrland, M., Lorentzen, L., and Baddeley, L.: Hyperspectral all-sky imaging of auroras, Opt. Express, 20, 27650–27660, 2012.

Zhang, Y. and Paxton, L. J.: An empirical Kp-dependent global auroral model based on TIMED/GUVI data, J. Atmos. Solar-Terr. Phys., 70, 1231–1242, 2008.

An autonomous adaptive low-power instrument platform (AAL-PIP) for remote high-latitude geospace data collection

C. R. Clauer[1,5], H. Kim[1,5], K. Deshpande[1,5], Z. Xu[1,5], D. Weimer[1,5], S. Musko[2], G. Crowley[3], C. Fish[4], R. Nealy[5], T. E. Humphreys[6], J. A. Bhatti[6], and A. J. Ridley[2]

[1]Center for Space Science Engineering and Research, Virginia Polytechnic Institute and State University, Blacksburg, VA, USA
[2]Space Physics Research Laboratory, University of Michigan, Ann Arbor, MI, USA
[3]Atmospheric & Space Technology Research Associates, Boulder, CO, USA
[4]Space Dynamics Laboratory, Utah State University, Logan, Utah, USA
[5]Bradley Department of Electrical and Computer Engineering, Virginia Polytechnic Institute and State University, Blacksburg, VA, USA
[6]Department of Aerospace Engineering and Engineering Mechanics, The University of Texas at Austin, Austin, TX, USA

Correspondence to: C. R. Clauer (rclauer@vt.edu)

Abstract. We present the development considerations and design for ground-based instrumentation that is being deployed on the East Antarctic Plateau along a 40° magnetic meridian chain to investigate interhemispheric magnetically conjugate geomagnetic coupling and other space-weather-related phenomena. The stations are magnetically conjugate to geomagnetic stations along the west coast of Greenland. The autonomous adaptive low-power instrument platforms being deployed in the Antarctic are designed to operate unattended in remote locations for at least 5 years. They utilize solar power and AGM storage batteries for power, two-way Iridium satellite communication for data acquisition and program/operation modification, support fluxgate and induction magnetometers as well as a dual-frequency GPS receiver and a high-frequency (HF) radio experiment. Size and weight considerations are considered to enable deployment by a small team using small aircraft. Considerable experience has been gained in the development and deployment of remote polar instrumentation that is reflected in the present generation of instrumentation discussed here. We conclude with the lessons learned from our experience in the design, deployment and operation of remote polar instrumentation.

1 Introduction

Improvement in the quality and resolution of data for use in scientific analysis and discovery is a major driver in the advancement of geophysical science. Improvements are accomplished by increasing the spatial distribution of measurements and by increasing the quality and temporal resolution of the samples. This is particularly true in space science and in the investigation of space weather – the dynamic variation of electrical currents and energetic charged particle populations around the Earth and in the upper atmosphere that are produced by interactions of the supersonic solar wind with the geomagnetic field. The solar wind flows around and distorts the geomagnetic field, forming the magnetospheric cavity in the flow. The magnetic field is compressed on the sunward side, and drawn out into a long comet-like tail on the nightside of the Earth. Various processes couple energy and momentum from the solar wind into the magnetosphere to drive electric currents and energize plasma.

Ground arrays of instruments at high latitudes are particularly advantageous for monitoring space weather phenomena because, due to the dipole nature of the main magnetic field, the entire outer magnetosphere maps to a relatively small region at polar and auroral latitudes in both hemispheres. Measurements by arrays of instruments in the polar regions can be used to provide context to better understand observations from satellites in space. Measurements from polar instruments are also vital to the validation of global numerical models that may be used to describe and forecast space weather phenomena. It is, therefore, increasingly important to deploy arrays of geophysical instruments in polar regions to advance our understanding of the complex electrodynamic interactions that comprise space weather. The examination of simultaneous data from both the northern and southern polar regions is also very important because of the considerable asymmetries between the two hemispheres. For example, solar illumination differences between the summer and winter hemisphere produce large asymmetries in the conductance in the two polar ionospheres. The magnetic field in the Southern Hemisphere is significantly weaker and therefore experiences larger amounts of energetic particle precipitation into the ionosphere, producing localized channels of higher conductivity.

While the Northern Hemisphere is relatively well instrumented, the southern polar region is not, primarily because of the extreme Antarctic climate and lack of manned facilities with infrastructure to support instrumentation. Due to the lack of manned infrastructure, measurements from the southern polar region have historically been very sparse. However, during the past decade, technical development has enabled the initiation of robust measurement programs using autonomous instrument platforms that can be deployed remotely and operate unattended for extended periods of time. Examples of such systems include the low-power magnetometer (LPM) platforms operated by the British Antarctic Survey (BAS) and the Automated Geophysical Observatory (AGO) program supported by the National Science Foundation (Lessard et al., 2009; Melville et al., 2014). The BAS LPMs are simple yet robust systems that use only solar power and batteries to support the data acquisition system (Kadokura et al., 2008). They store data in solid-state memory and must be visited each year to acquire the data. During these visits, the electronics box with memory and acquisition system are simply swapped with a new system, so the service is simple and rapid. The US AGOs are more complex systems supporting multiple instruments and utilizing both wind and solar power. They also utilize satellite communications (Iridium) to acquire data and monitor the health of the system. Nevertheless, they often require maintenance visits during the summer field season. In the Appendix, we provide Web resources for various autonomous polar system designs taken from the *Autonomous Polar Observing Systems Workshop Report* .

Another example of remote autonomous system development is the evolving low-power magnetometer platforms developed originally for use on the Greenland ice cap and later for deployment in Antarctica. We report here on the most recent development and deployment of this platform that now supports multiple low-power instruments and is designed to operate unattended for at least 5 years. Designated

as an autonomous adaptive low-power instrument platform (AAL-PIP), it is a relatively simple, yet robust, system using only solar power and storage batteries for winter operation. A two-way satellite data link provides data acquisition and system engineering data, and provides the ability to load new software to modify the operation of the system. While any cluster of low-power instruments can be supported, the present systems deployed on the East Antarctic Plateau support (1) a fluxgate magnetometer, (2) an induction magnetometer; (3) the Connected Autonomous Space Environment Sensor (CASES), a new software-defined, dual-frequency Global Positioning System (GPS) receiver; and (4) a high-frequency (HF) radio experiment.

We have a long history of developing and deploying small autonomous instrument systems, beginning in 1990 through the present. A review of this history is given by Musko et al. (2009). Here we will focus our attention on the specific design requirements and changes that have led to the present AAL-PIP. The system is designed for remote deployment by small aircraft onto the Antarctic Plateau and to operate unattended for at least 5 years. Data are stored in solid-state memory but are also transmitted via Iridium satellite telemetry to laboratories at the University of Michigan and Virginia Tech. During the summer, the photo voltaic (PV) panels provide abundant power. During the dark of winter, power is provided only by a bank of lead-acid absorbed glass mat (AGM) batteries. Therefore, during the winter, data are stored in memory and only engineering information is transmitted via the Iridium satellite so that, while using less power we can still monitor the health of the system. Since the Iridium communication link is two-way, it is possible to command the station to send winter data if there is a special event or time interval needed. When the sun comes up again and power becomes abundant, the current data plus the archived data are transmitted.

The CASES dual-frequency GPS receiver requires more power and produces more heat in the electronics box than the other instruments. Therefore, it is not run continuously but rather scheduled to run for specific time intervals. Thus, it does not provide continuous data. The GPS signals are being examined primarily to investigate signal scintillation (variations in amplitude and phase of the signal) with the goal of understanding the associated plasma irregularities that produce the scintillations. The two-way communication with the platforms using the Iridium link enables the capability to modify the operation of the system and schedule the operation of the instruments in different modes as desired. In general, the fluxgate and induction magnetometers are operated in a standard data collection mode, while the GPS and HF radio can be scheduled to operate during specific time intervals, or during specific conditions in the magnetometer data that can be monitored by the system itself, and/or in various modes of operation.

In the event that the system detects a sufficiently low charge state in the batteries during the winter operation, the

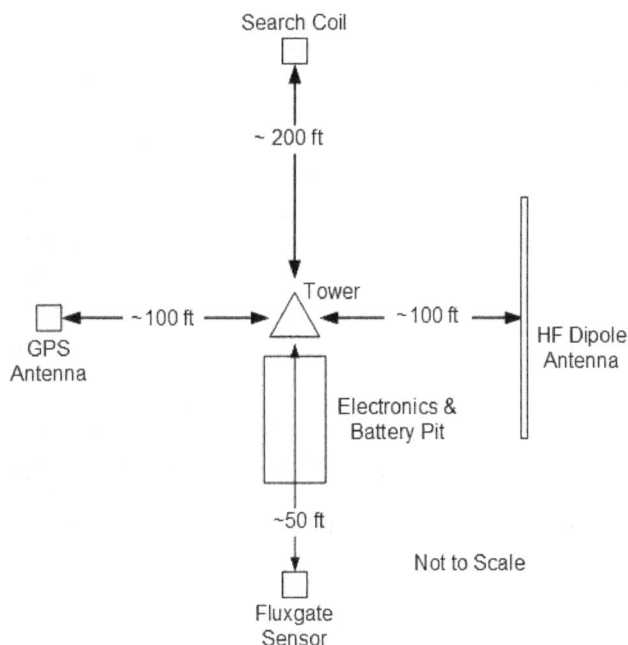

Figure 1. Schematic diagram of the site plan for the deployment of an AAL-PIP system supporting fluxgate magnetometer, search coil (or induction) magnetometer, CASES dual-frequency GPS receiver and HF transceiver experiment.

system will gracefully shut down and wait in this hibernation state until power is detected from the PV panels. When the sun comes up, power from the PV panels is initially used to heat the batteries to a temperature that will permit them to accept charging. The batteries are subsequently charged as the system resumes operation.

We are in the process of manufacturing 4 AAL-PIP systems for remote field deployment on the East Antarctic Plateau to form a chain along the 40° magnetic meridian. Prior to deployment in the remote location, each station is deployed at the South Pole to test for 1 year. Figure 1 shows a schematic drawing of the deployment plan for an AAL-PIP system. Figure 2 shows an AAL-PIP system during deployment at the South Pole.

2 Design considerations

The Antarctic Plateau is an extreme environment, having summer high temperatures that only reach −14 °C, while the winter lows can reach −70 °C. The infrastructure on the continent is limited in both availability and accessibility. The South Pole station is one of only a few manned stations that are located within the interior of the continent, and it is open for summer deployment teams only a few months of the year. It is, therefore, necessary to be able to deploy the system within a few days in the field by a small team of two or three people using small aircraft. Since accessing the stations is difficult and costly, once established, each station should

be able to operate unattended for several years. Finally, the sun is not available as a power source for a large portion of the year. Thus, our design requirements are that each system must

1. operate unattended at any location on the Antarctic plateau for at least 5 years with no maintenance;

2. instrument power requirements should be low, ≈ 1 W;

3. measure the magnetic field strength with 0.2 nT resolution in three orthogonal components and with 1 s cadence;

4. measure magnetic field waves in the ultra-low-frequency (ULF) range (0.1–5 Hz) with the resolution of tens of $pT/\sqrt{Hz}$ over the frequency response in two orthogonal horizontal components;

5. support a newly developed low-cost, software-defined, science-grade, dual-frequency GPS receiver for exploratory scintillation measurements;

6. be able to store at least 1 year of science and engineering data in internal nonvolatile memory;

7. transmit stored and near-real-time data to our laboratories at the University of Michigan and/or Virginia Tech via satellite communication link;

8. time-tag stored data with coordinated universal time (UTC) ±40 ms;

9. include an HF radio experiment to explore radio propagation characteristics between the deployed stations (over ranges of a few hundred to several hundreds of kilometers).

Considerable attention was also devoted to facilitating the deployment of these systems under difficult cold conditions. For example, most of the system can be assembled with gloves on. There are no nuts and bolts to be manipulated and no tools are required. Instead, pins are inserted to secure the tower sections, and T-bolts are used to secure the solar panels to the tower. This is illustrated in Fig. 3, showing a close-up view of the tower assembly with solar panels. The T-bolts secure the solar panels to top and bottom frames, and this assembly is winched to the top of the tower after the tower is raised and secured by guy wires.

Using the specifications described above, the present generation of AAL-PIP was designed as described in the following.

3 System and thermal design

The main consideration in the thermal design of the AAL-PIP is the desire to not have to revisit the site for maintenance. This consideration outweighs the desire for a 100 %

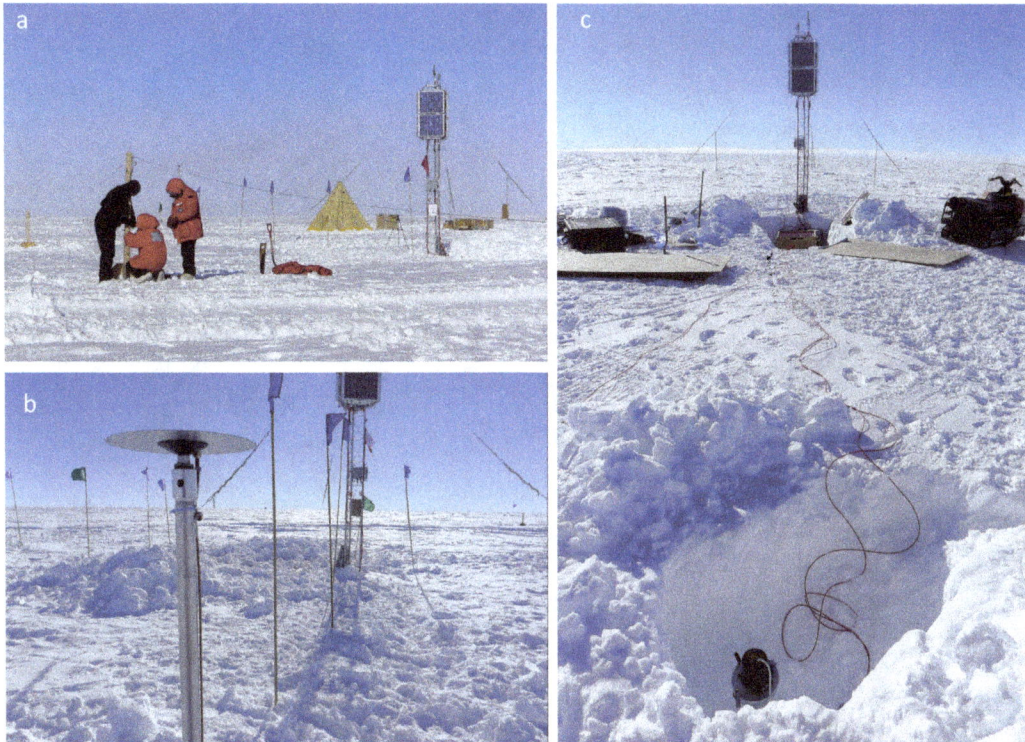

Figure 2. (a) shows HF radio antenna being installed next to system tower. **(b)** shows the GPS antenna for the dual-frequency CASES receiver mounted on a pole near the tower. **(c)** shows a view of the installation from the location of the fluxgate sensor pit. At the base of the tower is a pit containing the super-insulated electronics box and the battery box. The tower holds the photo voltaic (PV) panel array as well as antenna for the Iridium satellite.

duty cycle. This requires a system that can automatically shut down when conditions are poor and restart when conditions improve. The thermal design minimizes power usage during the austral winter. This, in turn, reduces the number of batteries that must be transported to the field site (a major consideration). The overall strategy is to prepare an electronics enclosure that is heavily insulated and minimizes the volume that requires heating for the electronics such that the electronics themselves provide sufficient heat. The measurement sensors are rated to operate at the ambient exterior temperature.

A separate battery box is insulated with 10.1 cm of polystyrene panels and heated to between -15 and $-20\,°C$ during the austral summer for battery charging. The battery box is not heated during the winter to reduce power consumption. This is a chosen design consideration between degrading the capacity of the batteries at cold temperatures and using power to heat the batteries to achieve greater capacity. In this simple design, as the batteries cool while the system continues to draw several watts, the voltage will gradually drop. When the battery voltage drops below 11.2 V, a low-voltage cutoff is activated and the system shuts down for the remainder of the winter. When, or if, this occurs depends upon the power used by the system (instruments, communications, and internal computer control) and the number of batteries. Increasing the number of batteries extends the

operation of the system, but it requires additional transport. Since lead-acid batteries are very heavy, this is a significant trade-off consideration because of weight limits for the small aircraft used in the deployment. As it is, for the more distant deployment sites, several flights are required and fuel caches must be deployed to support the AAL-PIP installation.

The electronics box, illustrated in Fig. 4, is insulated using a combination of 10.1 cm vacuum panels and 10.1 cm of polystyrene panels. The actual box is shown in Fig. 5. The insulated volume inside the electronics box is heated yearround to $-27\,°C$. The measured thermal resistance of the electronics box is $0.07\,W\,°C$, allowing the electronics box to maintain its interior temperature at $30\,°C$ above ambient by dissipating 2.1 W in the insulated volume.

The system utilizes six 40 W PV panels to charge the batteries during the period of daylight. When PV power is again available as the sun appears at the end of winter, the batteries are slowly heated to charging temperature. As the batteries warm up, their ability to deliver current increases and the remaining stored energy once again becomes available as the voltage rises. When the voltage rises to 11.8 V (typically around $-45\,°C$) the low-voltage detection circuit reconnects the load and the system resumes operation. In addition, a thermostat in the electronics box diverts power to a heater until the electronics reach operating temperature. In practice,

Figure 3. Tower close-up view showing orange pins used to secure tower sections and T-bolts at top securing the bottom of the PV panels.

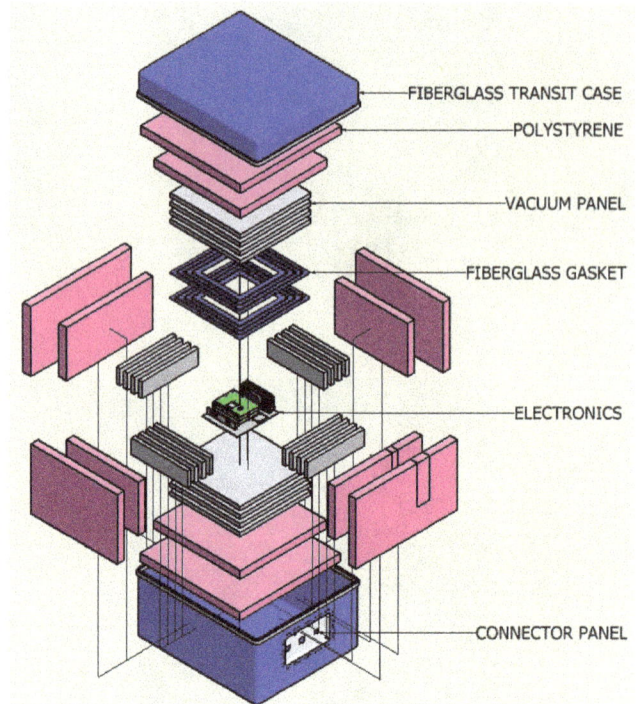

Figure 4. The super-insulated electronics box exploded view.

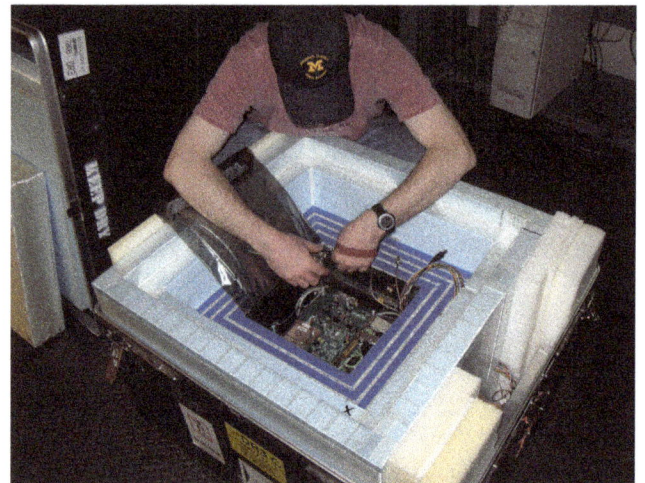

Figure 5. The super-insulated electronics box.

the heater is powered only once during initial start-up. After that, the Iridium modem is power-cycled to control the electronics temperature.

Figure 6 shows, from the top, measurements of the temperature in the battery box; the temperature in the electronics box and the battery voltage for the AAL-PIP system, designated as Sys4 during testing at South Pole; and PG2 after deployment at a remote location on the East Antarctic Plateau. The data run from January 2011 through February 2014, and the system was deployed to the East Antarctic Plateau during the Antarctic summer field season: December 2012–January 2013.

3.1   Power subsystem

In our earlier Greenland and Antarctic systems, we found that small wind turbines required yearly maintenance or replacement. This was mainly due to seizure of the bearings caused by grease loss and degradation. Since the AAL-PIP is required to run for at least 5 years without maintenance, the use of wind turbines was ruled out, even though wind power is available year round in Antarctica while solar power is not. While the AGO platforms mentioned earlier have had success using wind generators, the AGO platforms also require more frequent maintenance visits, and we are not yet convinced that wind generators can be relied upon for long-term reliable operation in remote Antarctic conditions.

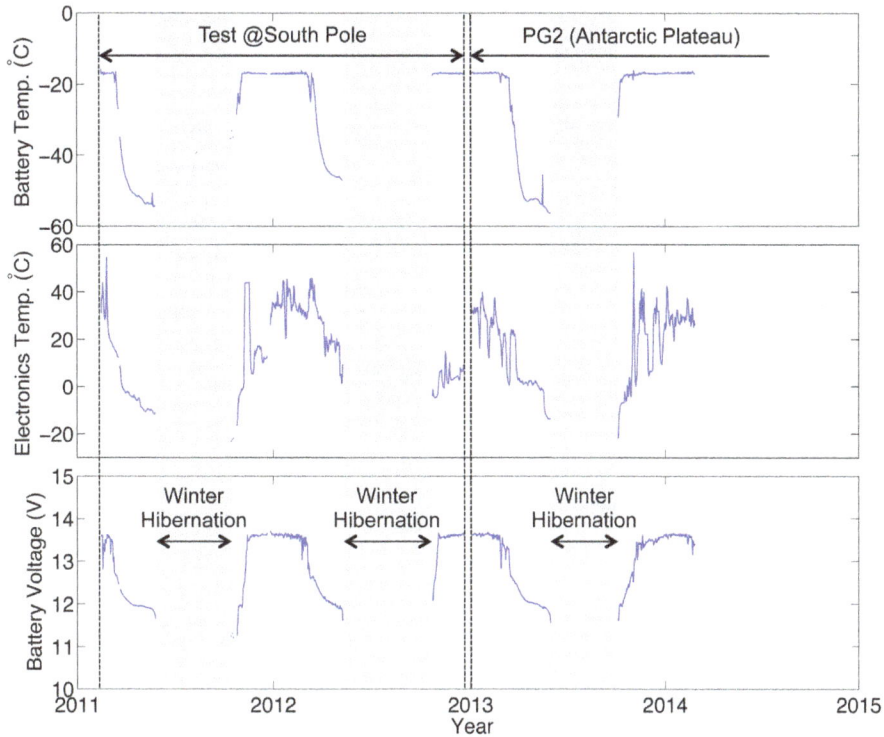

Figure 6. Measured temperature in the battery box (top), temperature in the electronics box (middle) and battery voltage (bottom) from 2011 to the end of February 2014. The AAL-PIP system was designated SYS4 during testing at South Pole station and renamed PG2 after deployment on the East Antarctic Plateau.

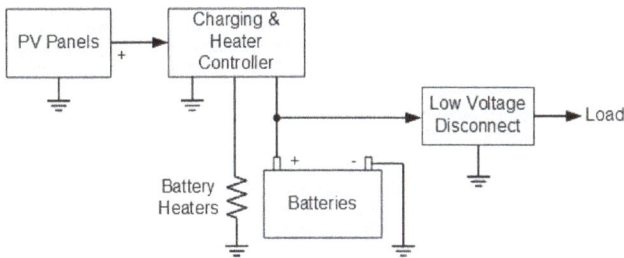

Figure 7. Power system block diagram.

Figure 7 shows a block diagram of the power system. The power system is designed around a bank of 16 100-amp-hour Power-Sonic absorbed glass mat (AGM) lead-acid batteries connected in parallel. Power-Sonic AGM batteries were selected because they have a proven track record of reliability and ruggedness in our earlier Greenland and Antarctica systems, and because they retain a significant amount of their rated capacity at low temperatures. In laboratory tests at the University of Michigan, a single Power-Sonic 100 amp-hour battery at $-55\,°C$ retained 48 % of its rated capacity when powering a 1.4 W load. This allows us to power the system using unheated batteries during the austral winter, when the batteries cannot be charged due to lack of sunlight. The batteries are charged during the austral summer when PV power is plentiful. Since the battery temperature must be higher than $-20\,°C$ to accept a charge, the batteries are heated to between -15 and $-20\,°C$ while charging during the summer. Because the batteries operate at low temperatures, and because they undergo only one full charge/discharge cycle per year, we expect battery life to exceed 10 years. Our earlier design LPM system, which uses the same batteries and charge/discharge strategy, is in its 9th year of Antarctic operation with no signs of battery degradation.

Figure 8 shows the top layer of two layers installed in the battery box. Battery handling under Antarctic conditions (high altitude and cold) can be hazardous due to battery weight and possibility of electrical shorts. We minimize the danger by using the following components and procedures. The batteries are shipped with Delphi automotive connectors prewired to the battery terminals. The battery terminals are also coated with a silicone caulk before shipment to prevent electrical shorts. At the field site, the batteries are moved one at a time from the battery shipping containers to the battery box. They are then connected to the power control electronics using a single wiring harness. The wiring harness is built using Arctic Ultraflex wire (Power Wire Products, Inc.) and is fitted with mating Delphi connectors. The connection is made by snapping the two connectors together. No tools are required for the battery installation or connection, and it can be accomplished wearing gloves. Figure 9 shows the power

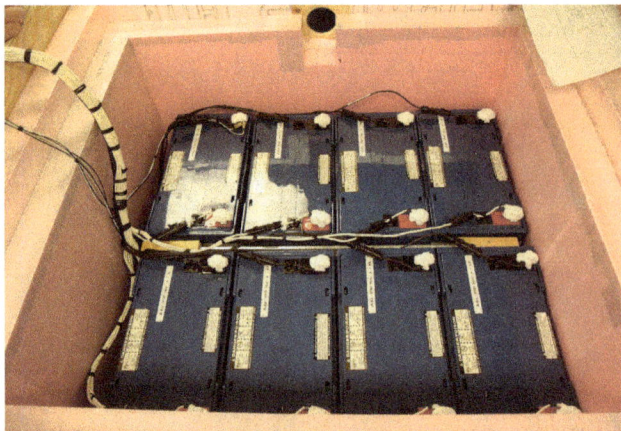

Figure 8. The top layer of two layers of batteries installed in the battery box. Note the aluminum battery heater plate between the two battery rows. Also note the custom wiring harness with Delphi connectors used to connect the batteries in parallel.

electronics box and fuse panel installed on top of the batteries in the battery box (left) and the power electronics printed circuit board inside the power electronics box (right).

Except for the PV panels and system power switch, the power system is contained entirely within the battery box, which is a heavy-duty plywood shipping crate lined with 10.1 cm of polystyrene insulation panels. To reduce the weight of the battery box for shipping and deployment, the batteries are shipped to the field site in separate foam-lined plywood shipping crates, each containing two batteries. For shipment, the battery box contains PV panels and other lightweight and fragile items. The batteries are moved from their shipping crates to the battery box at the deployment site. In addition to the batteries, the battery box contains the power electronics board, fuse panel and battery heaters. The battery heaters are commercial power resistors mounted on an aluminum panel that is located in the center of the battery bank.

The power electronics board is a custom designed-printed circuit board that contains the battery heating, charging and low-voltage disconnect (LVD) circuits. A custom design was selected over an off-the-shelf solution for three reasons: low parasitic power losses, the need to control battery temperature and the extremely low operating temperature range. Since the AAL-PIP system depends on stored battery power for up to 6 months per year, parasitic power losses must be minimized. Off-the-shelf battery charge controllers have significant parasitic power losses and cannot control battery temperature. They also cannot operate down to $-70\,^\circ$C.

The power electronics board performs the following functions:

1. sending all the PV power to the battery heaters when the batteries are too cold to charge

2. sending all the PV power to the batteries when the batteries are warm enough to charge and are not fully charged

3. disconnecting the PV panels when the batteries are warm and the batteries are fully charged

4. disconnecting the load (the electronics box) from the batteries when the battery voltage drops below 11.2 V and reconnecting the load when the battery voltage rises above 11.8 V.

We note also that the PV panels never power the load directly. PV power is only used to heat and charge the batteries. The AAL-PIP system does not use mechanical relays or any other moving parts. Solid-state switches are used throughout. The heater control, charge control and LVD circuits all employ hysteresis to eliminate unnecessary switching cycles when voltages and temperatures approach threshold values.

3.2 Computer control

The control subsystem consists of a three-board PC-104 stack, an Iridium modem and antenna. It is contained, except for the Iridium antenna, in the electronics box. Figure 10 shows the PC-104 stack that comprises the core of the control system.

The PC-104 stack consists of a Technologics TS-7260 single-board computer (SBC), a Technologics TS-Ser4 quad serial port board and custom I/O board. The TS-7260 SBC has many features, including a Cirrus 200 Mhz ARM9 32-bit CPU, 64 MB RAM, 128 MB flash memory, Ethernet, two USB ports, three serial ports, digital I/O, 12-bit A/D converter, watchdog timer and battery-backed real-time clock, and an operating temperature range of -40 to $+85\,^\circ$C. It runs the Linux operating system. A four-gigabyte flash drive connected to one of the USB ports provides mass storage for instrument and housekeeping data. A Garmin GPS receiver is used to provide an accurate time standard.

The only significant TS-7260 SBC deficiency for our application is inadequate system oscillator stability. This causes the system time to drift at a rate that varies strongly with the oscillator temperature. AAL-PIP is required to time-tag data with UTC $\pm$ 40 ms, but uncorrected system time can drift several seconds per hour. System time stability could be improved by continuously powering the Garmin GPS receiver and synchronizing frequently, but in order to save power, we use a different method. The Garmin GPS receiver is powered up once per hour to synchronize system time to UTC. While synchronizing the system time, the system time error and drift rate are measured using the GPS pulse-per-second (PPS) signal. The system clock rate is then adjusted, using Linux kernel functions, to compensate for the error and drift rate. Using this method, the system time is consistently maintained at UTC $\pm$ 5 milliseconds.

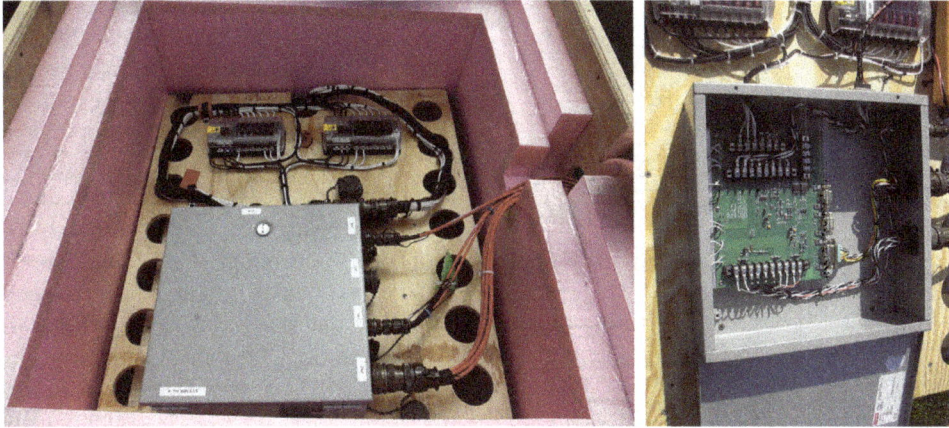

Figure 9. (left) Power electronics box and fuse panel installed on top of the batteries in the battery box. (right) Printed circuit board in the power electronics box.

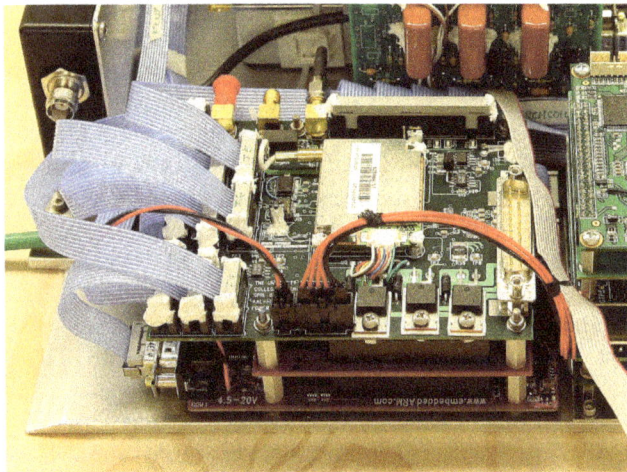

Figure 10. The PC-104 stack mounted on the electronics chassis. The custom I/O board is on the top.

The custom I/O board contains

– solid-state switches that control the Iridium modem, electronics box heater and instrument power.

– a Garmin GPS15H GPS receiver which transmits serial time messages to the TS-7260 SBC once per UTC second. It also transmits the digital PPS signal that is used to synchronize system time and measure time error and drift rate. The PPS signal is connected to one of the TS-7260 SBC interrupt inputs.

– temperature, current and voltage sensors that are used to monitor battery voltages, battery temperatures, electronics temperature and electronics current.

– a warm-up thermostat which diverts power to a 5 W heater until the PC-104 stack reaches operating temperature. The stack is then powered up and a second software-controlled heater is used as needed to maintain the electronics temperature.

– GPS antenna power and signal splitting circuitry that powers the antenna and routes its RF output to the Garmin and the CASES GPS receivers.

– connectors and onboard interconnections that connect off-board instruments and peripherals to each other and the PC-104 stack.

The TS-Ser4 board contains four RS-232 serial ports which are used to communicate with the Garmin GPS receiver, Iridium modem, CASES GPS receiver and HF transceiver.

The TS-7260 SBC runs version 2.6.29 of the Linux operating system. The operating system distribution was created by Todd Valentic of SRI International and is customized for the TS-7260 SBC. The AAL-PIP control software is written primarily in the Python language, with hardware interface code written in C. The software is structured as a set of independent processes. Each process is a separately running software program. Because Linux confines each process to its own memory space, if one process crashes, other processes are unaffected.

3.3 Measured power usage

The PC-104 stack, excluding the Garmin GPS receiver and antenna, requires 1.6 W. The Garmin GPS receiver requires 0.4 W. The Antcom dual-frequency GPS antenna requires 0.52 W. The Iridium modem and antenna require a total of 0.9 W in standby mode. While transmitting, they require an average of 3 to 4 W. The fluxgate magnetometer requires 0.8 W. The search coil magnetometer requires 1.6 W. The CASES GPS receiver requires 8.0 W. The HF transceiver requires between 8 and 10 W while transmitting.

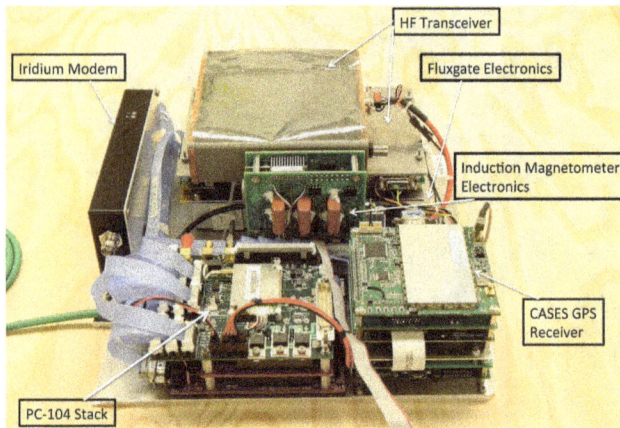

Figure 11. Electronics chassis.

3.4   Sensor array subsystem

The electronics for all of the sensors as well as the computer control and data handling are all enclosed in the super-insulated electronics box that is buried at the base of the tower during installation. Figure 11 shows the electronics chassis that is located in the electronics box.

3.4.1   Fluxgate magnetometer

The system supports a low-power, three-axis vector fluxgate magnetometer (LEMI-022AN) produced by the Laboratory for Electromagnetic Innovations (under the direction of Dr. Valery Korepanov), Lviv Centre of the Institute of Space Research, National Space Agency of Ukraine. It was specifically prepared to acquire stable three-component measurements of the Earth's magnetic field in the Antarctic region. The magnetometer consists of an electronic unit that is located in the super-insulated electronics box and a sensor with a connecting cable that is approximately 15 m in length. The main technical specifications of the instrument are given in Table A1.

3.4.2   Induction magnetometer

Induction magnetometers are widely used for space science research to measure magnetic field waves in space and on the ground. The magnetic field frequencies of interest in space physics typically fall into the ULF range (a few mHz to a few Hz). Induction magnetic sensors are copper wires wound around a highly permeable (μ) metal core. A voltage is induced across the coil when magnetic field through the sensor changes over time (dB/dt). Since induction magnetometers are sensitive to the magnetic field direction, it provides vector magnetic field information. The sensor is rated to operate at the ambient Antarctic temperature.

Each induction magnetometer deployed with the AAL-PIP system consists of two orthogonally mounted magnetic sen-

sors with preamps, a cable of between 60 and 200 m and a main analog circuit. The two-axis magnetic sensor configuration provides measurements of wave activity in the geomagnetic north–south and east–west directions. The sensor is located between 60 and 200 m away from the AAL-PIP system to minimize interference from the electronics. Signals detected by the magnetic sensors are filtered and amplified by the main analog circuit, and archived in the AAL-PIP data acquisition system with 12 bit digitization at the rate of 10 samples s^{-1} axis^{-1}. The main analog electronics include amplifiers and band-pass filters, so that the magnetic field variation in the ULF range of interest can be detected. The magnetometer measures magnetic field wave activity with a resolution of tens of pT/$\sqrt{\mathrm{Hz}}$ over the frequency response up to 2.5 Hz (-3 dB corner frequency). The power consumption of the induction magnetometer is 1.6 W.

3.4.3   GPS dual-frequency receiver

The Connected Autonomous Space Environment Sensor (CASES) receiver is a scientific grade, low-cost, dual-frequency, GPS receiver developed by Cornell University, the University of Texas at Austin and Atmospheric & Space Technology Research Associates (ASTRA). The development of CASES, its specifications and operation details are given by Crowley et al. (2011) and O'Hanlon et al. (2011). A space-qualified variant of CASES, called FOTON, has also been developed and is described by Lightsey et al. (2014).

This receiver is developed to be paperback-novel-sized, but its design was customized to a cubic form factor (10 cm × 10 cm × 10 cm) to fit inside the compact AAL-PIP electronics box. Furthermore, during the development, CASES passed the cold soak test below $-50\,^{\circ}$C and demonstrated reliable operation at about $-40\,^{\circ}$C before and after the test. The CASES receiver on an AAL-PIP system is connected to an Antcom GPS antenna. Power consumption is about 8 W.

For CASES, all the data acquisition and tracking operations as well as science- and navigation-related operations, such as TEC computations, scintillation monitoring and calculation of navigation solution and GPS observables are performed on a general-purpose digital signal processor (DSP). Based on the real-time data processing by a scintillation monitor called "SCINTMON", for each tracking channel, the DSP outputs scintillation parameters S4, σ_ϕ and τ_0 every 50–100 s; TEC, carrier phase and signal strength data at up to 10 Hz; and high-rate, in-phase and quadrature correlation products with an unambiguous carrier phase reference at up to 100 Hz (Dierendonck et al., 1993; Humphreys et al., 2010, 2009). The data are then routed to a single-board computer that runs a server program for data logging and remote monitoring. As an ionospheric scintillation monitor, CASES has many advantages, including GPS L2 civil code (L2C) tracking capability, incorporation of specialized tracking loops designed for operation in both weak-signal and scintillating

Table 1. Please provide a caption for Table 1.

Measurement range for each component	$\pm 65\,000$ nT
Cutoff frequency of analog low-pass filters (before digitization)	0.3 Hz
Output data sample rate formed by averaging 8 AD conversions	1/s
Resolution	0.01 nT
Temperature drift	< 0.5 nT $^{\circ}$C^{-1}
Spectral noise density at 1 Hz	< 10 pT
Sensor component orthogonality error	< 30 min of arc
Temperature operating range for electronics	-40 to $+50\,^{\circ}$C
Temperature operating range for sensor	-65 to $+50\,^{\circ}$C
Power supply voltage	12 V ± 2 V
Power consumption	< 0.5 W

environments, and connectivity via a number of different options. CASES on AAL-PIPs may be accessed through an Iridium link for remote logging, reconfiguration and reprogramming.

As a part of real-time processing, SCINTMON implements a scintillation-triggering algorithm. A trigger indicates the presence of a phase or amplitude fluctuation more vigorous than the user-specified triggering threshold. In response to a triggering event, CASES collects high-rate data only for that duration, whereas low-rate data are saved at all times whenever CASES is powered up.

Because the maximum Iridium data download rate is approximately 16 MB per day, data collected by all three space science instruments onboard – namely, the fluxgate magnetometer, the search-coil magnetometer and the CASES GPS receiver – need to be constrained. Each of the magnetometers collects about 1 MB of data per day. Thus, in normal daily operation, CASES collects low-rate data and triggered high-rate data until the daily memory limit for GPS data collection of about 12 to 13 MB is reached. Its scintillation-triggering strategy enables CASES to make maximal use of its allotted daily data limit by capturing and passing along high-rate data only spanning the most vigorous scintillation events.

Furthermore, a special AAL-PIP "storm mode" can be enabled in anticipation of an incoming solar storm. In storm mode, CASES is allowed to store much more than the standard daily data limit. Data stored locally in storm mode are retrieved over a period of several days post-storm.

Finally, CASES employs a temperature-compensated crystal oscillator (TCXO) as its reference oscillator. Although this feature causes CASES to be less expensive than most of the commercially available scintillation monitors, its usage introduces receiver clock errors which themselves can cause false detection of phase scintillation. However, receiver clock errors are removed in real time within SCINT-MON by differential processing before SCINTMON applies its scintillation-triggering tests, and they can be removed by post processing the data as described by Deshpande et al. (2012).

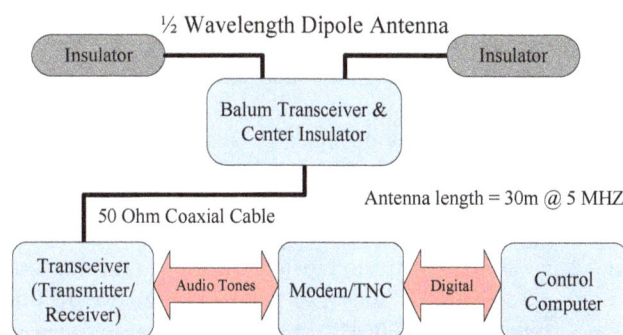

Figure 12. HF radio system.

3.4.4 HF radio experiment

At distances beyond a few kilometers HF radio propagation is an ionospheric phenomenon (Hunsucker and Bates, 1969; Blagoveshchensky et al., 2008). In the lower HF band a critical frequency can exist where a wave launched vertically is reflected. Frequencies near that critical frequency are useful for communication at distances up to several hundred kilometers. Under normal propagation conditions signal levels are typically strong and consistent. Propagation is enhanced by strong F region ionization. Strong ionization in the D region, however, attenuates the signal so that the optimum signal frequency varies toward higher frequencies during daylight hours but is reduced during night. This effect is further enhanced during geomagnetic storms when particle precipitation and X-ray flux levels increase (Warrington et al., 2012). An interesting aspect of this propagation mode is that it is enhanced when the antenna is designed to direct the maximum energy at high (nearly vertical) angles rather than toward the horizon. This mode is sometimes referred to as near vertical incidence sky wave (NVIS) propagation. High-latitude statistics on HF propagation over distances of hundreds of kilometers, particularly in the polar Antarctic region, are not well developed, so we include a HF radio experiment to collect information about propagation statistics between our array of stations (Collin, 1985).

The HF radio system (shown in Fig. 12) consists of three major modules: antenna, radio transceiver and modem. The antenna is a half-wave horizontal dipole antenna designed to operate at 5 MHz frequency. The antenna is constructed of wire supported approximately 1.5 m above the ice. Since the ice is relatively transparent to electromagnetic radiation at this frequency, the antenna height is not critical. Additionally, as the propagation is by way of sky wave (reflected from the ionosphere), the horizontal dipole best facilitates the high angle radiation direction required for relatively short distance HF propagation. The transceiver is a fixed frequency SSB radio which converts the radio frequency to and from the audio tones of the modem. The radio transmissions are very low power at approximately 3 W. The modem is a DSP-based device capable of numerous modes of operation. The modem interfaces with the AAL-PIP system control computer which controls radio operation and handles data through a serial data link.

The normal modulation mode is frequency-shift keying (FSK) at 100 Bd. The protocol is Simplex Teletype Over (RadioSITOR-B), which is a maritime HF communication standard using forward error correction (FEC). The relatively low data rate is designed to provide maximum effectiveness in the presence of HF noise and fading. Narrow-bandwidth filters (approximately 300 Hz) in the modem and transceiver provide a high signal-to-noise ratio which, along with the FEC, offsets much of the disadvantage of the low-power transmitter.

During operation of the HF antenna, the AAL-PIP computer applies a lookup table according to the system temperature to maintain the transceiver frequency within 50 Hz of the correct channel frequency. The frequency correction process is implemented by shifting the frequency of the modem audio tones since this may be done in small increments. The transceiver and modem have been tested for operation over a temperature range of -40 to $50\,^{\circ}$C with a cold soak below $-50\,^{\circ}$C.

While SITOR-B provides excellent communications in the radio system, FEC serves to mask some errors in the communication link. The modem provides several additional communication modes, some of which do not provide any error correction. The use of non-error-corrected modes can allow for experiments which better indicate the channel characteristics. Selection of mode, protocol and data rate and transmitted power level are under software control, so many experiments are possible.

The locations of the AAL-PIP systems in Antarctica as of January 2014 is shown in Fig. 13. In current experiments, the HF radio system is briefly turned on at 2 h intervals for test transmissions. Each station transmits in turn, with the other stations receiving. The transmission consists of several repetitions of a telegraphic test symbol (vvv) and the station identifier in plain text. Since the exact transmitted pattern is known, a character error rate can be found for each transmission. Initial results for tests between a site on the Antarctic

plateau and the South Pole station (a distance of approximately 600 km) indicate very high communication reliability except during periods of high D region absorption due to proton events. During significant proton events the signal reliability drops rapidly to zero as the 5 MHz signals are highly attenuated while passing through the D region. This suggests that the absence of signals is a good indicator of a significant event.

An example of D-region attenuation during geomagnetic storm periods in the 2011–2012 austral summer is shown in Fig. 14. Two major storms were evident during this season. The first occurred towards the end of January 2012, and the second occurred during the first part of March 2012. In both storm cases, the HF communication links between the installed systems were severely affected, with periods of total communication "blackout" witnessed during the height of the storms. There are also periods of communication disruption evident during minor storm events occurring in the same season. With the completion of the 2013–2014 summer season, many of the systems have been aligned along the geomagnetic field line, creating a much improved geometry for the study of HF communication impacts during geomagnetic storm time conditions in Antarctica.

4 Data handling

Because the AAL-PIP systems are extremely remote, maintaining Iridium communications is the highest software priority. Loss of communications means an expensive and time-consuming field trip. The communications software processes are well-tested, mature and independent of the other processes. Any of three different watchdog timers can force a hard reset if the Linux kernel crashes or communications are lost for a predetermined length of time. This scheme has proven reliable. We have never lost communication with an AAL-PIP system due to a software bug.

In previous systems, we used Iridium modem-to-modem connections for communications. We now use an Iridium protocol known as RUDICS (Router-based Unrestricted Digital Interworking Connectivity Solutions), which is a combination of a single Iridium modem and the Internet to provide communication between a remote system and an Internet-connected computer. RUDICS has several advantages over the modem-to-modem communications:

- fewer line drops because only one modem is used

- a single Internet-connected computer can handle many RUDICS connections simultaneously

- lower cost because only one modem per remote system is required.

Each remote AAL-PIP system initiates a RUDICS connection with the operations center computer located at the

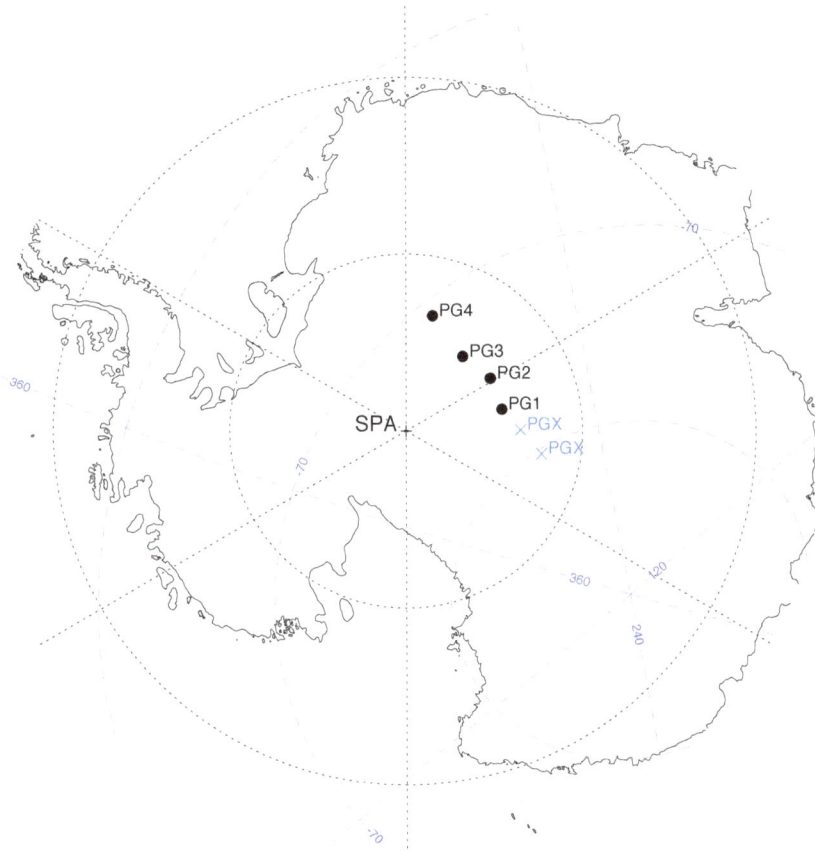

Figure 13. Map showing the locations of the AAL-PIP systems deployed in Antarctica since January 2014.

University of Michigan (UM) every 30 min. If the connection is idle for more than 5 min, the AAL-PIP terminates the connection to conserve power. Continuous contact with an AAL-PIP system can be maintained by executing a script on the operations center computer that pings the AAL-PIP system periodically. When the AAL-PIP system initiates a connection to the DOD Iridium gateway in Hawaii, the gateway then connects the AAL-PIP system to the operations center server using Internet and TCP/IP (Transmission Control Protocol/Internet Protocol) socket protocol. It typically takes about 40 s for an AAL-PIP system to connect to the server. One server can handle many RUDICS connections simultaneously. The RUDICS link provides about 2000 bits s^{-1} of half-duplex raw data throughput.

From the operations center server we can

- open one or more Unix Secure Shell (SSH) terminal sessions on the AAL-PIP

- automatically download data files for any set of instruments and dates.

We use SSH sessions to acquire diagnostic information and update AAL-PIP software. Updating the AAL-PIP software is done by copying two files to a particular AAL-PIP directory and rebooting the TS-7260 SBC. The remote SSH

session and software-updating capabilities have proven invaluable.

The operations center server automatically runs a data download program at 00:15 UTC daily. Typically, all of the data stored the day before are downloaded and then erased from the AAL-PIP flash drive. The download program can also be configured to download data for any subset of instruments and range of dates.

Iridium communications are subject to intermittent data corruption and dropped connections, so an error detection and retransmission protocol is required for automated file downloading. We are using a custom protocol to maximize performance over the relatively slow, unreliable Iridium link. Large blocks of data are transmitted to minimize protocol overhead and communications line turnarounds. Using this custom protocol, file download throughput approaches the link speed of 2000 bits s^{-1}.

Following the acquisition of the raw telemetry data by the operations center computer at UM, the data files are transferred to the Virginia Tech Web server that is hosted on a desktop computer. Here the files are converted into formats that are more readily accessible for scientific analysis, plotted, archived and made available on the Web and for file transfers. These tasks are accomplished by a combination of

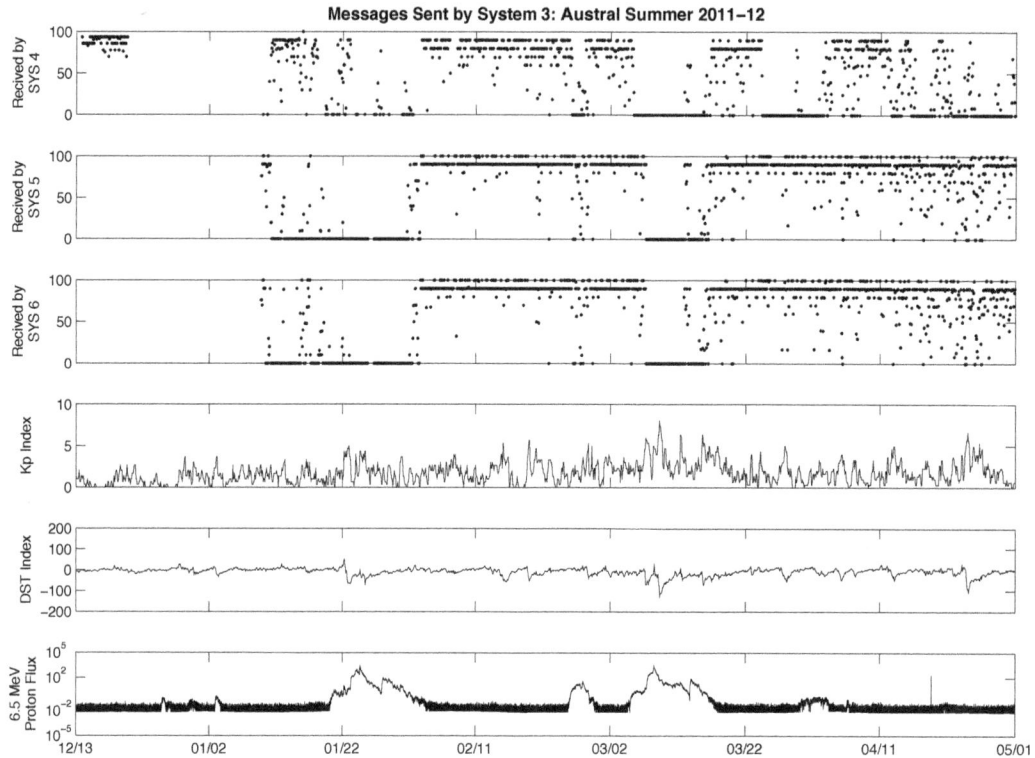

Figure 14. HF communication quality during the 2011–2012 summertime period. The top three panels indicate the percentage of characters received correctly as transmitted from system 3 to systems 4–6 (see Fig. 13 for a map of the location of the varying systems during this time frame) versus time. A level of 100 indicates that all receptions were error free. A level of 0 indicates that all of the receptions included a complete distortion of the transmitted file character set. The bottom three panels plot the K_p index, D_{ST} and proton flux $\left(\#/(s - cm^2 - sr) \right)$ over the same time period. All systems assumed nominal operations from 10 January 2012 until the end of the season (May 2012).

Unix Bash shell scripts, Python programs and IDL (Interactive Data Language) programs.

A shell script that starts the processing is executed automatically at fixed intervals, twice per day. A Python program connects to the SFTP server at UM, checks the data directory contents and downloads any new data files that are not already on the local file system. Sometimes there are no new files, such as when the data transfers are suspended during the Antarctic winters and no solar power is available. If new files are found then, following their download by SFTP (Secure File Transfer Protocol), the list is handed off for further processing by IDL programs.

The first step in the processing is conversion of the files into a "flat file" format that in principle can be accessed on any computing platform in any language. As the raw AAL-PIP data do not contain fixed time intervals, there are multiple files from each instrument for each day (approximately 24 per day), and these files need to be sorted and concatenated in the proper sequence to obtain a new, 1-day file.

Traditionally, flat files up through version 3 have used two separate files to hold an ASCII (American Standard Code for Information Interchange) text "header" and binary data. A newer, version 4 format is now used in which the text header

and binary data are combined into one file. The header contains a description of the data contents, in a standardized format that both is human readable and can be used by programs for further processing. The binary data begin at a fixed offset from the start of the file; this offset value is included in the text header but is normally 4096 bytes, which allows adequate space for most headers. The byte order of the data may be either big or little endian and is specified in the header. Programs that read flat files should be able to handle both formats. The flat files are stored in a database having subdirectories for each year and month, but with all magnetometer sites combined together.

Following the conversion to flat files, the new data are graphed, with the results saved in both PDF and PNG formats. The data are also converted to the Common Data Format (CDF). A Web page (http://mist.nianet.org) shows the most recent plots, and also has links to the archived plots and access to the CDF data files via HTTP. FTP access to the flat files is also available to the research group and collaborators. Due to the risk of hacker attacks that could result from "anonymous" FTP access, our FTP server requires a user name and password. These are made available to collaborators. Copies of the CDF data files are also available

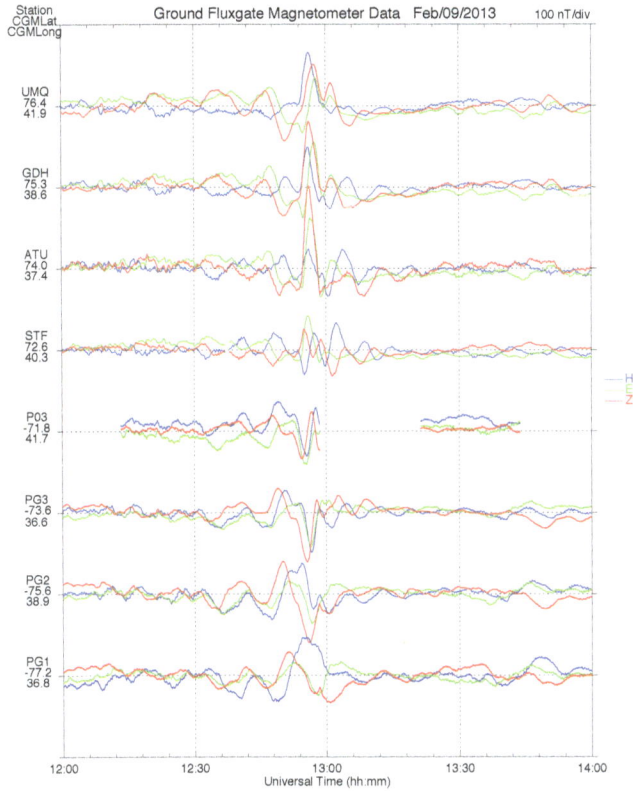

Figure 15. Geomagnetic impulse events observed by the fluxgate magnetometers at four station pairs in magnetically conjugate hemispheres (UMQ-PG1, GDH-PG2, ATU-PG3 and STF-P03) starting approximately at 12:45 UT on 9 February 2013.

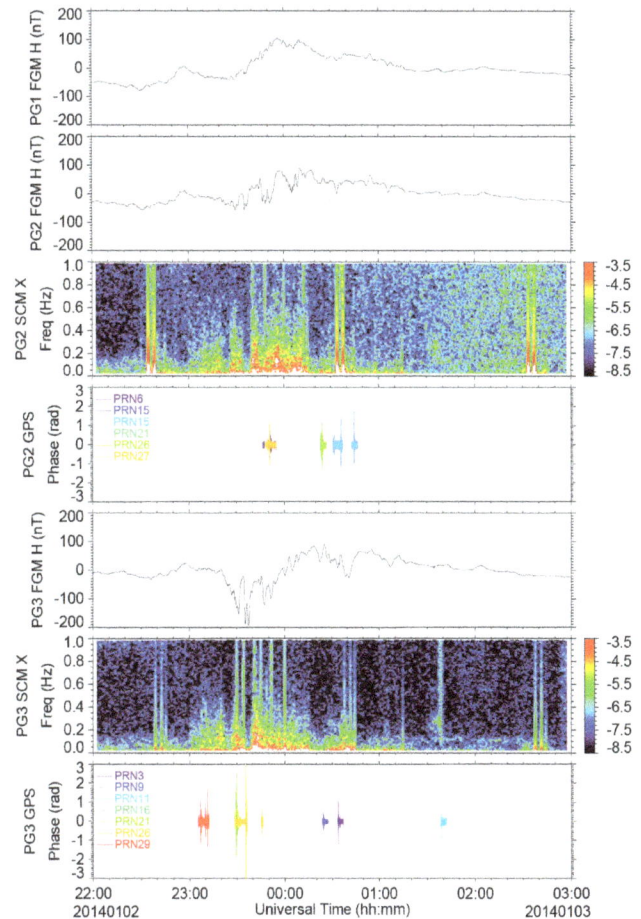

Figure 16. Measurements at the three remote field stations (PG1, PG2 and PG3) from 22:00 UT on 2 January 2014 to 03:00 UT on 3 January 2014 showing magnetic field disturbances and associated with GPS signal sintillations observed by **(a)** PG1 fluxgate magnetometer; **(b–d)** PG2 fluxgate and induction magnetometers, and GPS scintillation monitor; and **(e–g)** PG3 fluxgate and induction magnetometers, and GPS scintillations monitor. Note that the magnetic field data are measured from the H (fluxgate magnetometer) and X (induction magnetometer) components, both of which are aligned along the horizontal local magnetic field line.

in the THEMIS (Time History of Events and Macroscale Interactions during Substorms) database (Angelopoulos, 2008; Mende et al., 2008). The THEMIS project obtains these files automatically through FTP.

Backups of the data files and processing programs are maintained on a redundant, RAID system (redundant array of independent disks; the present capacity is 2 TB). All programs are also backed up on another system, normally through the use of DropBox. As the computer system that does the final processing and archiving of the data is accessible over the Internet for HTTP and FTP transfers, there is some exposure risk to malicious attacks (the firewall log shows multiple attempts daily). For this reason the RAID backup file system is not continuously connected to the host computer, ensuring that it would not be affected by a successful attack; backups to the RAID file system are done manually.

Figures 15 and 16 show example events from the data obtained by the AAL-PIP systems located in East Antarctica (PG1, PG2, and PG3). See the map in Fig. 13 for the locations of the remote field stations. Geomagnetic disturbances are clearly seen in Fig. 15, in which four station pairs in magnetically conjugate hemispheres (UMQ-PG1, GDH-

PG2, ATU-PG3 and STF-P03) observed sudden changes in magnetic signatures (called "magnetic impulse event") starting approximately at 12:45 UT. It is found, however, that there is asymmetry in the patterns as well as the amplitudes of the magnetic events. Kim et al. (2013) suggest that asymmetry in ground response patterns between the conjugate locations often shows little correlation with interplanetary magnetic field (IMF) orientation, season and ionospheric conductivity, indicating that a much more complex mechanism might be involved in creating interhemispheric conjugate behavior.

As shown in Fig. 16, geomagnetic field disturbances called "substorm" in association with ionospheric scintillations are observed by (a) the fluxgate magnetometer at PG1; (b–d) fluxgate and induction magnetometers, and GPS scintillation monitor at PG2; and (e–g) fluxgate and induction magnetometers, and GPS scintillation monitor at PG3. The substorm is a disturbance in the Earth's magnetosphere causing energy to be transferred from the tail side of the magnetosphere to the high-latitude ionosphere. This is accomplished through the precipitation of energetic particles along magnetic field lines into the atmosphere producing the aurora, and through an intensification of electrical current flowing in the highly conducting ionospheric auroral electrojet. GPS scintillations are often observed during geomagnetically disturbed times (Prikryl et al., 2011; Kinrade et al., 2012). Investigating interhemispheric comparisons of bipolar GPS scintillation maps, Prikryl et al. (2013) have reported that the scintillation occurrence is significantly higher in the southern cusp and polar cap compared to the northern regions. However, they also mention that the coverage of GPS receivers was insufficient to study the relation of IMF to the strong hemispherical assymmetry in the intensity of scintillations.

As shown here in the fluxgate magnetometer data in Fig. 16, the onset of the substorm event occurred at 23:20 UT. The spectrograms from the induction magnetometers (panels c and f), showing wave power as a function of frequency over time, display broadband pulsations during the substorm event. This is a very well known manifestation of substorms. The event also appears to be associated with the phase scintillations in the GPS signals (panels d and g). It should be noted that the GPS receivers were turned on from 22:00 to 03:00 UT. Similar results have been reported by Kim et al. (2014), showing a possible connection of ionospheric irregularities to a substorm event using simultaneous observations of ULF waves and GPS scintillations.

5 Conclusions

We have learned a number of lessons over the years through our engagement in remote polar measurements. With regard to the AAL-PIP system described here, the glove-friendly, tool-free deployment design has proven to be very well thought out and highly valued by the members of the deployment teams. This is one of the most innovative and valuable features of this new design.

The support of multiple instruments has increased the demand for power, with the consequence that the system spends a longer time in winter hibernation mode than with previous designs. New strategies for instrument operation or supplying additional batteries could help to reduce the hibernation period. The system has been designed such that the battery box is a modular unit and it is simple to add more battery boxes to the system. The limitation, however, is the ability for logistical support to accommodate the weight and transport of the batteries to the remote sites.

The system, as it is deployed now, provides no outputs for the deployment team to monitor, other than to see that the power is on. The orientation of the magnetic sensor is done by using an Iridium phone to talk with a person in the laboratory at Virginia Tech who is monitoring the data transmitted by the station. The magnetic sensor is rotated to place the x axis along the magnetic field by nulling the y axis (X and Y lie in the horizontal plane, and Z is positive vertical in the down directions). This has been satisfactory for the deployment of the magnetometer and gives a good end-to-end test of the system. However, if odd readings are obtained from any of the instruments or other components of the system do not operate as expected, it is difficult for the deployment team to undertake any simple diagnostic steps – checking cables, connectors, voltages, etc. It would be useful to have some type of diagnostic system output available to the deployment team for such purposes.

Our concern for operating in the cold environment using the super-insulated electronics box provided a surprise when the temperature of the electronics got above 40 °C during summertime with extended operation of the CASES GPS receiver. The result has been that we have needed to limit the operation of the receiver because of temperature considerations within the electronics box. We are able to heat the volume containing the electronics, but we have no way of actively cooling the volume. More thought must be given to this problem.

Appendix A

Table A1. Web resources for autonomous polar system design.

Wisconsin Automatic Weather Station Project	http://amrc.ssec.wisc.edu/aws/
McMurdo Long Term Ecological Research (LTER)	http://www.mcmlter.org/
IRIS/PASSCAL Polar Seismology	http://www.passcal.nmt.edu/content/polar/
Greenland Ice Sheet Monitoring Network	http://glisn.info/
UNAVCO Polar Geodetic Support	http://www.unavco.org/polartechnology
Power Systems for Polar Environments	http://www.polarpower.org
Polar Technology Conference	http://www.polartechnologyconference.org/
POLENET project site	http://www.polnet.org/
Antarctic PENQUIn Program	http://www.sos.siena.edu/antarctic/PENGUIn_Program
Augsburg College Space Physics	http://space.augsburg.edu/index.html
British Antarctic Survey Instrumentation	http://www.antarctica.ac.uk/bas_research/instruments/index.php
Virginia Tech Magnetosphere – Ionosphere Science Team	http://mist.nianet.org/
Autonomous Polar Observing Systems Workshop Report	http://www.iris.edu/hq/files/publications/other_workshops/docs/APOS_FINAL.pdf

Acknowledgements. Support for the development and testing of this system has been provided through Major Research Infrastructure (MRI) Grant ATM-922979 to Virginia Tech from the National Science Foundation. Additional support has been provided by the National Science Foundation for the operation and scientific investigation of data from the deployed AAL-PIP stations along the Antarctic 40° magnetic meridian by grants ANT-08398585 and PLR-1243398. Support at the University of Michigan has been provided by NSF grant ANT-0838861 and support at ASTRA has been provided by NSF grant PLR-1243225.

Edited by: L. Eppelbaum

References

Angelopoulos, V.: The THEMIS mission, Space Sci. Rev., 141, 5–34, 2008.

Blagoveshchensky, D. V., Kalishin, A. S., and Sergeyeva, M. A.: Space weather effects on radio propagation: study of the CEDAR, GEM and ISTP storm events, Ann. Geophys., 26, 1479–1490, doi:10.5194/angeo-26-1479-2008, 2008.

Collin, R. E.: Antennas and Radiowave Propagation, McGraw-Hill, New York, 1985.

Crowley, G., Bust, G. S., Reynolds, A., Azeem, I., Wilder, R., O'Hanlon, B. W., Psiaki, M. L., Powell, S., Humphreys, T. E., and Bhatti, J. A.: CASES: A novel low-cost ground-based dual-frequency GPS software receiver and space weather monitor, in: Proc 24th International Technical Meeting, Satellite Div. of the Institute of Navigation (ION GNSS 2011), 1437–1446, Inst. of Navigation, Manassas, VA, 2011.

Deshpande, K. B., Bust, G. S., Clauer, C. R., Kim, H., Macon, J. E., Humphreys, T. E., Bhatti, J. A., Musko, S. B., Crowley, G., and Weatherwax, A. T.: Initial GPS scintillation results from CASES receiver at South Pole, Antarctica, Radio Sci., 47, RS5009, doi:10.1029/2012RS005061, 2012.

Dierendonck, A. J. V., Klobuchar, J. A., and Huai, Q.: Ionospheric scintillation monitoring using commercial single frequency C/A code receivers, in: Proc of ION GPS-93, p. 1333, Inst. of Navigation, Manassas, VA, 1993.

Humphreys, T. E., Psiaki, M. L., Hinks, J. C., O'Hanlon, B., and Kintner, P. M.: Simulating ionosphere-induced scintillation for testing GPS receiver phase tracking loops, Selected Topics in Signal Processing, IEEE J., 3, 707–715, 2009.

Humphreys, T. E., Psiaki, M. L., and Kintner, P. M.: Modeling the effects of ionospheric scintillation on GPS carrier phase tracking, Aerospace and Electronic Systems, IEEE Trans., 46, 1624–1637, 2010.

Hunsucker, R. D. and Bates, H. F.: Survey of polar and auroral region effects on HF propagation, Radio Sci., 4, 347–365, doi:10.1029/RS004i004p00347, 1969.

Kadokura, A., Yamagishi, H., Sato, N., Nakano, K., and Rose, M.: Unmanned magnetometer network observation in the 44th Japanese Antarctic Research Expedition: Initial results and an event study on auroral substorm evolution, Polar Science, 2, 223–235, doi:10.1016/j.polar.2008.04.002, 2008.

Kim, H., Cai, X., Clauer, C. R., R. Kunduri, B. S., Matzka, J., Stolle, C., and Weimer, D. R.: Geomagnetic response to solar wind dynamic pressure impulse events at high-latitude conjugate points, J. Geophys. Res., 118, 6055–6071, doi:10.1002/jgra.50555, 2013.

Kim, H., Clauer, C. R., Deshpande, K., Lessard, M. R., Weatherwax, A. T., Bust, G. S., Crowley, G., and Humphreys, T. E.: Ionospheric irregularities during a substorm event: Observations of ULF pulsations and GPS scintillations, J. Atmos. Solar-Terr. Phys., 114, 1–8, doi:10.1016/j.jastp.2014.03.006, 2014.

Kinrade, J., Mitchell, C. N., Yin, P., Smith, N., Jarvis, M. J., Maxfield, D. J., Rose, M. C., Bust, G. S., and Weatherwax, A. T.: Ionospheric scintillation over Antarctica during the storm of 5–6 April 2010, J. Geophys. Res., 117, A05304, doi:10.1029/2011JA017073, 2012.

Lessard, M. R., Weatherwax, A. T., Spasojevic, M., Inan, U., Gerrard, A., Lanzerotti, L., Ridley, A., Engebretson, M. J., Petit, N., Clauer, R., LaBelle, J., Mende, S., Frey, H., Pilipenko, S., Rosenberg, T. J., and Detrick, D.: PENGUIn multi-instrument observations of dayside high-latitude injections during the March 23, 2007 substorm, J. Geophys. Res., 114, A00C11, doi:10.1029/2008JA013507, 2009.

Lightsey, E. G., Humphreys, T. E., Bhatti, J. A., Joplin, A., O'Hanlon, B. W., and Powell, S.: Demonstration of a Space Capable Miniature Dual Frequency GNSS Receiver, Navigation, 61, 53–64, 2014.

Melville, R., Stillinger, A., Gerrard, A., and Weatherwax, A.: Sustanable energy at the 100-W level for scientific sites on the Antarctic Plateau: Lessons learned from the PENGUIn-AGO project, Rev. Sci. Instrum., Rev. Sci.Instrum, 85, 045117, doi:10.1063/1.4871555, 2014.

Mende, S. B., Harris, S. E., Frey, H. U., Angelopoulos, V., Russell, C. T., Donovan, E., Jackel, B., Greffen, M., and Peticolas, L. M.: The THEMIS array of ground-based observatories for the study of auroral substorms, Space Sci. Rev., 141, 357–387, 2008.

Musko, S., Clauer, C., Ridley, A. J., and Arnett, K. L.: Autonomous low-power magnetic data collection platform to enable remote high latitude array deployment, Rev. Sci. Instrum., 80, 044501, doi:10.1063/1.3108527, 2009.

O'Hanlon, B. W., Psiaki, M. L., Powell, S., Bhatti, J. A., Humphreys, T. E., Crowley, G., and Bust, G. S.: CASES: A smart, compact GPS software receiver for space weather monitoring, in: Proc 24th International Technical Meeting, Satellite Div., 1745–2753, Inst. of Navigation, Manassas, VA, 2011.

Prikryl, P., Spogli, L., Jayachandran, P. T., Kinrade, J., Mitchell, C. N., Ning, B., Li, G., Cilliers, P. J., Terkildsen, M., Danskin, D. W., Spanswick, E., Donovan, E., Weatherwax, A. T., Bristow, W. A., Alfonsi, L., De Franceschi, G., Romano, V., Ngwira, C. M., and Opperman, B. D. L.: Interhemispheric comparison of GPS phase scintillation at high latitudes during the magnetic-cloud-induced geomagnetic storm of 5–7 April 2010, Ann. Geophys., 29, 2287–2304, doi:10.5194/angeo-29-2287-2011, 2011.

Prikryl, P., Zhang, Y., Ebihara, Y., Ghoddousi-Fard, R., Jayachandran, P. T., Kinrade, J., Mitchell, C. N., Weatherwax, A. T., Bust, G., Cilliers, P. J., Spogli, L., Alfonsi, L., Romano, V., Ning, B., Li, G., Jarvis, M. J., Danskin, D. W., Spanswick, E., Donovan, E., and Terkildsen, M.: An interhemispheric comparison of GPS phase scintillation with auroral emission observed at the South Pole and from the DMSP satellite, Ann. Geophys., 56, 2037–416X, 2013.

Warrington, E. M., Zaalov, N. Y., Naylor, J. S., and Stocker, A. J.: HF propagation modeling within the polar ionosphere, Radio Sci., 47, RSOL13, doi:10.1029/2011RS004909, 2012.

14

CLUSTER–STAFF search coil magnetometer calibration – comparisons with FGM

P. Robert[1], **N. Cornilleau-Wehrlin**[1,2], **R. Piberne**[1], **Y. de Conchy**[2], **C. Lacombe**[2], **V. Bouzid**[1], **B. Grison**[3], **D. Alison**[1], and **P. Canu**[1]

[1]Laboratoire de Physique des Plasmas, CNRS, Palaiseau, France
[2]LESIA-Observatoire de Paris, CNRS, Meudon, France
[3]Institute of Atmospheric Physics, Prague, Czech Republic

Correspondence to: P. Robert (patrick.robert@lpp.polytechnique.fr)

Abstract. The main part of the Cluster Spatio-Temporal Analysis of Field Fluctuations (STAFF) experiment consists of triaxial search coils allowing the measurements of the three magnetic components of the waves from 0.1 Hz up to 4 kHz. Two sets of data are produced, one by a module to filter and transmit the corresponding waveform up to either 10 or 180 Hz (STAFF-SC), and the second by the onboard Spectrum Analyser (STAFF-SA) to compute the elements of the spectral matrix for five components of the waves, $3 \times B$ and $2 \times E$ (from the EFW experiment), in the frequency range 8 Hz to 4 kHz.

In order to understand the way the output signals of the search coils are calibrated, the transfer functions of the different parts of the instrument are described as well as the way to transform telemetry data into physical units across various coordinate systems from the spinning sensors to a fixed and known frame. The instrument sensitivity is discussed. Cross-calibration inside STAFF (SC and SA) is presented. Results of cross-calibration between the STAFF search coils and the Cluster Fluxgate Magnetometer (FGM) data are discussed. It is shown that these cross-calibrations lead to an agreement between both data sets at low frequency within a 2 % error. By means of statistics done over 10 yr, it is shown that the functionalities and characteristics of both instruments have not changed during this period.

1 Introduction

Data calibration of spectra and waveforms issued from a search coil magnetometer is not a new problem. Among previous space physics missions using search coil magnetometers, let us mention GEOS-1 and GEOS-2 as the first ESA spacecraft dedicated to the study of waves and particles in the magnetosphere (Knott, 1975; Jones, 1977). The GEOS wave consortium (S300 experiment) comprised a tri-axis search coil magnetometer built by the predecessors of the spatial team of the Laboratoire de Physique des Plasmas (LPP). The technology used in CLUSTER–STAFF experiments has been substantially upgraded since this epoch, but the principle remains the same: how to calibrate magnetic waveforms issued from a search coil rotating across a high ambient DC field, knowing that the transfer function varies with the frequency? This kind of problem has been solved in this epoch for time–frequency studies (Robert et al., 1978, 1979). Nevertheless, since the creation of the CLUSTER Active Archive (Perry et al., 2005), the need to have a continuously calibrated waveform became essential, and a dedicated method, detailed in this paper, was deployed.

To calibrate a set of search coil data is one thing, to be sure that the calibration is right is another thing. It is true for the calibration of any instrument, but particularly important for search coil calibration where the solution is not unique. In fact, it depends on calibration parameters, themselves depending of the frequency band of the signal (see Sect. 6.4). Using the amplitude of the spin tone measured in the spin plane by the search coil, it is possible to compute the two DC field components in this plane, and so compare them with a fluxgate magnetometer instrument. This was done in the GEOS epoch, where the agreement found was $\sim 4\%$ in magnitude and $\sim 4°$ in direction (Robert, 1979b).

As the CLUSTER data are archived and will be used a long time after the lifetimes of the instrument and the corresponding PI and engineers, it was about time to do this work now to get the best possible calibration and to do necessary checks of cross-calibration. Special care has therefore been devoted to the calibration and cross-calibration of the magnetic wave measurements before launch on the ground, then onboard during the commissioning phase, and throughout the mission. A special effort to compare STAFF-SC data with the FGM onboard flux gate magnetometer data (Balogh et al., 1997, 2001) has been undertaken from the beginning of the mission until now. This was encouraged by the Cluster Active Archive (CAA) activities and particularly by the organisation of regular cross-calibration meetings. The present paper, after a short reminder of the STAFF experiment (Sect. 2), presents the instrument transfer functions determined on the ground, followed by the in-flight verification (Sect. 3). A comparison of the sensitivity of the instrument both on the ground and in space is then discussed (Sect. 4). The transformation of the raw data that are acquired in a spinning system (Cluster spacecraft are spin stabilised) into a fixed physically meaningful reference frame needs the series of successive coordinate transformations described in Sect. 5. The conversion of telemetry data into physical units, that is to say the calibration process itself, is then presented for the waveform data (Sect. 6) and for the Spectrum Analyser data (Sect. 7). The continuity of spectral values as well as the similarity in the wave characterisation between the two STAFF experiments are presented in Sect. 8. Section 9 is devoted to the comparison between STAFF and FGM data and to a discussion of the obtained results, followed by the conclusion.

2   Instrument description

STAFF is one of the five experiments of the Wave Experiment Consortium (WEC); see Pedersen et al. (1997). The optimisation of the analysis of the five components of the electromagnetic waves is among the objectives of the WEC. The STAFF experiment comprises a boom-mounted three-axis search coil magnetometer, a preamplifier and an electronics box that houses the two complementary data-analysis packages: the digital Spectrum Analyser, and an onboard waveform unit. The experiment is briefly described below, with some emphasis on elements of interest for further wave characteristic determination and the comparison between the four spacecraft. For a detailed description of the experiment, see Cornilleau-Wehrlin et al. (1997, 2003). The search coils are mounted at the extremity of a radial boom to avoid interferences from the spacecraft. Figure 1 gives a schematic

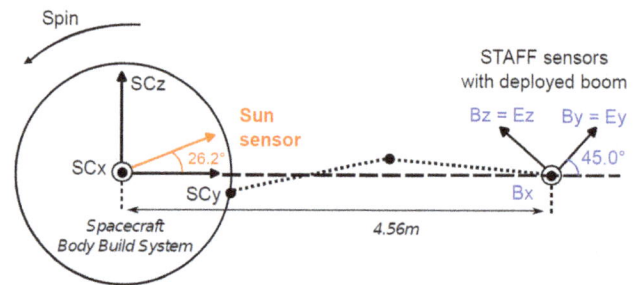

Figure 1. Position of STAFF search coil antennas on the Cluster spacecraft, with respect to the spacecraft and EFW antennas.

of the position of the STAFF antennas with respect to the spacecraft body axis.

The frequency range of the search coil measurements is 0.1 Hz to 4 kHz. The signals go to the preamplifiers, which incorporate a first-order high-pass filter at 0.3 Hz in order to diminish the spin signal. The three output signals then enter the waveform unit, where the analogue waveform signal is sent to different Cluster experiments. First, inside the waveform unit, it is filtered and digitised before being sent to DWP, the Digital Wave-Processing experiment (Woolliscroft et al., 1997) interfacing between wave experiments and the spacecraft. Second, it is sent to the STAFF Spectrum Analyser (STAFF-SA) and to other experiments (see the STAFF block diagram in Fig. 2). These are the Electron Drift Instrument (EDI) (Paschmann et al., 1997), the Wideband (WBD) Plasma Wave (Gurnett et al., 1997) and the Electric Field and Wave experiment (EFW) (Gustafsson et al., 1997). The internal memory of EFW permits, among different possibilities, to get small snapshots of the five-component waveform up to 4 kHz.

The magnetic waveform unit comprises low-pass filters of the seventh order, at either 10 or 180 Hz, selected by telecommand in accordance with the telemetry rate, giving a 42 db attenuation per octave. The sampling rates are 25 and 450 Hz respectively. The output signals are digitised in a real 16-bit analogue-to-digital converter. The 96 dB dynamic range allowed by the 16-bit digitalisation permits the analysis of simultaneously natural waves of a few 10^{-5} nT Hz$^{-1/2}$ and the large signal induced by the rotation of the spacecraft in the environmental DC field, up to some 2000 nT at 0.25 Hz. With such a dynamic range we can get accurate measurements, even at the inversion of the DC magnetic field, e.g. at the magnetopause crossing. The experiment had been designed for its initial orbit, for which the perigee was at a radial distance of 4 Earth radii from the Earth's centre. During the prolongation of the Cluster mission, due to mechanical laws, the perigee has decreased a lot, and there are periods around the perigee where the waveform does saturate. While the data are useless for these periods, this has not induced a degradation of the experiment's capabilities. Owing to telemetry limitations, a reduction in the dynamic data range from 16 to

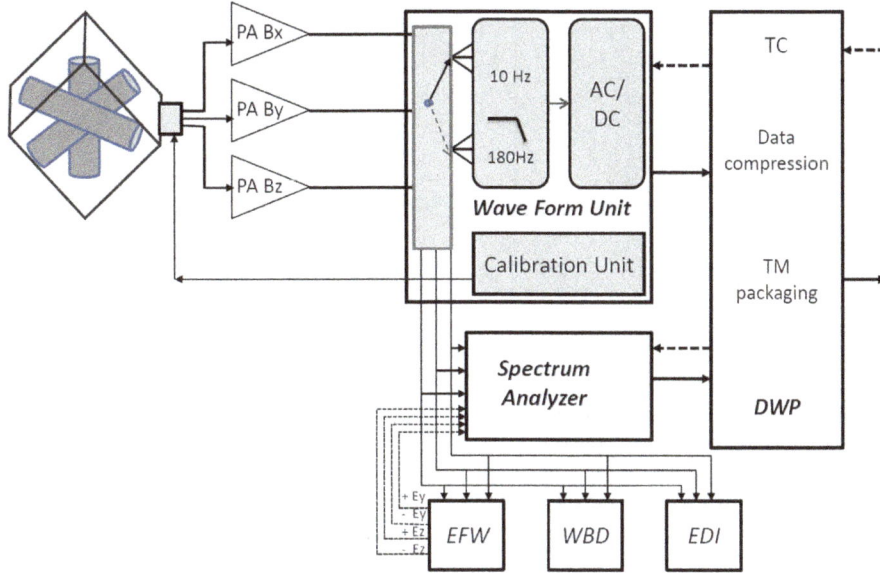

Figure 2. Block diagram of the STAFF experiment and its links to other Cluster experiments.

12 bits is performed inside DWP. The principle is to transmit the full 16-bit word at the beginning of each telemetry packet, and later the difference between the successive samples coded on 12 bits, in such a way that the dynamics of the experiment is preserved even at boundary crossings. Conservative back-up solutions can be selected by telecommand, being either a cruder compression, or having no compression at all. The back-up compression is used during three hours around the perigee, where the spin signal can be above some 200 nT, and at high telemetry rates where the waveform is acquired up to 180 Hz. The three modes have been tested successfully during the commissioning phase.

At higher frequencies for which the telemetry does not permit one to get the waveform, the onboard Spectrum Analyser is part of the STAFF experiment. In addition to the three search coil output signals, the Spectrum Analyser receives the signals from the four electric field probes of the EFW experiment. These are used to form a pair of orthogonal electric field dipole sensors. All five inputs ($2 \times E + 3 \times B$) are used to compute in real time the 5×5 Hermitian cross-spectral matrix at 27 frequencies distributed logarithmically in the frequency range 8 Hz to 4 kHz. The components in the spin plane are despun onboard. All channels are sampled simultaneously, and the integration time for each channel is the same as the overall instrument time resolution, which can be commanded to values between 125 ms and 4 s. The five autospectral power estimates are obtained with a dynamic range of approximately 100 dB and an average amplitude resolution of 0.38 dB. The ten cross-spectral power estimates are normalised to give the coherence. The precision of the phase depends upon the magnitude of the coherence: for a signal with magnitude in the highest bin, it is approximately 5°

close to 0, 180, and ±90°, increasing to about 10° midway between these angles.

The STAFF waveform box also houses an onboard calibration unit that permits one to detect a potential failure of a part of the experiment and to recalculate the transfer function in case of any variation in the experiment response, which is crucial for the comparison between the four spacecraft. The calibration sequence, run once every orbit, consists in sending successively a white noise and fixed-frequency sine waves (~ 7 and 100 Hz), the intensity of which is diminished step by step. The calibration signals are sent at the input of the search coil through the feedback wiring (Fig. 7 of Cornileau-Wehrlin et al., 1997).

3 STAFF experiment transfer functions

3.1 Initial transfer functions

The magnetic sensors together with their associated preamplifiers were first calibrated on the ground. There is a special facility at a quiet site in the Forest of Orléans which is located at the Chambon-la-Forêt Observatory. The calibration facility was built by previous members of the laboratory in the 1960s. The facility consists of a set of three 1 m diameter Helmholtz loops orthogonally mounted to generate a magnetic field. At the centre is a table on which the sensors to be calibrated are put. This table can move around a central axis and is carefully graduated. This facility is also equipped with big loops that were intended to compensate for the Earth's magnetic field. In fact this is not used, as for search coils we are only concerned with rapid (faster than 10 s) variations in the magnetic field. Free space field and stimuli are used, to get respectively the instrument sensitivity and transfer

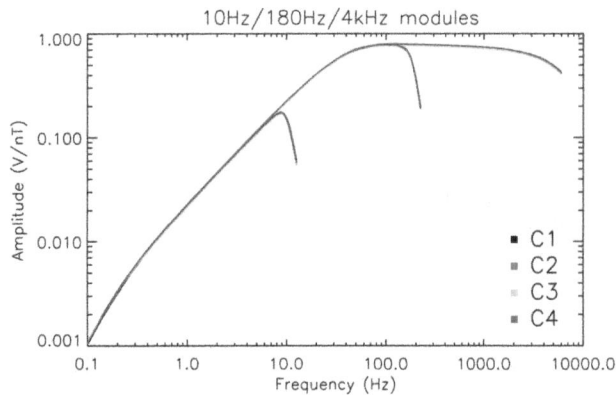

Figure 3. Amplitude of the transfer functions as a function of the frequency, at the output of the preamplifier for the whole frequency range 0.1 Hz–4 kHz, and at the output of the 180 Hz and the 10 Hz filters respectively, for the Bx component. Data for the four spacecraft are overplotted.

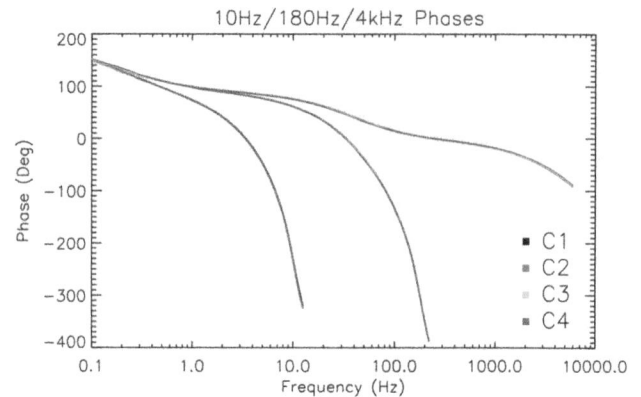

Figure 4. Phase of the transfer functions in degrees as a function of the frequency, at the output of the preamplifier for the whole frequency range 0.1 Hz–4 kHz, and at the output of the 180 Hz filter and the 10 Hz filter respectively, for the Bx component. Data for the four spacecraft are overplotted.

function. We measured there the transfer functions of the antennas and of their respectively associated preamplifiers. The 10 and 180 Hz filter transfer functions have been established in the laboratory, not at the Chambon-la-Forêt Observatory. The deduced combined transfer functions at the output of the antennas and preamplifiers, and at the output of two-range high-pass filters 10 and 180 Hz, are plotted in Figs. 3 and 4. The transfer functions of the four spacecraft are overplotted, for one component, Bx, and do not show significant differences (see Sect. 3.3 below).

Another quantity has been measured in Chambon-la-Forêt: the angle between the mechanical axis of the search coil antennas and the magnetic axis. This is obtained by rotating the antennas on the table and knowing the mechanical axis, and by looking at the antenna response at each angle, we determine the antenna diagram. The angles are small, and the axis can be assumed to be aligned, within an error smaller than $0.3°$. The orthogonality of the three axes has also been verified. By the way, from the spin signal seen on the Bz axis (parallel to the spin axis), it has been shown that the angle between the spacecraft spin axis and the Z antennas is of the order of $0.5°$. As it has been decided not to take into account this small misalignment, it has been also decided to neglect the very small non-orthogonality of the sensors. Note that $\sin(0.5°) = 0.0087$, which is close to 1 %. As we will see in Sect. 9, 1 % is also the best agreement found between STAFF and FGM, with all sources of errors. Another work could take into account these small errors, but it should be done at once for STAFF and FGM. It could be done in the future.

3.2 Corrections applied to the initial transfer functions

While measurements during the commissioning phase showed that the sensitivity and transfer functions were as expected from ground measurements (see Fig. 2 of

Cornilleau-Wehrlin et al., 2003), it appeared during scientific operation that we observed a systematic underestimation of SC1 measurements of about 10 % at low frequencies, in particular at the spacecraft spin frequency. Moreover, comparisons of STAFF waveform data with FGM data evidenced another 10 % underestimation. The reasons for these differences have been studied and explained, leading to a correction of the transfer functions. It is those corrected transfer functions that are given in Figs. 3 and 4. Let us explain the different issues. First we look at the 10 % difference between SC1 and the others. Going back to records of measurements in Chambon-la-Forêt, we found that the current loop axis we used for SC1 was not the same as for the other spacecraft. In addition, the big loops aimed at compensating for the DC Earth magnetic field, which was aligned with the current loop axis used for SC1 antennas, but which was not used for the other spacecraft search coils, disturbed the magnetic field by means of an induced magnetic field opposite to the one produced by the current loop even at very low frequencies. This explained the differences between SC1 and the others. Second, looking very carefully at each of the three current loops, we then found that their structures were no longer perfect plane circles. All this has been verified by means of a reference search coil (that has been tested by different Helmholtz coil facilities and compared with predicted measurements). From this a corrective transfer function has been established:

$$\text{FT\_CORR} = 1.1 \times \frac{\left(1 + j\frac{f}{\text{fch}}\right)}{\left(1 + j\frac{f}{\text{fcb}}\right)},$$

with fch $= 85$ Hz and fcb $= 102$ Hz.

The transfer functions of SC 2, 3 and 4 are corrected by this formula. The corrected transfer function of SC 1 is computed by averaging the other three complex transfer functions. The new transfer functions thus obtained have

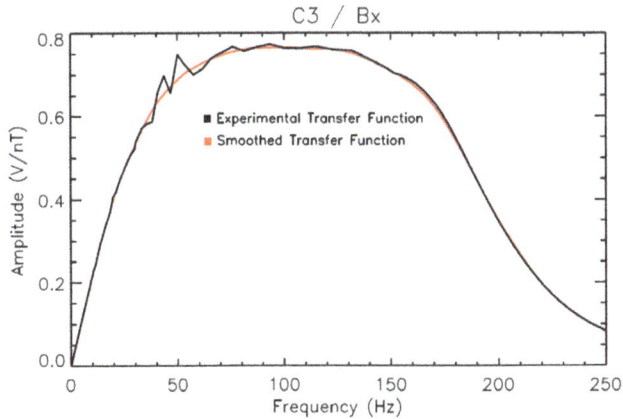

Figure 5. Example of a smoothed transfer function, here the Bx component for SC3, at the output of the 180 Hz filter. The influence of power lines at 50 Hz is clear; this has been smoothed.

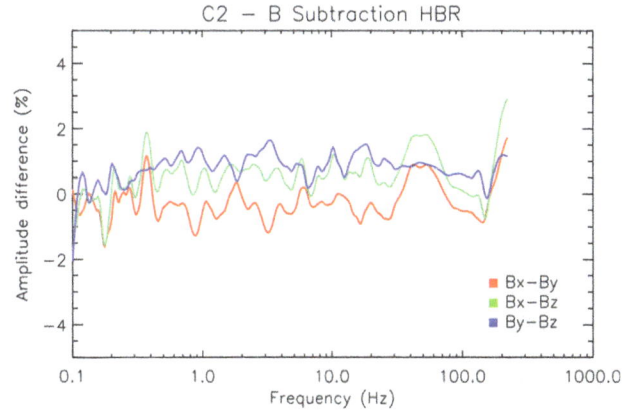

Figure 6. Example for SC2 of the normalised difference in the answer in amplitude of the three components of the magnetic waveform data on the spacecraft.

been applied to STAFF data and compared again with FGM data, as will be shown later on in this paper, giving satisfactory results. It was then estimated that we had found the error sources, and that we could not go further. This comparison seems to show that the facility loops had already undergone the deformation at the time of Cluster STAFF search coil calibration, in 1999, observed some years previously. Since then, a new set of lops, as circular, planar and orthogonal as possible, has been installed at Chambon la Forêt Oservatory.

As the site magnetic quietness is not perfect, there remain some variations in the transfer function which were attributed to the environmental electric array, namely at 50 and 150 HZ. This lead to a smoothing of the new transfer function, as can be seen in Fig. 5.

3.3 Similarity of the search coils, between the three components and between the four spacecraft

As the aim of the Cluster mission is to perform three-dimensional measurements, this implies the ability to combine the data of the different spacecraft either to derive quantities as a curl to get e.g. small-scale currents or to apply the so-called K-filtering method (Pinçon and Lefeuvre, 1991) to disentangle possible different waves modes in turbulent spectra, it was a requirement to produce four experiments as similarly as possible (see e.g. Fig. 4 in Cornilleau-Wehrlin et al., 1997). An example is given in Fig. 6 below, where it can be seen that the relative difference in response in amplitude of the transfer function in the frequency range 0.1–180 Hz is less than 2 %. The normalised differences $Bx-By$, $Bx-Bz$ and $By-Bz$ are overplotted in red, green and blue respectively. For other spacecraft and for 10 Hz filter output, the normalised differences have the same order of magnitude. Note that the differences start to increase around 180 Hz, where the low-pass filters start to be efficient.

3.4 In-flight calibration

As mentioned above (Sect. 2), in order to verify the health of the experiment in operation, the in-flight calibration mode is run once per orbit. A systematic check is done as soon as the data arrive in the lab to verify that data remain within given limits. This did not lead to any unexplained alarm. More detailed analysis can be done – and will be done – to analyse how the experiment behaves after 12 yr (or more) of operation, being built for 2 yr. An example of such a check is given in Fig. 7. This is the result of the analysis of the white noise which is sent by the onboard cal-box to the search coils, using the feedback wires of the search coils. In this step of the calibration mode, the strongest signal is sent. After a Fourier transform of the signal, it is averaged in successive frequency bands to facilitate the verification. The power as a function of the frequency reflects the combined transfer functions of the search coils and the 10 Hz filter. In the figure, two data sets are superimposed, one obtained at the beginning of the mission in 2001 and the other recently in 2012, in the same period of the year (same region of the magnetosphere). One can see the stability of the experiment's behaviour with time.

4 STAFF sensitivity

The determination of the instrument's magnetic sensitivity is an important issue in what concerns the validity of the scientific data analysis. As mentioned in the previous section, the instrument's sensitivity is determined on the ground at the quiet site of Chambon-la-Forêt. An example is given in Fig. 8, onto which have been superimposed to the ground-determined sensitivity the results of measurements in space when in a region with no wave activity (lobes of the magnetosphere). The in-flight data are for one spacecraft (SC4) on one day. Data from STAFF-SA and from waveforms in the 10 and 180 Hz bands are overplotted. The

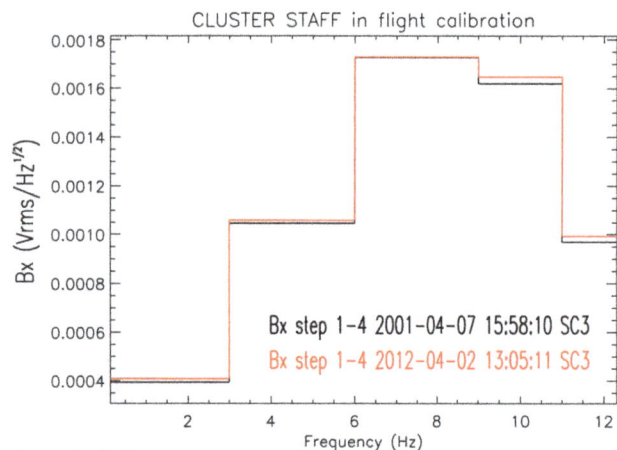

Figure 7. Superimposition of the result of two periods of in-flight calibration in 2001 and 2012. Here a pseudo white noise sent to the search coil antennas is measured at the output of the 10 Hz filter.

10 and 180 Hz filter data cannot be simultaneous, but are close in time. Bz is plotted on top and Bx at the bottom. One sees that the in-flight STAFF-SC data are as good as on the ground. For STAFF-SA output, the in-flight sensitivity is better than the ground sensitivity; indeed, the in-flight experimental noise is below the ground sensitivity curve. This could be explained by the absence in flight of the 50 Hz power line signal and its harmonics seen on the ground. Nevertheless, a few interferences are seen, at 70, 140 and 280 Hz, internal to STAFF-SA, and at 900 Hz, coming from the DWP clock. When looking at the background noise for the 180 Hz waveform, one sees some thin interference lines, the frequencies of which vary from time to time and from one spacecraft to another. This may limit the sensitivity of the measurements in the higher frequency range. The increase in the noise level at and above 10 Hz (180 Hz) comes from the cut-off frequencies of the filters. Due to the effect of the spin signal (see below), the noise level is higher on Bx than on Bz (parallel to the spin axis) at low frequencies.

Figure 9 intends to show the possible evolution of the noise level with time. The spectra are obtained up to 9 Hz by 10 Hz low-pass filter waveform data, and above 9 Hz, they come from STAFF-SA. The above-mentioned interferences are seen clearly on Bx (and By), mainly on Cluster 3 and Cluster 2. Data are averaged over one hour, taken in quiet periods in the same region for the four periods, in the Earth's lobes. Data for spacecraft 1 to 4 are plotted from top to bottom, with Bx components on the left and the Bz ones on the right. Being similar to Bx, they are not shown. One can notice the rather stable level with time, with nevertheless some increase for the A frequency band of STAFF-SA (8–64 Hz). The higher level at low frequencies ($f < 0.3$ Hz) is due to an effect of the local spin signal (high level of DC magnetic fields).

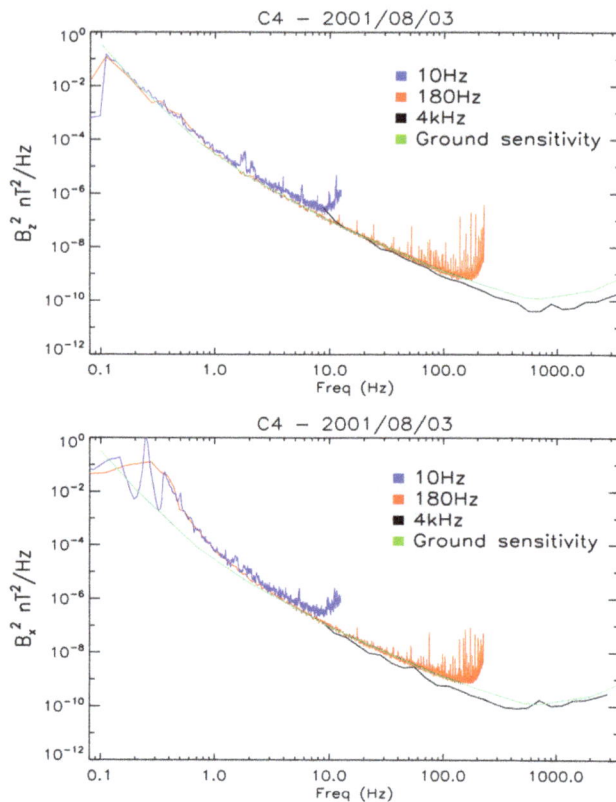

Figure 8. Example of comparison of the sensitivity measured on the ground (green line) at the quiet site of Chambon la Forêt and in flight during a quiet period, for SC4 Bz (top panel) and Bx (bottom panel) components (By is identical to Bx). Outputs of Spectrum Analyser (black line) of 180 Hz filter (red) and 10 Hz filter (blue) are superimposed (see text).

5   Sensor rotation and coordinate systems

To transform telemetry data into significant physical units, we need to convert the data from the sensor coordinate system into one or another system, and in particular to transform from the spinning system into a fixed one, with respect to the Sun and the Earth, for instance. For the waveform data, all transformations are done on the ground, whereas for STAFF-SA data, part of the transformation is done on board. The following sections are dedicated to defining all intermediate coordinate systems required for this operation. Notice that these definitions can be used for other experiments of the same type, on any other mission.

All transformation matrices are named as A_to_B, where A and B are two different coordinate systems. To convert a vector given in the A system to the same vector expressed in the B system, the following expression is used:

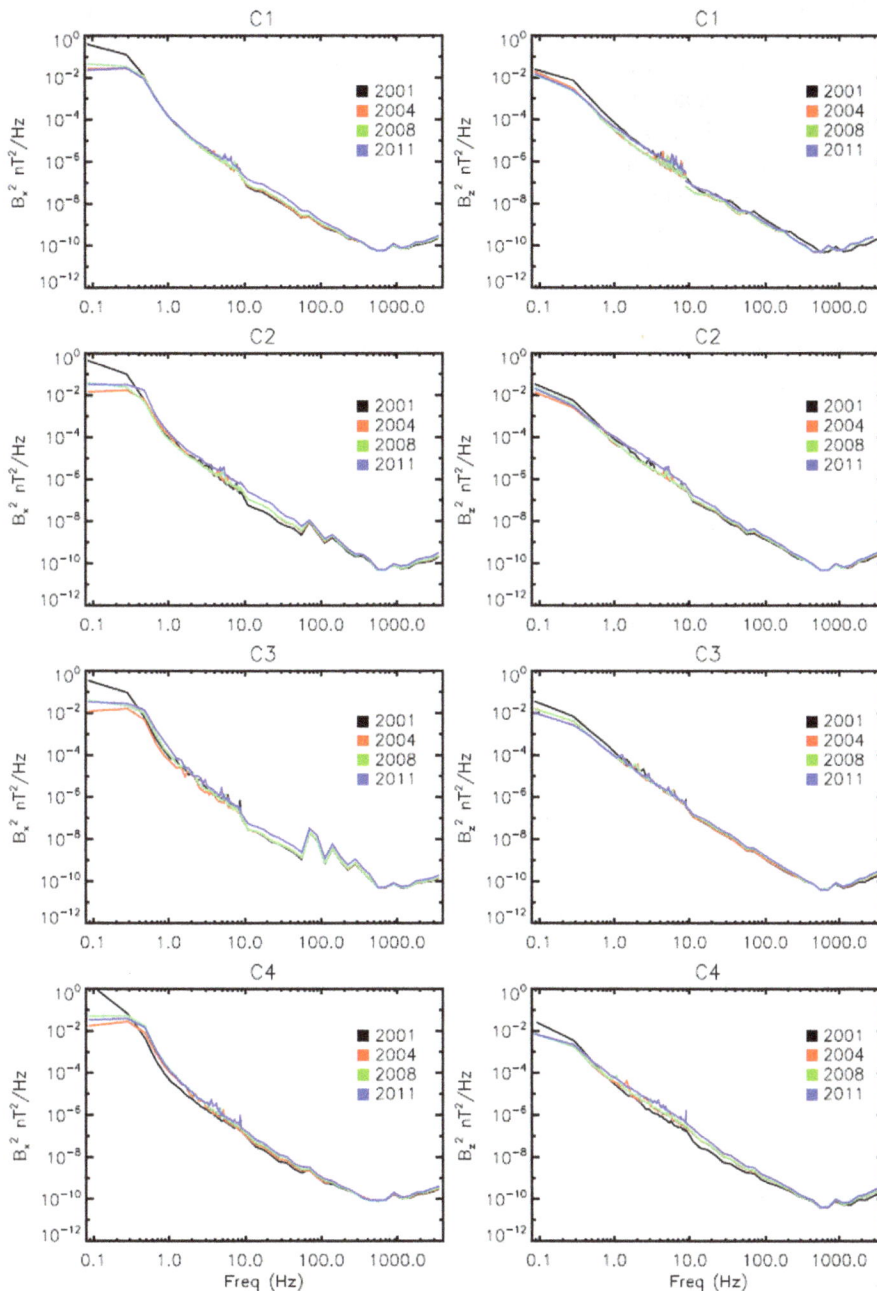

Figure 9. Evolution of the sensitivity with time, for four different years, in the lobes (quiet region) for the four spacecraft and for the Bx and Bz components (By, not shown, is similar to Bx) for both parts of the experiment (waveform up to 9 Hz, Spectrum Analyser above 9 Hz). The chosen time intervals are 3 August 2001 12:00–13:00 UT, 12 August 2004 11:00–12:00 UT, 12 August 2008 19:00–20:00 UT and 19 August 2011 00:10–01:10.

$$\begin{pmatrix} x \\ y \\ z \end{pmatrix}_B = A\_\text{to}\_B \begin{pmatrix} x \\ y \\ z \end{pmatrix}_A .$$

For a general computation of this kind of matrix, see Robert (1993, 2003, 2004).

5.1 The Sensor Coordinate System (SCS)

This is the system where the original signal is measured (see Fig. 1). This system could be a non-perfect orthogonal system (see Sect. 3.1).

5.2   The Orthogonal Sensor System (OSS)

This is a Cartesian orthogonal coordinate system. The original sensor system can be a non-orthogonal system. The first step is to transform the data vector into an orthogonal coordinate system, the Z axis being the reference of the new orthogonal sensor system. The corresponding matrix, called "SCS_to_OSS", close to a unit matrix, is required and must be applied; values are supposed to be constant in time. Nevertheless, for the first time, taking into account the low deviation of the sensor to an orthogonal system for CLUSTER–STAFF ($\sim 0.2°$, see Sect. 3.1), this correction is not applied and the matrix is set to the unity matrix.

$$\text{SCS\_to\_OSS} \cong \begin{pmatrix} 1 & 0 & 0 \\ 0 & 1 & 0 \\ 0 & 0 & 1 \end{pmatrix}$$

5.3   The Data Sensor System (DSS)

The Body Build System (BBS, see next section) is a system fixed to the geometry of the spacecraft, and is used as the spacecraft system reference for all the experiments. Generally, for most of the spacecraft missions, the Z axis is close to the maximum principal inertia axis also called the spin axis (for spin-stabilised spacecraft). Nevertheless, for CLUSTER, this axis has been defined as the X axis (see Fig. 1).

In all our data, the convention taken is that Z is the spin axis. It means that we have an intermediate coordinate system, called the Data Sensor System (DSS), which corresponds to the previous OSS, but where the axes are permuted to make Z close to the spin axis.

With respect to Fig. 1, X_{OSS}, Y_{OSS}, and Z_{OSS} become Y, Z, and X in DSS.

This permutation is obtained by the following matrix:

$$\text{OSS\_to\_DSS} = \begin{pmatrix} 0 & 1 & 0 \\ 0 & 0 & 1 \\ 1 & 0 & 0 \end{pmatrix}.$$

5.4   The Body Build System (BBS)

In the case of CLUSTER, the Z axis of the Data Sensor System is close to the X axis of the BBS, but the misalignment angle is not easy to determine. It is also true for the small angle between this X_{BBS} and the true spin axis (precession and nutation motions). Nevertheless, an estimate of the cumulative angle is made in Sect. 5.5. Here, we neglect this small misalignment and assume that $Z_{\text{DSS}} = X_{\text{BBS}}$. In all cases, two other axes may be rotated by an important angle (see Fig. 1). The corresponding matrix is required, called "DSS_to_BBS"; values are supposed to be constant. Practically, for the STAFF search coils of CLUSTER, this matrix is a rotation matrix of $\alpha = 45°$.

$$\text{DSS\_to\_BBS} = \begin{pmatrix} 0 & 0 & 1 \\ \cos\alpha & -\sin\alpha & 0 \\ \sin\alpha & \cos\alpha & 0 \end{pmatrix}$$

5.5   The Spin Reference System (SRS)

The Spin Reference System has its Z axis parallel to the spin axis. This is a spinning system, rotating at the spin frequency. As mentioned above, there is a small misalignment between the X_{BBS} axis and the Z_{SCS} axis, as there is another slight misalignment between the X_{BBS} axis and the Z_{DSS} axis. It is not easy to separate the two angles, but it is possible to estimate the small angle between the Z_{SCS} axis and the true spin axis which defines Z_{SRS}. This angle θ could be estimated by the measurement of the low spin signal on the Z_{SCS} component.

If B_{xs}, B_{ys}, and B_{zs} are the amplitudes in nT of the spin sine on the three x, y, and z components of the SCS, this angle is estimated by

$$\tilde{\theta} = \frac{B_{zs}}{\sqrt{B_{xs}^2 + B_{ys}^2 + B_{zs}^2}}.$$

This angle could be constant, but can also have small variations during operations on the spacecraft (trajectory modifications, etc.). It has been estimated to an average value of $\sim 0.5\%$, and, for the first time, has not been taken into account, so the "BBS_to_SRS" matrix is set to

$$\text{BBS\_to\_SRS} \cong \begin{pmatrix} 0 & 1 & 0 \\ 0 & 0 & 1 \\ 1 & 0 & 0 \end{pmatrix}.$$

This is a simple circular permutation.

5.6   The Spin Reference2 (SR2) system

The SR2 system, also called "SSS" for Spacecraft-SUN System, or "DS" for despun, is derived from the SRS by a *despin* operation. The spinning spacecraft is "stopped" just at the time where the X axis is in the plane containing the Z spin axis and the direction of the Sun. The rotation angle required is derived from the Sun pulse, which gives the time where the Sun sensor is in the plane defined by the spin axis and the direction of the Sun. Knowing the position of the Sun sensor onboard the spacecraft (see Fig. 1) and the time of each telemetry point, we can deduce the spin phase angle φ_s. This angle, and the corresponding time measurement, is required to build the "SRS_to_SR2" matrix. The terms of this matrix are fast varying with time. f_s is the spin frequency given in the auxiliary data. The phase angle φ_s is calculated for each time tag of the data thanks to the Sun pulse signal. This gives

$$\text{SRS\_to\_SR2} = \begin{pmatrix} \sin(2\pi f_s t + \varphi_s) & \cos(2\pi f_s t + \varphi_s) & 0 \\ \cos(2\pi f_s t + \varphi_s) & -\sin(2\pi f_s t + \varphi_s) & 0 \\ 0 & 0 & 1 \end{pmatrix}.$$

5.7   The Geocentric Solar Ecliptic (GSE) system

The GSE system is a well-known system, with the Z axis perpendicular to the ecliptic plane, and the X axis toward the

Sun. To do the transformation of the SSS to the GSE, the direction of the spin axis in the GSE system is required. Due to the gyroscopic effect of a spinning spacecraft, the spin axis is approximately constant in an inertial system, and so has a yearly variation in the GSE system, except during spacecraft manoeuvres.

SR2 to GSE transformation is done using the module "tsr2gse" routine of the ROCOTLIB software (see Robert, 1993, 2003, 2004). The Cartesian GSE coordinates of the direction of the spin axis are required as the corresponding time measurement. To transform spin right ascension and the spin declination angle given in the STAFF-SC CAA data in the Geocentric Equatorial Inertial (GEI) system, routine "tgeigse" can be used. These angles are also available in the auxiliary files available at CAA (latitude and longitude angles of the spin axis direction in GSE).

Note that in the GSE system, each component mixes both parallel and perpendicular components with the spin axis. Because sensitivity is strongly different at low frequencies in the parallel and perpendicular components in the SR2 system, it is recommended to filter the data below $\sim 0.6\,\text{Hz}$ before coordinate transformation. This is done for CAA Complex Spectra products.

5.8 The Inverse SR2 (ISR2) system

This is equivalent to the SR2 system (or SSS), where the Z and Y axes have inverse signs. This system is useful for CLUSTER, where the Z axis of the ISR2 system is close to the Z axis of the GSE system, so ISR2 is a rather good approximation of the GSE system, and does not require spin direction in the GSE system.

$$\text{SR2\_to\_ISR2} = \begin{pmatrix} 1 & 0 & 0 \\ 0 & -1 & 0 \\ 0 & 0 & -1 \end{pmatrix}$$

5.9 Simplification of the cumulative matrix products

The cumulative matrix product requested to transform original data given in SCS coordinates into a fixed coordinate system such as SR2 can be greatly simplified if we neglect all the small misalignment angles mentioned above. By the way, the first mass processing on the STAFF-SC data was to produce a database for the level 1 data (telemetry data) in the DSS, which is delivered to the CAA. The only difference between the DSS and the SCS sensor coordinate is a circular permutation of the components to get the Z axis close to the spin axis, since we assume that the SCS is orthogonal and equal to the OSS (see Sect. 5.3).

So, to transform data expressed in the DSS into the "fixed" SR2, we have to apply the cumulative matrix product:

$$\begin{pmatrix} x \\ y \\ z \end{pmatrix}_{\text{SR2}} = [\text{SRS\_to\_SR2}][\text{BBS\_to\_SRS}][\text{DSS\_to\_BBS}] \begin{pmatrix} x \\ y \\ z \end{pmatrix}_{\text{DSS}}.$$

Assuming all small misalignment angles to be close to zero, we get

$$[\text{BBS\_to\_SRS}][\text{DSS\_to\_BBS}] = \begin{pmatrix} \cos\alpha & -\sin\alpha & 0 \\ \sin\alpha & \cos\alpha & 0 \\ 0 & 0 & 1 \end{pmatrix}.$$

Using the expression "SRS_to_SR2" given in Sect. 5.6, with $\omega_{\text{s}} = 2\pi f_{\text{s}}$ after some calculus we get

$$\begin{pmatrix} x \\ y \\ z \end{pmatrix}_{\text{SR2}} = \begin{pmatrix} \sin(\omega_{\text{s}}t + \varphi_{\text{s}} + \alpha) & \cos(\omega_{\text{s}}t + \varphi_{\text{s}} + \alpha) & 0 \\ \cos(\omega_{\text{s}}t + \varphi_{\text{s}} + \alpha) & -\sin(\omega_{\text{s}}t + \varphi_{\text{s}} + \alpha) & 0 \\ 0 & 0 & 1 \end{pmatrix} \begin{pmatrix} x \\ y \\ z \end{pmatrix}_{\text{DSS}}.$$

By neglecting all the small misalignment angles, the transformation from the Data Sensor System to the fixed SR2 system is simply reduced to a rotation in the spin plane of the fast varying angle $\psi = (\omega_{\text{s}}t + \varphi_{\text{s}} + \alpha)$.

This simplification is used for CLUSTER–STAFF calibration, but cannot be used for spacecraft or rockets having precession or nutation, or a non-constant direction of the spin axis. In this case, the full computation must be done.

6 STAFF-SC calibration method

6.1 Spectrum calibration in sensor frame

The STAFF-SC experiment is a waveform unit which delivers magnetic waveform $x(t)_{\text{Volt}}$ in the SCS sensor reference frame. The transfer function being frequency dependent but not proportional, all components of this signal at various frequencies must be corrected both in amplitude and phase, so the signal delivered by the search coil is

$$x(t)_{\text{Volt}} = \int_{-\text{fs}/2}^{+\text{fs}/2} X(f)_{\text{nT}}\alpha(f)e^{2i\pi ft}\mathrm{d}f, \tag{1}$$

where $X(f)$ is the spectrum of the true ambient signal, in nT, $\alpha(f)$ is the complex transfer function of the sensor in $\text{V}\,\text{nT}^{-1}$, and fs/2 is the upper detectable frequency in Hz.

For the first time, let us consider the calibration of a single spectrum.

After digital processing, fs being then the sampling frequency, Eq. (1) becomes

$$x_{k(\text{Volt})} = \sum_{-N/2}^{N/2} X_n\alpha_n e^{2i\pi nk}. \tag{2}$$

X_n is the Fourier transform of the real signal x_k, in nT, to be estimated. The spectral resolution is $\delta f = f_{\text{s}}/N$.

So, to retrieve the original spectra (in nT), a simple Fourier transform is required:

$$X_n = \frac{1}{\alpha_n}\frac{1}{N}\sum_{-N/2}^{N/2} x_k e^{-2i\pi nk}. \tag{3}$$

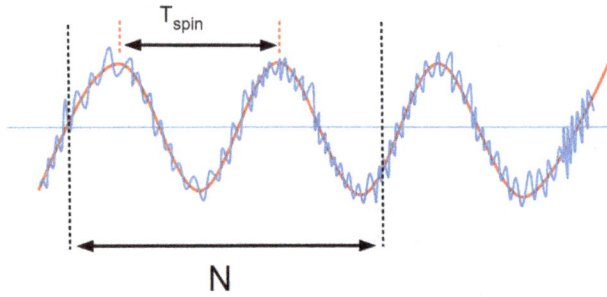

Figure 10. Spin tone superimposed onto rapid variations of the magnetic field that the search coils intend to measure.

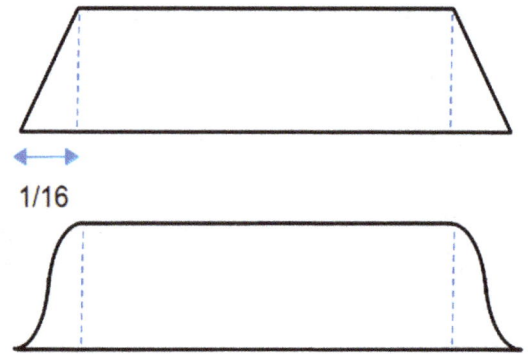

Figure 11. Trapeze and round trapeze used as weighting functions.

This is the theory. In practice, the original x_k signal is formed by a large sine signal at spin frequency fspin (~ 0.25 Hz), due to the rotation of the sensors into a large DC field (~ 100 nT), and the fluctuations (a few nT) are superimposed, so the amplitude of the "useful" frequency range is ~ 100 times less intense than in the DC field (spin signal at ~ 0.25 Hz, see Fig. 10). Furthermore, a Fourier transform assumes a periodic signal of period N, and thus introduces large discontinuities on the edges of the window which generate meaningless high-frequency components (see Robert et al., 1978, Robert, 1979a).

Thus, the first step is to remove this large sine signal with dedicated software which computes the amplitude and phase of the sine for a given spin frequency and removes it. Note that this measurement of amplitude and phase on the two Bx and Bz DSS spinning components allows us to compute the two components in the spin plane of the DC field, by applying the complex coefficient of the transfer function at the spin frequency, taking into account the phase angle given by the Sun pulse time to convert results into a non-spinning frame system (see later). Then, to avoid discontinuities on the edge of the window, the second step is to apply a weighting function on the signal after centering on zero. The weighting function must preserve the shape of the signal, but must also ensure that the weighted signal is periodic, so that its edges fall to zero. By experience, the choice of a very long trapezoid works well, as shown in Fig. 11.

So, the estimate of the original spectrum (in nT) is computed by

$$\tilde{X}_n = \frac{1}{\alpha_n} \frac{1}{N} \sum_{-N/2}^{N/2} x_k W_k e^{-2i\pi nk}.$$

Before computing the estimate of the calibrated waveform $\tilde{x}_{k(nT)}$, we now have to study the successive coordinate systems used to convert the signal recorded by the sensors into a useful coordinate system.

6.2 Computing calibrated waveforms in the SR2 system

From Eq. (3) we can estimate the calibrated waveform in the SCS by an inverse FFT, as

$$\tilde{x}_{k(nT)} = \sum_{-N/2}^{N/2} \tilde{X}_n e^{2i\pi nk}.$$

Afterward, the calibrated waveform in the SR2 system is computed by applying the successive coordinate system matrix defined in Sects. 5.1 to 5.6; practically, for CLUSTER–STAFF, we use the simplified equation given in Sect. 5.9, and we get

$$\begin{pmatrix} \tilde{x}_k \\ \tilde{y}_k \\ \tilde{z}_k \end{pmatrix}_{SR2} = \begin{pmatrix} \sin(\omega_s t_k + \varphi_s + \alpha) & \cos(\omega_s t_k + \varphi_s + \alpha) & 0 \\ \cos(\omega_s t_k + \varphi_s + \alpha) & -\sin(\omega_s t_k + \varphi_s + \alpha) & 0 \\ 0 & 0 & 1 \end{pmatrix} \begin{pmatrix} \tilde{x}_k \\ \tilde{y}_k \\ \tilde{z}_k \end{pmatrix}_{DSS}.$$

One result is that the SCS and the DSS differ only by a circular permutation (see Sects. 5.2 and 5.3).

6.3 Computing the calibrated spectrum in the SR2 system

The previous waveform being calibrated and expressed in the SR2 system, the complex spectra is simply given by the FFT of this calibrated waveform, as

$$\tilde{X}_n = \frac{1}{N} \sum_{-N/2}^{N/2} x_k W_k e^{-2i\pi nk}.$$

The weighting function can be chosen freely. For CLUSTER–STAFF CAA products, we chose a trapeze function as described in Fig. 11. For other applications, an alternative could be a "rounded trapeze", by replacing the edges with a $\sin^2$ function rather than a line. The same operation is of course done for $\tilde{Y}_n$ and $\tilde{Z}_n$.

6.3.1 Window effect

Due to the weighting function, the previous calibrated waveform is significant only around the centre of the window. To enlarge this part, we generally use a weighting function, the shape of which is a trapeze or a rounded trapeze represented in Fig. 11. We can see that only $\sim 7/8$ of the waveform is significant.

In normal use, the length of the window is about a few spin periods. At least one spin period is required to estimate the amplitude and phase of the sine, and two or three spin periods allow a best estimate. Beyond a few spin periods, the DC field could change significantly, and the spin tone estimate will be less accurate. A good compromise is between one and four spin periods to estimate properly the DC field and to remove spin tone.

6.3.2 Summary of the various steps done during spectrum calibration

Calibration of the telemetry data is done in successive steps, described below. At this level, the calibration is done on successive data windows to obtain the calibrated spectra as described above.

Get waveform in volts

Telemetry data are given in integer values (called TM counts) within a $[0, 65\,535]$ interval corresponding to a volt range of $[-5, +5\,\text{V}]$. The value of $65\,535$ comes from the used sample words of 16 bits in length. This step does the conversion simply as

$$x\,(t_n)_{\text{Volt}} = \left[x\,(t_n)_{\text{TM}} \times 10/65\,535 \right] - 5.$$

Selecting the time length of the windows determines the Δt. Δf is one resolution of each spectrum.

This step is named "Calibration step #1: Volts, spinning sensor system, with DC field".

Cleaning raw waveforms

This step consists in removing the high-amplitude signal at spin frequency due to the SC rotation into a high DC field. Indeed, the wave useful signal is very low (a few nT) compared with the high spin tone (a few nT, up to ~ 5–$600\,\text{nT}$). Even if the transfer function coefficient in amplitude is small at the spin frequency, the spin tone in volts remains too high to do a correct fast Fourier transform (see Robert, 1979a).

This step is named "Calibration step #2: Volts, spinning sensor system, without DC field".

The independent calibration of the spin tone, both in amplitude and phase, on the two Bx and By components allows the determination of the two components in the spin plane, which can be compared with the same components measured by the FGM experiment.

Calibration of each component within a given window

For the first time the signal is centered, and then a light trapezoidal windowing is applied to reduce edge effects before applying the FFT. Next, in the frequency domain, for each frequency, the complex spectrum is divided by the complex transfer function to get a calibration in amplitude and phase. Since the transfer function is close to zero for frequencies close to zero, a cut-off frequency is applied, generally fixed at 0.1 Hz. Lastly, an inverse Fourier transform is performed to return in the time domain and to get a calibrated waveform in the given window, always in the Sensor Spinning System.

This step is named "Calibration step #3: nTesla, spinning sensor system, without DC field".

Get the calibrated waveform in the fixed SR2 system

By applying the appropriate matrix given in Sect. 6, which requires spin phase computation, one gets the calibrated waveform in the SR2 system.

This step is named "Calibration step #4: nTesla, fixed SR2 system, without DC field".

Add DC field values on X and Y

This is an optional step which allows comparison with FGM data, because one obtains, for the two X and Y components in the spin plane, both the DC field and the fluctuation.

This step is named "Calibration step #5: nTesla, fixed SR2 system, with xy DC field".

Get the calibrated waveform in the GSE system or others

This is an optional step. From step 4, waveforms can easily be converted in the GSE system or other geocentric systems (GSM, MAG, GEO ...) by using the ROCOTLIB software (see Robert, 1993, 2003, 2004).

Remark: this method provides a calibrated waveform *which is only significant in the central part of the window*, and produces discontinuities at the edges of each window, so this method cannot be used to produce continuously calibrated waveforms (see Robert, 1979a, 2009). The method used to produce continuously calibrated waveforms is described in the next section.

Nevertheless, this method can be used to produce the estimate of the calibrated spectrum $\tilde{X}^{\text{SR2}}_{n(\text{nT})}$ in the SR2 system (see Sect. 6.3) by applying a simple Fourier transform. The main advantage is the low CPU time consumption.

6.4 Waveform continuous calibration method

6.4.1 Method chosen for CLUSTER

To obtain a continuous waveform, we have to repeat the previous operation by overlapping successive windows and keeping the central points, as illustrated in Fig. 12. The calibration is done on a window of Nkern telemetry (TM) points, which determines the frequency resolution of the intermediate calibrated spectrum, that is the accuracy of the calibration. The calibration window is then shifted by Nshift points.

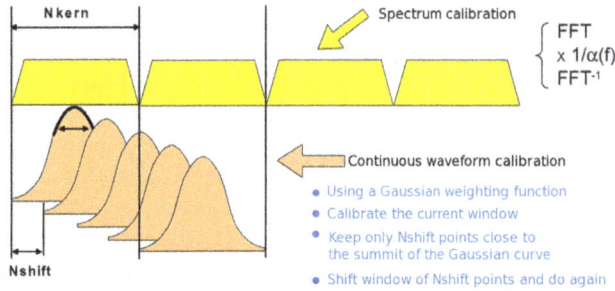

Figure 12. Illustration of the continuous calibration method (bottom) as compared with spectrum calibration.

Table 1. Parameters chosen for the routine production at CAA for the continuous calibration of the waveform as a function of the sample frequency.

	Sample freq. (Hz)	NKern	Window duration (s)	Spin period (s)	Nshift
NBR	25	1024	40.95	~ 10	2
HBR	450	4096	9.10	~ 2.3	2

The Nkern number must be optimised in order to

– Do a correct despin: the window duration must be long enough to have a good estimate of the spin tone, but not too long, because the amplitude (and phase) of the spin tone varies with time; the DC field could change in both direction and amplitude. A good compromise is between 2 and ~ 10 spin periods. One period is the minimum to run the despin algorithm.

– Have a high enough frequency resolution: the window duration must be long enough to get a significant sampling of the transfer function and to get a good accurate calibration.

The number Nshift must be optimised in order to

– Be the shortest possible, for instance two points corresponding to the summit of the weighting function in the window.

– Reduce CPU time by increasing Nshift, but quality will be reduced too. If the weighting function is not constant during the Nshift point centered on the window, a parasite line appears on the spectrograms at fsr/Nshift frequency (fsr being the sampling rate).

Tests show that the best result is obtained by using a Gaussian weighting function, and by shifting the windows from one to the next one by two points, saving each time the two central points at the summit of the Gaussian.

For the CLUSTER–STAFF-SC CWF CAA production, the chosen values for the Nkern and Nshift parameters are given in Table 1. These values have been chosen to process a calibration which works whatever the amplitude and the time variation of the DC field, under normal conditions. This is important, because we can see that by the choice of the calibration parameters, *the solution for the calibration data is not unique*. For very particular conditions, these parameters can be adjusted to get the best quality of the calibration. For instance, if low frequencies are not interesting, it is preferable to filter the data above twice the spin frequency, to avoid undesirable spin effects. The length of the calibration window

also plays a role in the calibration, depending on whether the covered period is stationary or not. In a general way, a long window leads to a more accurate calibration, but is time consuming, and enlarges the data gap. However, a long window could also lead to a non-perfect cleaning of the large sine signal due to the DC field, especially if it is fast varying. In any case, the best compromise must be sought in relation to the nature of the data (stationary waves or magnetopause crossing can require different calibration parameters). This is why we keep available in the Cluster Archive the level 1 waveform (uncalibrated). It is the expertise of the experimenter that will lead to better results.

6.4.2 Other calibration methods

The previous method, which is a deconvolution in the frequency domain, can be summarised by

$$x_{k(\mathrm{nT})} = \mathrm{FFT}^{-1}\left\{\frac{1}{\alpha_n}\mathrm{FFT}\left(x_{k(\mathrm{Volt})}w_k\right)/\right\}.$$

More recently, the THEMIS mission has also included search coils (Roux et al., 2008), and for the data processing we used the deconvolution in the time domain (Le Contel et al., 2008), which can be summarised by

$$x_{k(\mathrm{nT})} = \left(x_{k(\mathrm{Volt})}w_k\right) \times A_k,$$

where "$\times$" is the convolution operator and A_k the impulse function of $\frac{1}{\alpha_n}$, i.e. $\mathrm{FFT}^{-1}\left\{\frac{1}{\alpha_n}\right\}$.

From Plancherel's theorem, notice that the two expressions are equivalent:

$$\begin{cases} x(t_k) \times A(t_k) \Leftrightarrow & X(f_n) \cdot \alpha(f_n) \\ x(t_k) \cdot w(t_k) \Leftrightarrow & X(f_n) \times W(f_n) \end{cases}.$$

We find again in this method the concept of Nkern and Nshift, with the same meaning, as the weighting function, and the need of "cleaning" the waveform by removing the spin tone before any other processing.

Comparison of the two methods has been done by applying the two different software packages to the same data set, and concludes with a good agreement. Details of these comparisons will be done in another paper. Note that the calibration software used for CLUSTER is written in Fortran90,

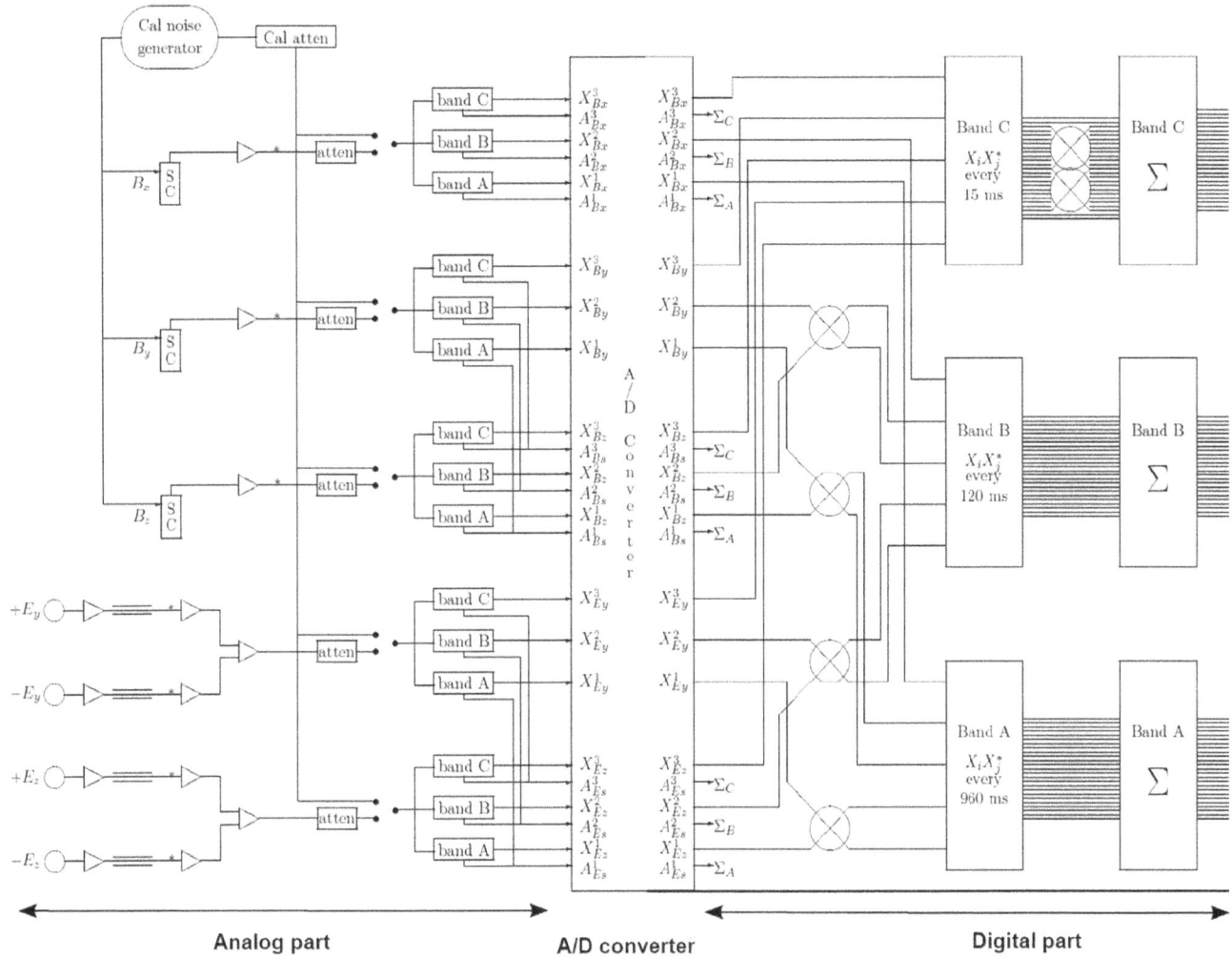

Figure 13. The different parts of the STAFF Spectrum Analyser instrument (STAFF-SA). From left to right: analogue part, A/D converter, digital part.

while the calibration software used for THEMIS is written in IDL. In this last case, the convolution operation has been done by the built-in "CONVOL" IDL function, which is very efficient. So, the good agreement of the two results is a proof of the validity of these two programs, which is very important for validating archive databases.

7 STAFF-SA Spectrum Analyser

STAFF-SA has 5 input channels connected to 5 sensors: 3 magnetic and 2 electric. An overview of the instrument is given in Fig. 13. It makes estimations of the auto- and cross-spectral power density at 27 frequencies, arranged into 3 bands, A, B, and C, which have their own automatic gain control.

7.1 Onboard calculations

The separation into three bands is performed by the analogue part of the receiver. The digital part performs the despin for the spin-plane components and makes a filtering in nine narrower frequency bands. It then calculates the cross-spectral matrix of the five components in amplitude and in phase. The AGC are fixed during the time of an analysis, time controlled by telecommand. The different steps of the onboard calculation can be seen in Fig. 14, and for more details on the onboard calculation, see Sect. 4.3 of Cornilleau-Wehrlin et al. (1997).

In order to calculate the cross-spectral matrix, the components that are in the spin plane are despun:

$$B_u = B_y \cos(m) + B_z \sin(m)$$
$$B_v = B_z \cos(m) - B_y \sin(m),$$

where m is the instantaneous angular position of the spacecraft as derived from the onboard Sun reference pulse (SRP), and u and v are the fixed coordinates corresponding to the position of the STAFF search coil (SCS) antennas at the time of the Sun pulse (i.e. when the SRP sees the Sun, see Fig. 1). The angle between this reference frame and the SR2 reference frame is then $45 - 26.2 = 18.8°$.

7.2 Routine on-ground calibration

The calibration model that is applied on the ground to the raw spectral matrix data is a combination of mathematical algorithms and tables of coefficients (S(AGC), D(AGC), $\tilde{S}$ (Freq), $\tilde{D}$ (Freq)) used by these algorithms. The set of coefficients comes from measurements performed in the laboratory, including the inverse transfer function of the sensors, STAFF search coils and EFW antennas. The spin-plane receivers are strapped together in pairs with common AGC outputs. The calibration model will treat the sum and the difference of both the receivers. It is convenient to assume that the frequency-dependent and AGC-dependent variations of the analogue transfer function can be separated.

Functions S and D take account of the variation in the AGC level, assumed to be the same for all frequencies within any given band. The parameters S characterise the analogue receivers; the mean spectral noise density in the overall pre-converter passband of analogue receiver m is a function of the corresponding AGC output A.

Functions $\tilde{S}$ and $\tilde{D}$ allow for the variation with frequency within each digital input channel. The parameters $\tilde{S}$ characterise the bandpass of the digital spectrum analyser and also the variation with frequency of the analogue receivers. They allow independent auto- and cross-spectral estimates to be obtained in both amplitude and phase at each of the nine frequencies f.

To calculate $\tilde{S}$ and $\tilde{D}$, we have chosen a reference noise level corresponding to AGC = 80 (AGC = 0 to 255). To this ideal calibration we had foreseen to apply a small correction in two different cases: first if the spin is not nominal, i.e. is different from 4 s, this parameter being routinely provided by the spacecraft auxiliary data; second if the spin-plane receivers are not identical. In this later case, the correction parameters have been identified during ground tests of the instrument before launch (Harvey et al., 2004).

To calculate the spectral matrix, we use an iterative numerical method, with a convergence test to stop the calculation. Note that we have assumed the variation of autocorrelations with the AGC level to be the same for all frequencies, within any given band; this is not entirely true for some frequencies, and it can explain some small anomalies. The last operation is to transform the data that are in a non-standard fixed reference frame into SR2. For the components that are in the SC spin plane, a rotation of delta $\varphi = -18.8°$ has to be applied, as well as a BBS-to-SRS rotation matrix to have Bz parallel to the spin axis (see Sect. 5).

This overall treatment of STAFF-SA data gives the complete complex spectral matrix (SM), the diagonal coefficients of the matrix being the power spectral density (PSD) for the five components, in physical units. These PSD diagonal coefficients are kept at a better time resolution than the overall SM. To obtain from the spectral matrix the polarisation and propagation parameters, one can use the PRASSADCO program that has been specifically developed for use by Cluster STAFF-SA (Santolik, 2003), as can be seen in the following section.

8 STAFF-SC/STAFF-SA continuity and other cross-checks

Once the transfer function is calculated, one of the first checks is to compare the results of the data analysis by the two STAFF sub-experiments. Both continuity in the spectra and similarity in the wave characterisation results have been checked.

8.1 Power spectra

For this purpose we have used a special mode of operation which allows the maximisation of the frequency overlap between the two experiments, between 8 and 180 Hz. Examples of different kinds of wave data are given below in Fig. 15. Two kinds of wave fluctuations are shown, for two different spacecraft. Only the Bx and Bz components are shown. Whereas the overlap and continuity are rather good, one can be aware of some effect on STAFF-SA at lower frequencies, as reported in the CLUSTER–STAFF CAA calibration report (Robert et al., 2012).

8.2 Wave characteristic determination

It is important to determine the characteristics of measured waves, which can be done by means of the three magnetic orthogonal sensors. The parameters that can be obtained at one point of measurement are, in particular, wave planarity, ellipticity, sense of polarisation, and propagation angle with respect to the main magnetic field.

These quantities are obtained on the ground for the waveform data up to either 10 or 180 Hz, depending on the telemetry mode, whereas the coefficients of the complex spectral matrix are calculated onboard by the STAFF-SA for the frequency range 8 Hz–4 kHz. Figure 16 gives an example of such an analysis performed by data coming from both the waveform (left panel) and from SA (right panel), observed by Cluster 1 close to the magnetopause on 26 March 2007. The colour scale is the same for both data set analyses, but the frequency resolution is different: linear for waveform data and log for SA. The top panel shows the dynamic spectra. Different polarisation and propagation parameters are then plotted in the time–frequency plane for amplitudes greater than a minimum level only (for PSD larger than 10^{-6} nT^2 Hz^{-1}). These parameters are the ellipticy of the emission, the propagation angle theta (k, B), and the

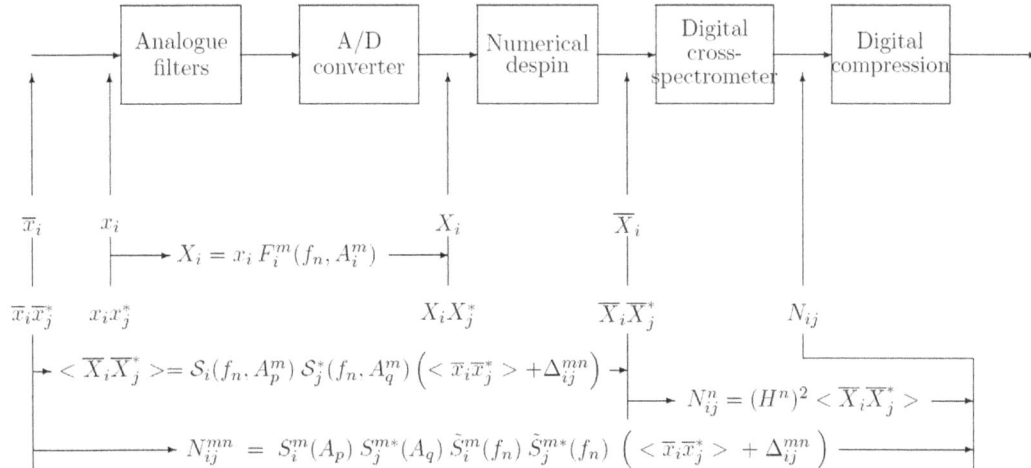

Figure 14. Relationships between the different signals in STAFF-SA.

azimuth angle ($\phi = 0$ indicates the spacecraft Sun direction projected in the plane perpendicular to B). The three polarisation parameters are computed through the singular value decomposition method (Santolik et al., 2003) using the PRASSADCO tool (Santolik, 2003).

The common frequency band between STAFF-SA and STAFF-SC is about 60–225 Hz. The main emission observed during that time interval is detected between 07:22:45 and 07:23:30 UT around 50 Hz, and up to more than 200 Hz in the middle of the time interval. One can remark that the polarisation and propagation parameters calculated from the two data sets give the same results, in the limit of their respective frequency and time resolutions. During this event, the magnetic emissions were clearly right-hand polarised: the theta angle was most of the time less than 45°, and the phi angle displays a nice rotation (from -90 to -180, 180 then to 90°). The similar values of the different parameters obtained from both parts of the experiment, together with physically "reasonable" results, give some confidence in the validity of the data processing performed. This kind of test was performed at the beginning of the mission in order to find an error in the rotation matrix, which has been solved since.

9 STAFF-SC/FGM comparisons

9.1 Interest of such a study

The possibility of recovering two DC magnetic field components by the search coil experiment is particularly useful, because it allows a comparison with the result of the Fluxgate Magnetometer experiment. It has already been done for the GEOS mission (see Jones, 1977); here the agreement found was $\sim 4\,\%$ in magnitude and $\sim 4°$ in direction (Robert, 1979b).

For CLUSTER, the two STAFF and FGM experiments have run successfully since the beginning of the mission. In the framework of the CAA cross-calibration meeting, it was obvious to look after a comparison of the result of the two instruments in their common frequency range. This is not only useful for data validation, but this also permits the clarification of the respective roles of the two instruments.

So, during all calibration meetings, from 2006 until now, the STAFF-FGM comparisons were day by day in progress. Thanks to this kind of comparison, we realised that the transfer function of the search coils was underestimated (Robert, 2nd Cross Calibration Workshop, 2006). After investigation, it was shown that the shape of the calibration device (Helmholtz coils) was slowly distorted with time (see Sect. 3.2 and Robert, 14th Cross Calibration Workshop, 2011). New calibration tables were used, and the STAFF-FGM agreement improved.

9.2 Data origin

FGM data are issued from CAA, in "full" resolution mode, in the GSE system. They are converted into the SR2 system to compare the spin plane components with STAFF-SC data.

Search coil data are calibrated following the process described in Sect. 6.4, in normal bit rate mode (NBR) sampled at 25 Hz mode, and of course in the SR2 system. Step 5, "nTesla, fixed SR2 system, with xy DC field", is used.

9.3 Direct waveform comparison

9.3.1 A typical event studied on various scales

The following results have been shown in different cross-calibration meetings. The 24 February 2001 case is interesting, because we can look at it on various timescales. Figure 17 shows a waveform comparison between 21:00 and 22:00 UT. The $B_\perp$ component is computed from B_{xs} and B_{ys} components in the spin plane as

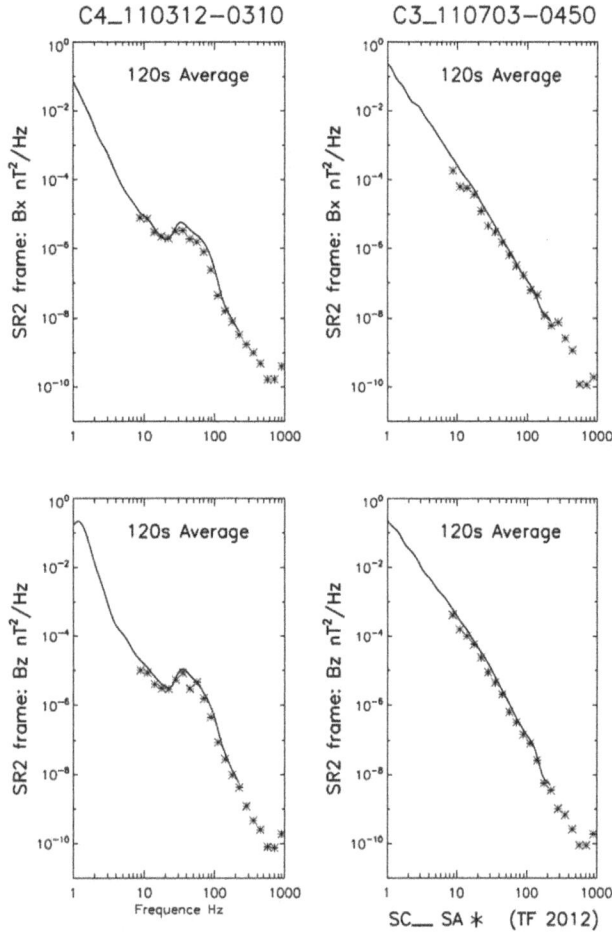

Figure 15. Examples of a comparison between a wavelet spectrum of the waveform data (black line) and a STAFF-SA spectrum (cross). A special operation mode has been used to maximise the overlap between the two experiments, between 8 and 180 Hz. Here are the Bz and Bx components for two different events (left and right sides respectively).

$$B_\perp = \sqrt{B_{xs}^2 + B_{ys}^2},$$

while the direction or phase φ is computed from

$$\sin\varphi = B_{ys}/B_\perp$$
$$\cos\varphi = B_{xs}/B_\perp.$$

On the $B_\perp$ component, at a first glance, the agreement is good, with

$$\frac{\Delta B_\perp}{B_\perp} = \frac{\left(B_\perp^{STA} - B_\perp^{FGM}\right)}{\left(B_\perp^{STA} + B_\perp^{FGM}\right)/2} \sim 4\%,$$

while the phase angle difference is

$$\Delta\varphi = \left(\varphi^{STA} - \varphi^{FGM}\right) \sim 5°.$$

After re-sampling the data to obtain the same sample rate on the two data sets, we can compute the mean difference point to point, and the result is much better:

$$\frac{\Delta B_\perp}{B_\perp} = 0.77\%$$
$$\sigma B_\perp = 0.84\%.$$

If we zoom in on the narrow spike between 21:56 and 21:58 UT (Fig. 18), we can see that the agreement is also good on a short timescale ~ 2 mn. We find as previously that

$$\frac{\Delta B_\perp}{B_\perp} < 1\%$$
$$\Delta\varphi \cong 3°.$$

9.3.2 Statistic over 10 yr on spin plane DC field

Figure 19 shows statistics performed over 10 yr of STAFF-FGM DC field comparisons. The 58 events altogether have been chosen, under four various conditions each year:

- low DC field, low ULF activity,

- low DC field, high ULF activity,

- high DC field, low ULF activity,

- high DC field, high ULF activity.

Figure 19a shows the relative difference $\Delta B_\perp/B_\perp$ in %, where we can see that this difference is roughly constant for each spacecraft during the 10 yr studied.

Figure 19b shows the standard deviation of $\Delta B_\perp/B_\perp$, which is between 0.5 and 5 %, except for one point at 12 %, but which corresponds to a very low $B_\perp$, so $\Delta B_\perp/B_\perp$ become relatively high, taking into account the accuracy of the measurement.

Figure 19c shows the amplitude of the $B_\perp$ DC field for each event, from a few nT to 500 nT.

Lastly, panel d gives the phase difference of the $B_\perp$ component in the SR2 system.

Concerning the relative stability of $\Delta B_\perp/B_\perp$, we can see that it is independent of the magnitude of the DC field, whatever the level of ULF activity. Furthermore, for each spacecraft, this difference remains constant all over the ten-year study. This is an important result, because it shows that the transfer function remains constant from the beginning of the mission until 10 yr after. This result could be confirmed by a dedicated study of the onboard calibration signals.

Another important result is the difference from one spacecraft to another: in fact, the best result seems be obtained for SC1, where the transfer function has been obtained by the averaging of the three others (SC2, SC3 and SC4). This result is thus directly directed by the estimate of the transfer function on the ground, and gives an estimate of their accuracy (see Sect. 3). The choice has been made to keep each of the 3×4 transfer functions slightly different, but, as these tables should theoretically all be identical, another choice could have been to set all tables to the SC1 average table. This could be done in a future work.

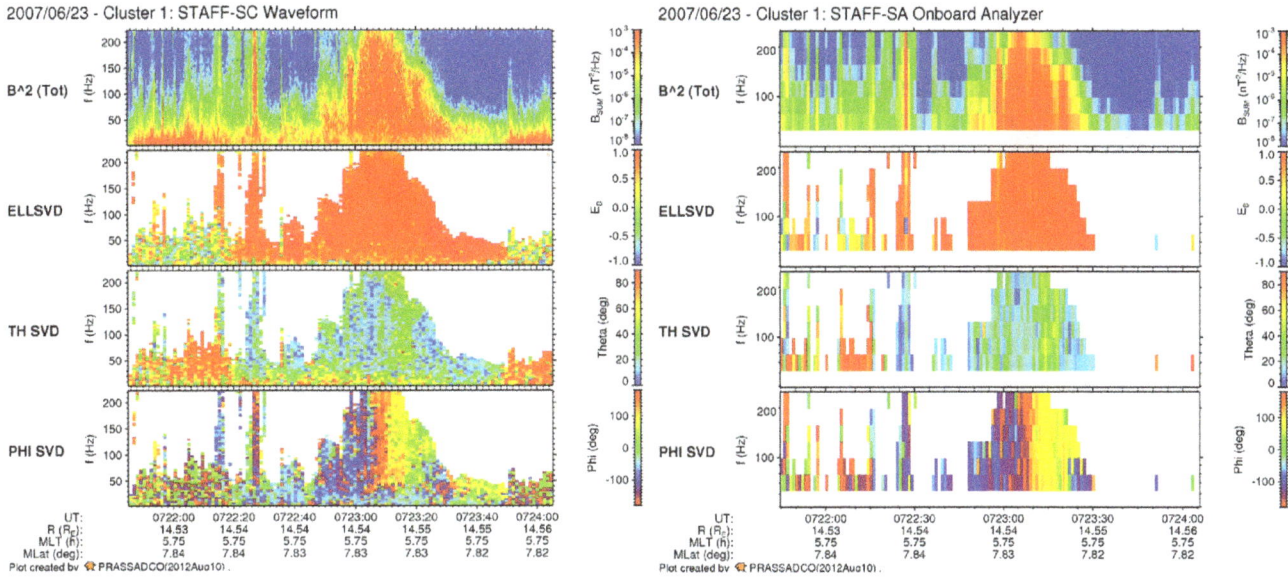

Figure 16. For each instrument (Waveform data – STAFF SC, and Spectrum Analyser – STAFF SA), the same polarisation and propagation characteristic quantities are plotted in a frequency–time diagram: from top to bottom, the total magnetic PSD (power spectral densities), the ellipticy, the propagation angle $\theta(k, B)$ and the azimuth angle (ϕ). To highlight the polarisation of the intense emissions, parameters are plotted only for PSD values, above a threshold. Note that the frequency scale is linear for STAFF-SC and log for STAFF-SA.

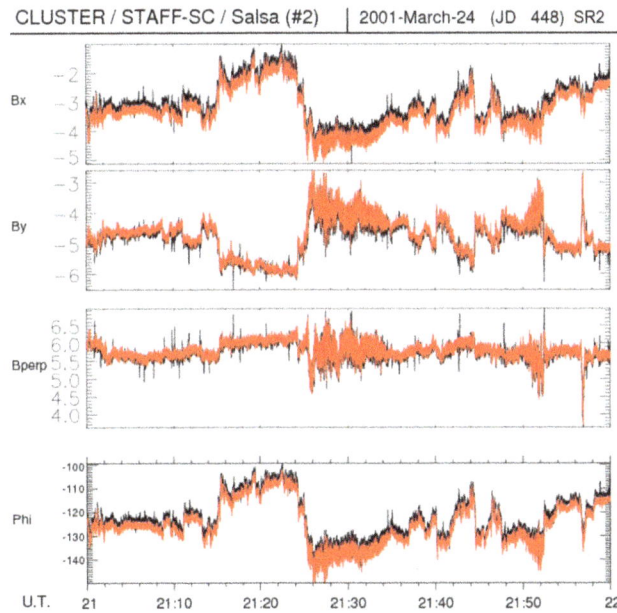

Figure 17. Direct waveform comparison on a large scale (15th Cross-Calibration Workshop, 17–19 April 2012). Black: STAFF, red: FGM.

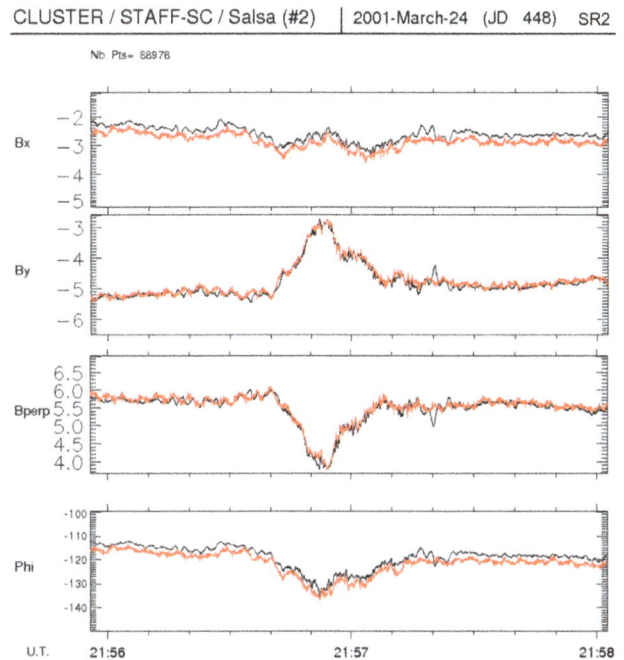

Figure 18. Zoom on 2 min of data around the spike seen at 21:57 UT of Fig. 17. (15th Cross-Calibration Workshop, 17–19 April 2012). Black: STAFF, red: FGM.

Concerning the direction, most of the time this $\Delta\varphi$ difference is between 2 and 4°. Nevertheless, for some cases, this difference changes in sign, and is between −2 and −4°. This change has not been explained up to now.

9.3.3 Comparison at 1 Hz

Figure 20 shows an event with an almost monochromatic wave at low frequency (~ 1 Hz) superimposed onto a low DC

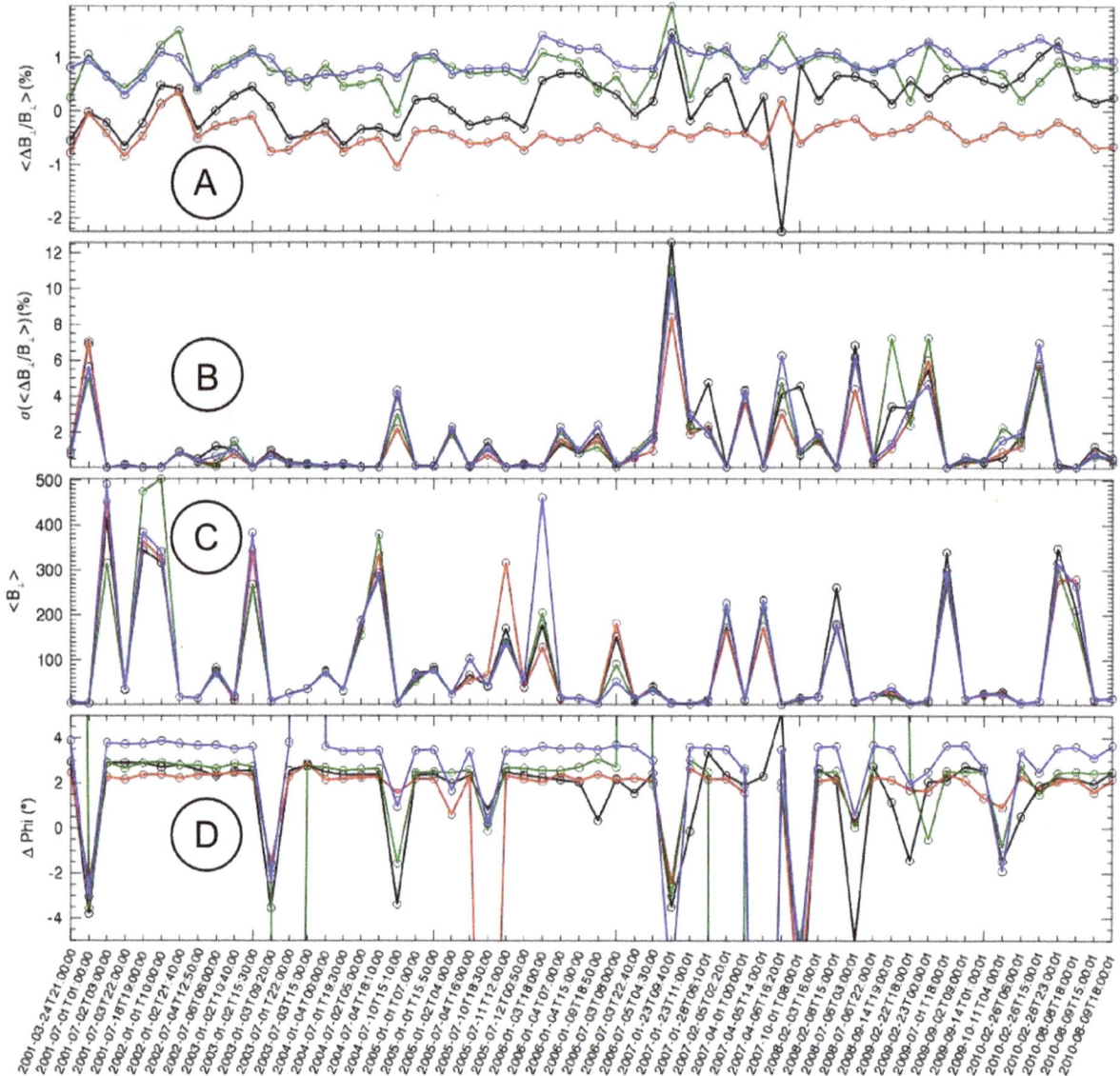

Figure 19. Statistics over 10 yr of STAFF-FGM spin plane DC field comparison for the four spacecraft (black, red, green, and blue for spacecraft 1, 2, 3, and 4, respectively).

variation. In the left part, one can see a constant difference of $\sim 1\,\%$ in the $B_\perp$ component, as expected, and a phase difference of $\sim 4°$. The zoom (in the right part) still shows the same agreement on the DC part, both in amplitude and phase. To see a more precise comparison for the component at 1 Hz, we shift the FGM data of 3.3 nT (1.1 %) to have a better superimposition of the two curves (Fig. 21). The result is rather satisfying, a good fit being found at a first glance, but a spectral analysis is required to get a best estimate of the difference (see Sect. 9.4.1).

9.3.4   Comparison at 6 Hz

The following example corresponds to another almost monochromatic wave at $\sim 6\,\text{Hz}$, always superimposed

onto a low DC variation (Fig. 22). The wave occurs at $\sim 09{:}39\,\text{UT}$ on By. Agreement on DC fields remains the same ($\Delta B / B < 1\,\%$, $\Delta\phi \sim 3°$).

By zooming on the wave (Fig. 23), we can identify a $\sim 6\,\text{Hz}$ wave whose amplitude and phase seem to be in good agreement, but as previously, a spectral analysis is required to get more details (see Sect. 9.4.2).

9.4   Spectrum comparison

9.4.1   STAFF-SC/FGM sensitivity

Figure 24 shows a spectrum of STAFF and FGM done during a very quiet period, which means that these two curves can be considered to be the sensitivity of the two instruments. The

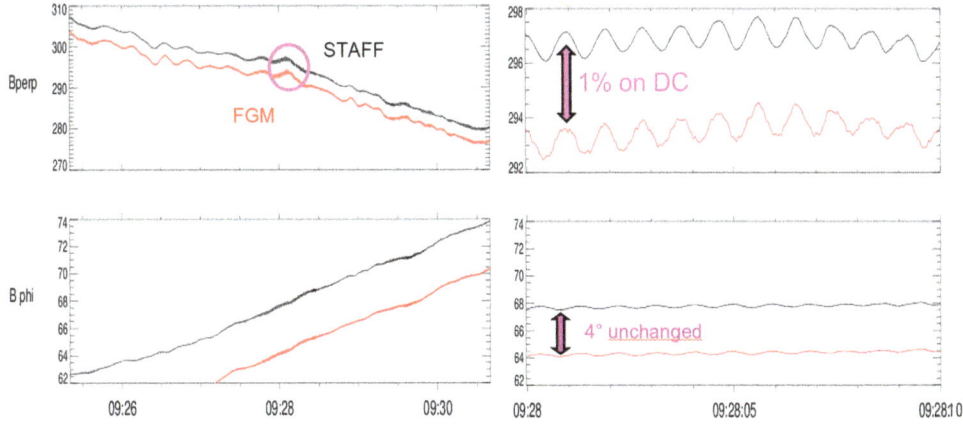

Figure 20. Wave comparison at 1 Hz (CLUSTER–Tango (#4), 23 September 2001).

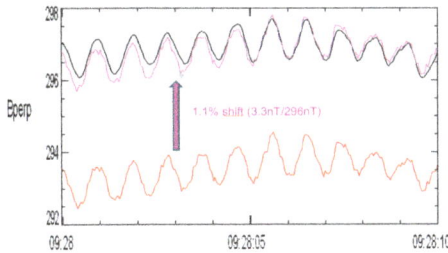

Figure 21. Wave at 1 Hz, STAFF-FGM superimposed.

two curves cross at ∼ 0.7 Hz, that is to say that at this frequency the two instruments have the same sensitivity. Below 0.7 Hz, FGM is not only more sensitive, but of course gives the three components of the DC field contrary to STAFF. Above 0.7 Hz, the search coils are more sensitive and can detect events of smaller magnitude. This leads to the choice of the one experiment rather than the other, according to whether you look at DC or at waves, and for waves, to which frequency range you want to focus on. In fact, the two experiments are quite complementary.

9.4.2 1 Hz event

Figure 25 shows the FGM and STAFF spectra corresponding to the waveform event of Fig. 20. The strong peak at 1 Hz spreads from 0.5 to 1.5 Hz, and the agreement between the two instruments is very good, even for the second peak at ∼ 2.5 Hz. To quantify the exact difference, a dedicated study should be done, requiring filtering of the high frequencies, spike removal and Shannon interpolation for the SATFF-FGM resampling. The noise above 3 Hz is higher for FGM, as expected; nevertheless, it is above the sensitivity shown in Fig. 24.

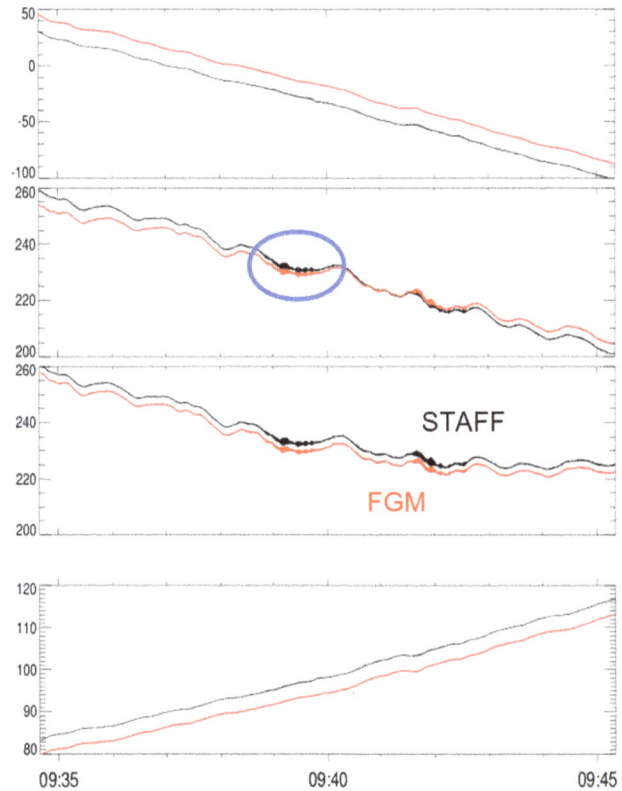

Figure 22. Wave comparison at 6 Hz (CLUSTER–Tango (#4), 23 September 2001. 14th CAA cross-calibration meeting, York, 5–7 October 2011).

9.4.3 6 Hz event

Figure 26 shows the spectra corresponding to the waveform event at 6 Hz of Figs. 22 and 23. As above, the strong peak at 6 Hz spreads from ∼ 4.5 to 6.5 Hz, and shows very good agreement between STAFF and FGM. Nevertheless, the second peak at ∼ 7.75 Hz is not recorded by FGM, its sensitivity not being sufficient at this frequency. On the other hand, a

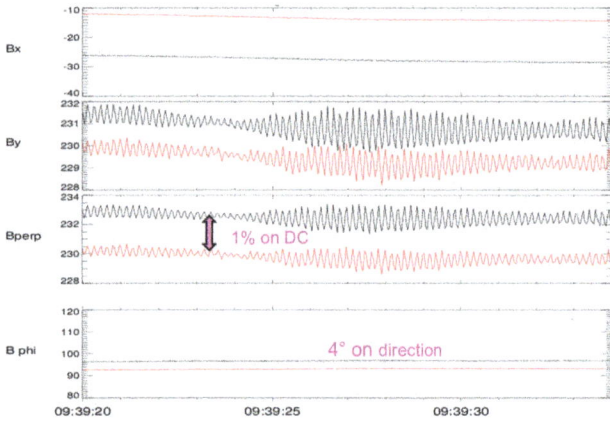

Figure 23. Zoom on wave comparison.

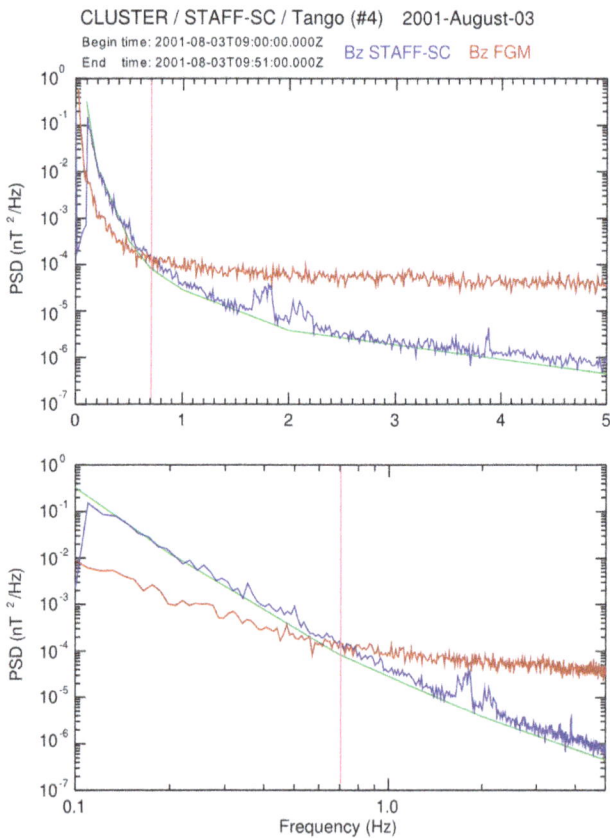

Figure 24. STAFF-FGM spectra comparison for a very low power event, to show respective sensitivity of the two instruments. Blue: STAFF, red: FGM. Top panel: lin–log scale, bottom panel: log–log scale. The green line is the STAFF ground measurement sensitivity, measured before launch, and corresponds roughly to the sensitivity observed in flight.

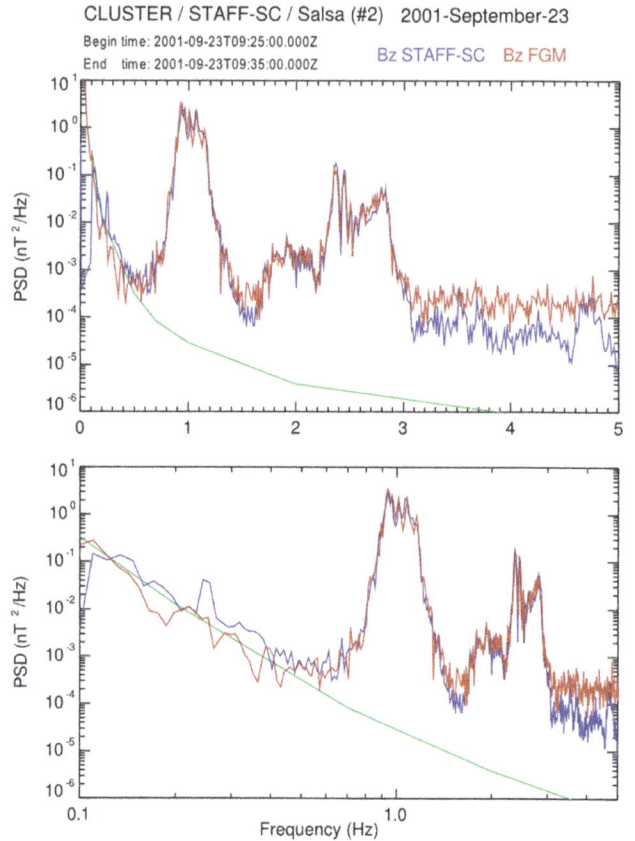

Figure 25. STAFF-FGM spectra comparison for events at 1 Hz.

low frequency below 0.4 Hz is not recorded by STAFF. This example is also a good illustration of the respective interest of the two instruments.

9.4.4 Wide frequency band event

Figure 27 shows a strong signal over the whole frequency bandwidth. The agreement is very good between 0.1 and ∼4 Hz. Above 4 Hz, the power spectral density ($nT^2 Hz^{-1}$) of STAFF and FGM differs by a factor of nearly 2. Since the event is strong, the two instruments are widely above their sensitivity (the green line corresponds to the STAFF-SC sensitivity). Furthermore, this is STAFF, which is above FGM. A deeper study must be done to explain this. The effect of fall-off on the FGM frequency response at this frequency could be studied in a future work.

Figure 28 (from Nikiri et al., 2006) shows another example of power spectra comparison, in the cusp region. One can notice good agreement in the overlapping frequency bands, between 0.5 and 2 Hz, as for the previous 1 Hz event, so it confirms that below ∼4 Hz, the agreement is very good. It shows in the complementarity of both experiments, permitting analyses of a wide frequency range with good precision.

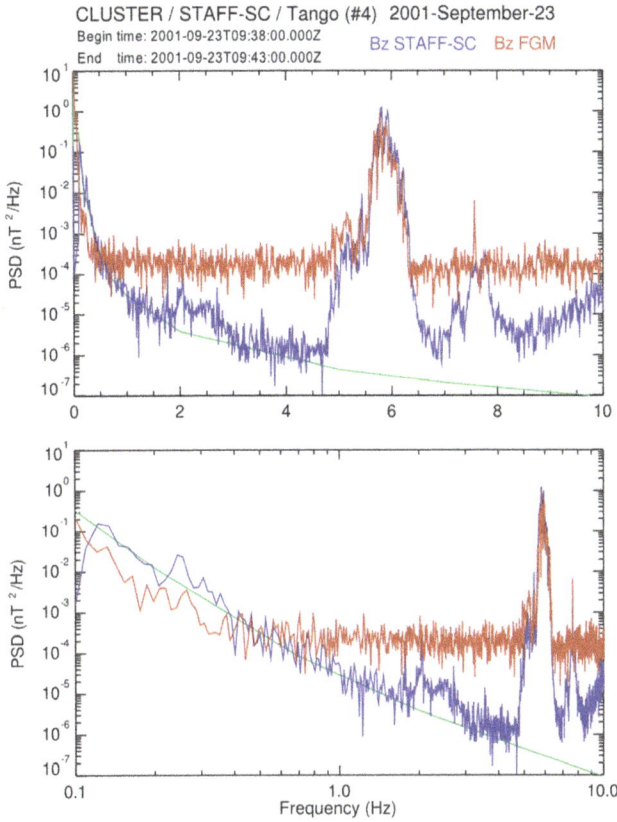

Figure 26. STAFF-FGM spectra comparison for events at 6 Hz.

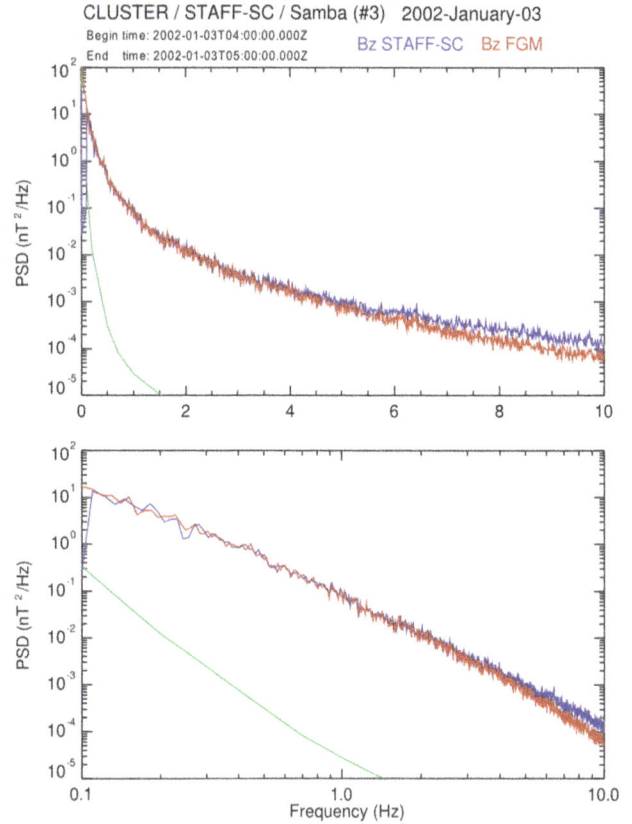

Figure 27. STAFF-FGM spectra comparison for a large frequency band event.

9.4.5 Doppler effect on sensitivity

Classical spectra are computed in the fixed (not spinning) system, as shown in Fig. 29, where we chose an event corresponding to the background noise. Ground sensitivity is plotted in black. Only the Z component (blue curve) is not disturbed by the spin, but the X and Y components (red and green curves respectively), in the spin plane, suffer from the Doppler effect: the large peak at spin frequency (0.25 Hz) corresponds to the 0 Hz frequency in the spinning system of the sensor, and thus could lead to an infinite value, since the transfer function at this frequency is null. Practically, to avoid an undefined value, one chooses a very low value rather than 0.

The two holes on X and Y at 0.25 ± 0.1 Hz correspond to the cut-off frequency chosen during the calibration (see Sect. 6). Furthermore, the spin effect added to the Doppler effect leads to a decrease in sensitivity in the X and Y components.

In other words, because the transfer function is close to zero at $f = 0$, this means that a right-handed polarised wave at spin frequency in the spinning coordinate system *cannot be recorded* by the STAFF sensor (see Fig. 30). Indeed, it is seen at $f = 0$ by the sensor, so the search coils do not provide any signal at this frequency. For this right wave,

$$f_{SR} = f_{SR2} - f_{spin}.$$

On the other hand, a left-handed polarised wave at any frequency, including DC, *is recorded* by the STAFF sensor. For this left wave,

$$f_{SR} = f_{SR2} + f_{spin}.$$

This also means that, at a low frequency, we *cannot* expect full agreement between STAFF and FGM, except for left-handed polarised waves, but for frequencies $f \gg f_{spin}$, as we found previously, there is good agreement.

To understand the consequences of the Doppler effect with a non-linear transfer function well, Fig. 31 shows this spectrum in the SR spinning coordinate. The peak at the spin frequency now corresponds to the DC magnetic field, and the cut-off frequency looks the same for the 3 XYZ components. Moreover, we can see that beyond twice spin frequency, the sensitivity becomes the same on the three components, which is not the case in the fixed SR2 system.

For more details on the Doppler effect on detected waves, see Robert et al. (1978, 1979) and Robert (1979a).

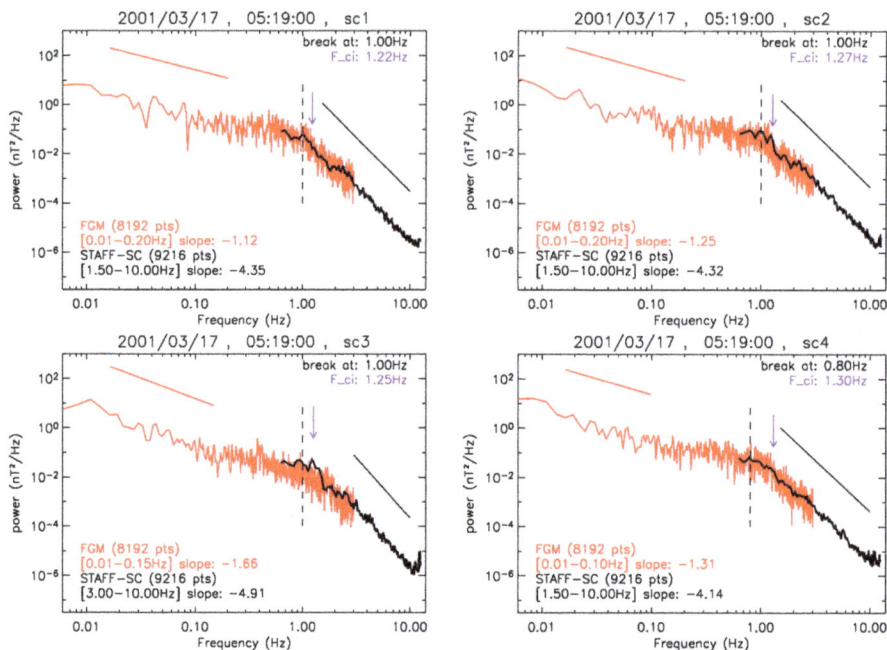

Figure 28. An example of comparison between FGM (in red) and STAFF (in black) power spectra, from Nikiri et al. (2006), in the cusp region. One can notice the good agreement in the overlap frequency bands, between 0.5 and 2 Hz. It shows that one can use a combination of the two complementary experiments, for instance to calculate the spectral power law index and the frequency break.

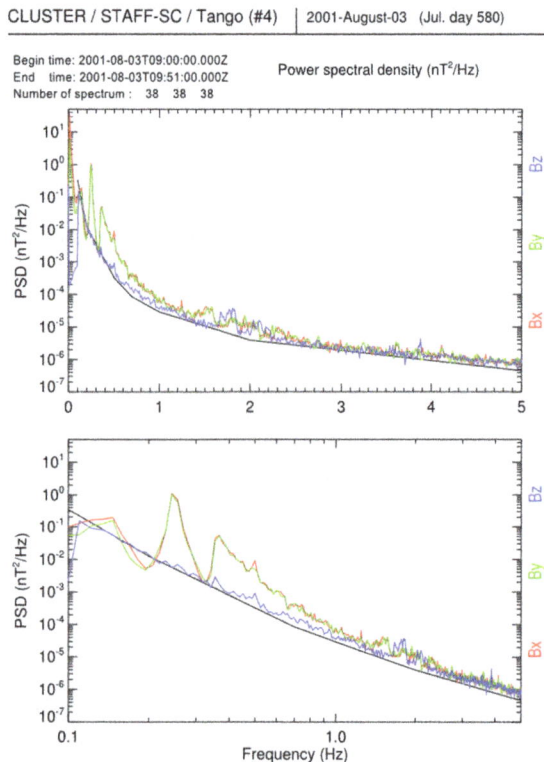

Figure 29. STAFF Power spectral density of background noise in the SR2 system.

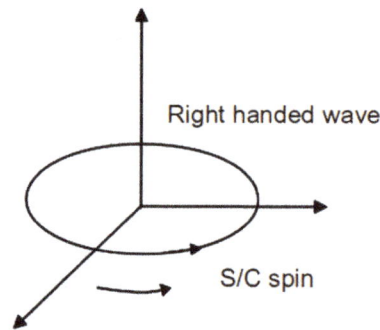

Figure 30. Schematic diagram illustrating the Doppler effect on the recorded waves.

10   Conclusions

A method to calibrate the waveform delivered by a rotating search coil has been proposed, and used for CLUSTER–STAFF-SC data. It has been shown that the solution to waveform calibration data is not unique and depends on the signal itself. Various coordinate systems required to transform telemetry data into a fixed and known coordinate system have been defined. The method and coordinate systems defined here can be used for another mission of the same kind. This paper also shows that the quality of the transfer functions of the instrument is a key to getting the best accurate calibrated waveforms and spectra. Sampling and transfer function determine the sensitivity of the instrument, which has

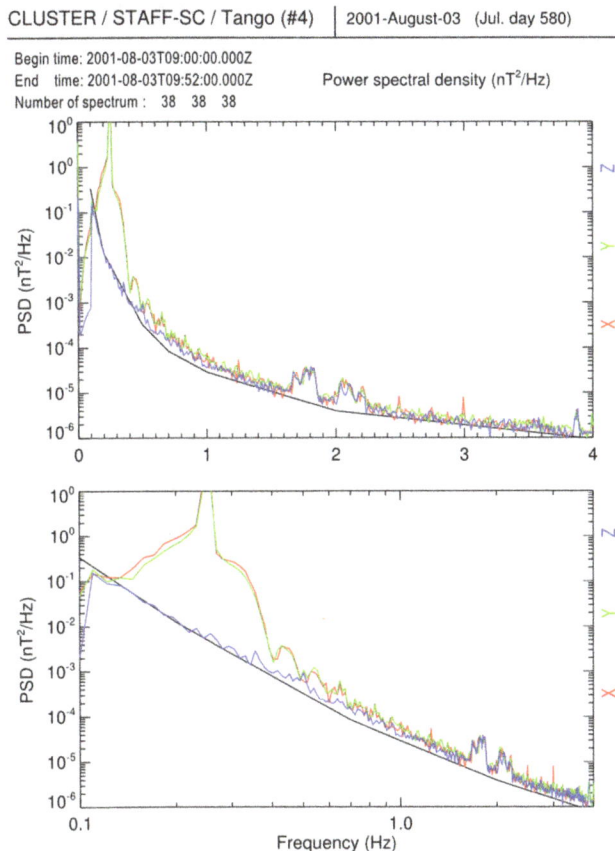

Figure 31. STAFF Power spectral density of background noise in the spinning system.

been established both from the calibrated waveform issued from STAFF-SC and for the calibrated spectra issued from STAFF-SA. It has been shown that we get good continuity between the two sub-instruments, into their common frequency range.

Cross-calibration between STAFF-SC in normal mode (0–10 Hz) and FGM in the same frequency range leads to an agreement of $\sim \pm 1\%$ on the DC field in the spin plane and within a few % between 0.5 and 10 Hz. The respective sensitivity of the two instruments, deduced from observations done during a period of a very quiet magnetic activity, shows the two curves crossing at 0.7 Hz. Below this frequency, the fluxgate is more sensitive and gives the three components of the DC field. Above this frequency, the search coil is more sensitive.

Statistical study of the DC field, as measured by FGM and by STAFF at spin frequency over ten years, shows a constant difference between the two instruments, and so demonstrates the stability of the quality of measurements performed by both instruments.

Acknowledgements. The authors deeply thank A. Balogh, first PI of the FGM experiment, and the following ones, E. A. Lucek and C. Carr, and the whole FGM team, who contributed to delivering to the CAA an invaluable set of data available to the scientific community.

All engineers who have been working on the STAFF data since the beginning of the mission, too numerous to be mentioned, are fully thanked. The authors are indebted to Christophe Coillot for the establishment of the corrective transfer function formula, Bertrand de la Porte for the test on the transfer function, and Kateryna Musatenko for processing the in-flight calibration data.

Many thanks to the CAA itself for its action, and for the organisation of a long series of cross-calibration meetings from which a large part of this work is issued.

STAFF instrumentation and parts of data analysis have got support from ESA and CNES.

Edited by: H. Laakso

References

Balogh, A., Dunlop, M. W., Cowley, S. W. H., Southwood, D. J., Thomlinson, J. G., Glassmeier, K. H., Musmann, G., Lühr, H., Buchert, S., Acuña, M. H., Fairfield, D. H., Slavin, J. A., Riedler, W., Schwingenschuh, K., and Kivelson, M. G.: The cluster magnetic field investigation, Space Sci. Rev., 79, 65–91, 1997.

Balogh, A., Carr, C. M., Acuña, M. H., Dunlop, M. W., Beek, T. J., Brown, P., Fornacon, K.-H., Georgescu, E., Glassmeier, K.-H., Harris, J., Musmann, G., Oddy, T., and Schwingenschuh, K.: The Cluster Magnetic Field Investigation: overview of in-flight performance and initial results, Ann. Geophys., 19, 1207–1217, doi:10.5194/angeo-19-1207-2001, 2001.

Cornilleau-Wehrlin, N., Chauveau, P., Louis, S., Meyer, A., Nappa, J. M., Perraut, S., Rezeau, L., Robert, P., Roux, A., de Villedary, C., de Conchy, Y., Friel, L., Harvey, C. C., Hubert, D., Lacombe, C., Manning, R., Wouters, F., Lefeuvre, F., Parrot, M., Pinçon, J. L., Poirier, B., Kofman, W., and Louarn, P.: The CLUSTER Spatio-Temporal Analysis of Field Fluctuations (STAFF) Experiment, Space Sci. Rev., 79, 107–136, 1997.

Cornilleau-Wehrlin, N., Chanteur, G., Perraut, S., Rezeau, L., Robert, P., Roux, A., de Villedary, C., Canu, P., Maksimovic, M., de Conchy, Y., Hubert, D., Lacombe, C., Lefeuvre, F., Parrot, M., Pinçon, J. L., Décréau, P. M. E., Harvey, C. C., Louarn, Ph., Santolik, O., Alleyne, H. St. C., Roth, M., Chust, T., Le Contel, O., and STAFF team: First results obtained by the Cluster STAFF experiment, Ann. Geophys., 21, 437–456, doi:10.5194/angeo-21-437-2003, 2003.

Cross Calibration Workshops: The purpose of the cross-calibration workshops is to provide a forum where observations from different Cluster instruments are compared in detail, presentations of organized workshops between September 2005 and March 2013, available at: ftp://ftp.lpp.polytechnique.fr/robert/keep/Biblio_et_CV/ESA_CrossCal_Meetings, 1st CrossCal Workshop, ESTEC, 2–3 February2006/CrossCal01-STAFF_Robert.ppt, 2nd CrossCal Workshop, ESTEC, 16 May 2006/CrossCal02-STAFF_Robert.ppt, 8th CrossCal Workshop, Kinsale, Co. Cork, Ireland, 28–30 October 2008/CrossCal08-STAFF_Robert.ppt 9th CrossCal Workshop, Jesus College, Cambridge, 23–27 March 2009/CrossCal09-STAFF_Burlaud.ppt, 10th CrossCal

Workshop, L'Observatoire de Paris, Paris, 3–4 November 2009/CrossCal10-STAFFActions_Burlaud.ppt/CrossCal10-STAFF-SA_De_Conchy.ppt/CrossCal10-STAFF-SC_Robert.ppt, 11th CrossCal Workshop, Hotel die Tanne, Goslar, 7–9 April 2010/CrossCal11-STAFF_Burlaud.ppt,/CrossCal11-STAFF_Robert.ppt, 12th CrossCal Workshop, CESR, Toulouse, 26–28 October 2010/CrossCal12-STAFF_Cornilleau.ppt, /CrossCal11-STAFF_Robert.ppt, 13th CrossCal Workshop, IRF, Uppsala, Sweden, 13–15 April 2011/CrossCal13-STAFF_Robert.ppt, 14th CrossCal Workshop, York, UK, 5–7 October 2011/CrossCal14-STAFF_Robert.ppt/CrossCal14-STAFF_Santolik.ppt, 15th CrossCal Workshop, University College of London, London, 17–19 April 2012/CrossCal15-STAFF_Robert_Piberne.pptx, 16th CrossCal Workshop, Toulouse, France, 6–9 November 2012/CrossCal16-STAFF_Robert_Piberne.ppt, 17th CrossCal Workshop, ESOC, Darmstadt, Germany, 25–27 March 2013/CrossCal17-STAFF_Piberne.ppt, 18th CrossCal Workshop, Cosener's House, Abingdon, UK, 23–25 October 2013/CrossCal18-STAFF_Piberne.pptx.

Gurnett, D. A., Huff, R. L., and Kirchner, D. L.: The wide-band plasma wave investigation, Space Sci. Rev., 79, 195–208, 1997.

Gustafsson, G., Boström, R., Holback, B., Holmgren, G., Lundgren, A., Stasiewicz, K., Ahlén, L., Mozer, F. S., Pankow, D., Harvey, P., Berg, P., Ulrich, R., Pedersen, A., Schmidt, R., Butler, A., Fransen, A. W. C., Klinge, D., Thomsen, M., Fälthammar, C. G., Lindqvist, P.-A., Christenson, S., Holtet, J., Lybekk, B., Sten, T. A., Tanskanen, P., Lappalainen, K., and Wygant, J.: The electric field and wave experiment for the Cluster mission, Space Sci. Rev., 79, 137–156, 1997.

Harvey, C. C., Belkacemi, M., Manning, R., Wouters, F., de Conchy, Y.: STAFF Spectrum Analyzer, Conversion of the Science Data to Physical Units, Technical Report OBSPM-0001, issue 7, rev. 5, Observatoire de Paris Meudon, Paris, 2004.

Jones, D.: Introduction to the S-300 wave experiment onboard GEOS, Space Sci. Rev., 22, 327–332, 1977.

Knott, K.: Payload of the GEOS scientific geostationary satellite, ESA Sci. Tech. Rev., 1, 173–196, 1975.

Le Contel, O., Roux, A., Robert, P., Coillot, C., Bouabdellah, A., de la Porte, B., Alison, D., Ruocco, S., Angelopoulos, V., Bromund, K., Chaston, C. C., Cully, C., Auster, H. U., Glassmeier, K. H., Baumjohann, W., Carlson, C. W., McFadden, J. P., and Larson, D.: First results of the THEMIS Search Coil Magnetometers, Space Sci. Rev., 141, 509–534, doi:10.1007/s11214-008-9371-y, 2008.

Nykyri, K., Grison, B., Cargill, P. J., Lavraud, B., Lucek, E., Dandouras, I., Balogh, A., Cornilleau-Wehrlin, N., and Rème, H.: Origin of the turbulent spectra in the high-altitude cusp: Cluster spacecraft observations, Ann. Geophys., 24, 1057–1075, doi:10.5194/angeo-24-1057-2006, 2006.

Paschmann, G., Melzner, F., Frenzel, R., Vaith, H., Parigger, P., Pagel, U., Bauer, O. H., Haerendel, G., Baumjohann, W., Sckopke, N., Torbert, R. B., Briggs, B., Chan, J., Lynch, K., Morey, K., Quinn, J. M., Simpson, D., Young, C., McIllwain, C. E., Fillius, W., Kerr, S. S., Mahieu, R., and Whipple, E. C.: The electron drift instrument for cluster, Space Sci. Rev., 79, 233–269, 1997.

Pedersen, A., Cornilleau-Wehrlin, N., de la Porte, B., Roux, A., Bouabdellah, A., Décréau, P. M. E., Lefeuvre, F., Sené, F. X., Gurnett, D., Huff, R. R., Gustafsson, G., Holmgren, G., Woolliscroft, L. J. C., Thompson, J. A., and Davies, P. H. N.: The Wave Experiment Consortium (WEC), Space Sci. Rev., 79, 93–106, 1997.

Perry, C., Eriksson, T., Escoubet, P., Esson, S., Laakso, H., McCaffrey, S., Sanderson, T., and Bowen, H.: The ESA Cluster Active Archive, in: Proceedings of the Cluster and Double Star Symposium 5th Anniversary of Cluster in Space, ESTEC, Noordwijk, 2005.

Pinçon, J. L. and Lefeuvre, F.: Local characterization of homogeneous turbulence in a space plasma from simultaneous measurements of field components at several points in space, J. Geophys. Res., 96, 1789–1802, 1991.

Robert, P.: Cluster software tools – Part I: Coordinate transformations library, Document de Travail, DT/CRPE/1231, July, available at: ftp://ftp.lpp.polytechnique.fr/robert/keep/Biblio_et_CV/Working_documents/1993_Robert_DTCRPE1231_ROCOTLIB.pdf (last access: 3 December 2013), 1993.

Robert, P.: Intensité et polarisation des ondes UBF détectées à bord de GEOS-1, Méthode d'analyse numérique du signal et production en routine de sommaires expérimentateurs, Problèmes rencontres et solutions pratiques, Note Technique CRPE/ETE/71, May, available at: ftp://ftp.lpp.polytechnique.fr/robert/keep/Biblio_et_CV/Working_documents/1979_Robert_NTCRPE71_Intensite_et_Polarisation_des_ondes_UBF.pdf (last access: 4 December 2013), 1979a.

Robert, P.: Measurement by the S-300 experiment of two components of the DC magnetic field, in the X–Y plane of the satellites GEOS-1 and GEOS-2, Comparison with the results of the S-331 magnetometer, Document de Travail CRPE/ETE, July, available at: ftp://ftp.lpp.polytechnique.fr/robert/keep/Biblio_et_CV/Working_documents/1979_Robert_DTCRPE_Measurement_by_S300_of_2comp_DC_field.pdf (last access: 3 December 2013), 1979b.

Robert, P.: ROCOTLIB: a rather complete suite of coordinate transformation routines, Issue 1, Rev. 8, November 2003, available at: http://cdpp.eu/index.php/Scientific-libraries/rocotlib.html (last access: 3 December 2013), 2003.

Robert, P.: ROCOTLIB: a coordinate Transformation Library for Solar-Terrestrial studies, & french version, ROCOTLIB: une bibliothèque de changement de coordonnées pour les études Soleil-Terre Le Bulletin du Centre de Données de la Physique des Plasmas Num. 8t, CNES, 2004.

Roberg, P.: Cross Calibration Workshops, document "CrossCal10-STAFFSC_Robert.ppt", L'Observatoire de Paris, 3–4 November 2009, Paris, 2009.

Robert, P., Kodera, K., Perraut, S., Gendrin, R., and de Villedary, C.: Polarization characteristics of ULF waves detected onboard GEOS-1, Problems encountered and practical solutions, XIXth URSI General Assembly, Helsinki, Finland, 31 July–8 August, available at: ftp://ftp.lpp.polytechnique.fr/robert/keep/Biblio_et_CV/Publications/1978_Robert_URSI_Helsinki.pdf (last access: 3 December 2013), 1978.

Robert, P., Kodera, K., Perraut, S., Gendrin, R., and de Villedary, C.: Amplitude et polarisation des ondes UBF détectées à bord du satellite GEOS-1, Méthodes d'analyse, problèmes rencontrés et solutions pratiques, Ann. Télécommun., 34, 179–186, 1979.

Robert, P., Burlaud, C., Maksimovic, M., Cornilleau-Wehrlin, N., and Piberne, R.: Calibration Report of the STAFF Measurements in the Cluster Active Archive (CAA), Doc. No. CAA-STA-CR-002, issue 3.0, 16 May 2012, available at: http://caa.estec.esa.int/documents/CR/CAA_EST_CR_STA_v30.pdf (last access: 3 December 2013), 2012.

Roux, A., Le Contel, O., Robert, P., Coillot, C., Bouabdellah, A., de la Porte, B., Alison, D., Ruocco, S., and Vassal, M. C.: The search coil magnetometer for THEMIS, Space Sci. Rev., 141, 265–275, doi:10.1007/s11214-008-9455-8, 2008.

Santolik, O.: Propagation Analysis of STAFF-SA Data with Coherency Tests (a User's Guide to PRASSADCO), LPCE/NTS/073.D, Lab. Phys. Chimie Environ./CNRS, Orleans, France, available at: http://aurora2.troja.mff.cuni.cz/~santolik/PRASSADCO/tc2/doc/guide.pdf (last access: 3 December 2013), 2003.

Santolik, O., Parrot, M., and Lefeuvre, F.: Singular value decomposition methods for wave propagation analysis, Radio Sci., 38, 1010, doi:10.1029/2000RS002523, 2003.

Woolliscroft, L. J. C., Alleyne, H. S. C., Dunford, C. M., Sumner, A., Thompson, J. A., Walker, S. N., Yearby, K. H., Buckley, A., Chapman, S., Gough, M. P., and the DWP Co-investigators: The digital wave processing experiment on Cluster, Space Sci. Rev., 79, 209–231, 1997.

15

Calibration of QM-MOURA three-axis magnetometer and gradiometer

M. Díaz-Michelena[1], **R. Sanz**[1,*], **M. F. Cerdán**[1,2], **and A. B. Fernández**[1]

[1]National Institute of Aerospace Technology, Ctra. de Ajalvir, km 4, 28850 Torrejón de Ardoz, Madrid, Spain
[2]Departamento de Física de la Tierra, Astronomía y Astrofísica I, Universidad Complutense de Madrid, Pza. de las Ciencias, s/n, 28040 Madrid, Spain
[*]now at: CNR-IMM MATIS at Dipartimento di Fisica e Astronomia, Università di Catania, Via S. Sofia 64, 95123 Catania, Italy

Correspondence to: M. Díaz-Michelena (diazma@inta.es)

Abstract. MOURA instrument is a three-axis magnetometer and gradiometer designed and developed for Mars MetNet Precursor mission.

The initial scientific goal of the instrument is to measure the local magnetic field in the surroundings of the lander i.e. to characterize the magnetic environment generated by the remanent magnetization of the crust and the superimposed daily variations of the field produced either by the solar wind incidence or by the thermomagnetic variations. Therefore, the qualification model (QM) will be tested in representative scenarios like magnetic surveys on terrestrial analogues of Mars and monitoring solar events, with the aim to achieve some experience prior to the arrival to Mars.

In this work, we present a practical first approach for calibration of the instrument in the laboratory; a finer correction after the comparison of MOURA data with those of a reference magnetometer located in San Pablo de los Montes (SPT) INTERMAGNET Observatory; and a comparative recording of a geomagnetic storm as a demonstration of the compliance of the instrument capabilities with the scientific objectives.

1 Introduction

MOURA is a three-axis magnetometer and gradiometer instrument, to be included in the Spanish payload for the Finnish-Russian-Spanish Mars MetNet Precursor Mission (MMPM, 2014), rescheduled for 2018. The mission concept of MMPM is to deploy the first lander of a net of meteorological stations based on the penetrator concept over the surface of Mars. One of the targeted measurements of MOURA instrument will be to measure the change in remanent magnetization of Mars lithospheric minerals. We will measure the thermoremanent behaviour of surface rocks and search for temperature transitions for the compositional analysis of the crust (Sanz et al., 2011; Fernández et al., 2013) aiming to explain its local magnetic anomalies, with intensities several orders of magnitude higher than the Earth ones. The second scientific objective is to measure the variations of the field related to the solar wind effects. The intensities corresponding to these magnetic fields are summarized in Table 1 for the Earth and Mars.

Due to the limited development time (2 years), mass and energy constraints of the mission (150 g and > 0.5 W for the three Spanish payloads), and the Martian environment envelope (temperatures ranging from -90 to $20\,^{\circ}$C in operation and from -120 to $125\,^{\circ}$C storage, and a total irradiation dose up to 15 krad Si^{-1}), MOURA development has singular characteristics, which have an impact on its performances. MOURA consists of a double design: one compact sensor with macroscopic front-end electronics including many COTS and PEMS components up-screened for the mission, and a second with a mixed applied specific integrated circuit (ASIC) based front-end (Sordo-Ibáñez et al., 2014). This work focuses on the former one. MOURA instrument is located on top of the inflatable structure of the lander (Fig. 1) to provide a certain distance from the penetrator, avoiding

Table 1. Main field intensities of the magnetic sources on Earth and Mars.

Source	Feature	Earth	Mars (estimated)	Main temporal and spatial characteristics
Interior	Palaeodipole	10^{22}–10^{23} Am^2	$< 2 \times 10^{18}$ Am^2	Inversions
Crust	Strong, stable, remanent magnetization	0.1–1 Am^{-1}	12–20 Am^{-1}	Tens of km thick magnetized layer
Surface	Surface rocks	100 nT	10^5 nT	
	Local strong magnetic fields	10–10^4 nT	2200 nT	Earth: attributed to some geological structures (volcanoes, magnetite outcrops …) Mars: mainly in the highlands in the Southern Hemisphere
Ionosphere and magnetosphere	Daily variation	20 nT (200 nT in the equator)	0.5–5 nT in the Equator	
	Sun storms/ substorms	100–1000 nT	Few nT	Earth aurora: Auroral oval surface, 80–500 km height Mars aurora: 30 km long – 8 km height
	Elongated shape density bulges near the surface	–	Few nT	Local magnetospheres
	Pulses	0.1–100 nT	–	
	Flux ropes	20 nT	< 5 nT	
	Interplanetary – planetary transition	30 nT	< 5 nT	Earth: solar wind– magnetosphere, Mars: solar wind–ionosphere
	Several layers	≈ 2500 nT (IGMR)	Few nT	From 80 to 200 km height
	Plasma current	10–20 nT	< 5 nT	

any extra mass for a deployment system. Therefore, apart from the two magnetometers (for close gradiometry) and the compensating temperature sensors, it has a tilt angle detector to determine the relative position with respect to the horizontal.

Because of the above-mentioned mission constraints, both the sensing and the electronics suppose a trade-off between performances under the expected environment, power and mass budget. In addition, the magnetic signal of the electronic components was also carefully measured, in order to improve the magnetic cleanliness of the compact instrument. As a result, the parts list for the electronics was restricted according to their magnetic signal. Under the mentioned demanding criteria, the selected sensing element was the triaxial HMC1043 magnetic sensor by Honeywell (Honeywell Magnetic Sensors, 2014). The HMC1043 sensors belong to a family of sensors based on anisotropic magnetoresistance (AMR) effect (Freitas et al., 2007) which has been exhaus-

tively up-screened (temperature, thermal shock, life cycle, and radiation) by INTA (Sanz et al., 2012) and successfully used in previous space missions (D. Michelena et al., 2010; D. Michelena, 2009; DTUsat, 2014). Although the selection of HMC1043 for the two magnetic sensing elements presents advantages in terms of weight and power consumption, the AMR technology based sensors present several drawbacks like their resolution (lower compared with other sensing technologies, like the fluxgates), or an important dependence of their response (gain and offset) with temperature (Ripka et al., 2013; Díaz-Michelena et al., 2014). This point is particularly challenging because MOURA is expected to be allocated outside the lander (Fig. 1) and thus be exposed to Mars surface thermal fluctuations. As one of the main objectives of MOURA is to measure the thermal variation of Martian magnetic minerals magnetization, this thermal characterization of the instrument becomes critical. Due to the necessities of the project, and after the successful qualification (mechan-

Table 2. MOURA characteristics summary.

Characteristics	Conditions	Min	Typical	Max	Unit
Sourcing voltage (V_{dd})		4.5	5	5.5	V
Set/reset voltage ($V_{s/r}$)		10	12	15	V
Sourcing current (I_{dd})	@ 5 V, RT, stand-by	81		86	mA
Set/reset current ($I_{s/r}$)	@ 12 V, RT		< 2		mA
Operating temperature		−100		70	°C
Storage temperature		−130		125	°C
Field range	Nominal		±65		μT
Extended range	Auto-offset compensation		±130		μT
Linearity error	Nominal range		< 0.5		% FS
Hysteresis error	Nominal range		< 0.1		% FS
Repeatability error	Nominal range		< 0.1		% FS
Sensitivity			0.45		Cts nT^{-1}
Resolution			2.2		nT
PSD	@ 0.5 Hz samples averaged: 1		0.85		nT $\sqrt{Hz}^{-1}$
	@ 0.5 Hz samples averaged: 10		0.42		nT $\sqrt{Hz}^{-1}$
	@ 0.5 Hz samples averaged: 100		0.28		nT $\sqrt{Hz}^{-1}$
Mass			72		g
Box dimensions			$150 \times 30x \times 15$		mm

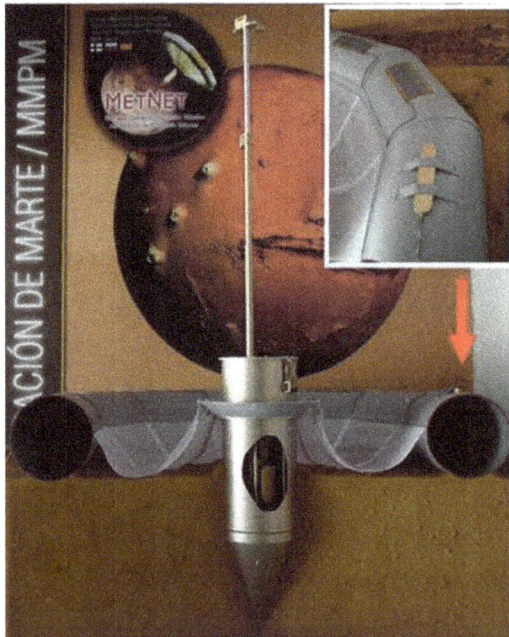

Figure 1. MMPM mock-up. The inset shows the position of MOURA sewn to the inflatable structure of the lander (red arrow).

ical shock, vibration, thermal and vacuum), the qualification model (QM) of the magnetometer was slightly modified, and therefore should be strictly denominated engineering qualification model (EQM). This EQM is still fully representative (electric and functional) of the flight model (FM) but not me-chanically representative. This fact will have implications in the calibration with temperature of the instrument.

In the present work we focus on the first calibrations performed to MOURA EQM (MOURA from now on) as is (Fig. 2), which involves: magnetic, tilt angle detector, including gravity measurements characterization, and thermal behaviours. The purpose of this calibration is to demonstrate the capability of MOURA instrument to fulfil the above-mentioned scientific objectives on Mars by means of measurements on Earth. For this reason, the field range has been increased to ±65 000 nT (extendable to ±130 000 nT, see Table 2).

On Earth the contrast in magnetic field intensity in on-ground prospections is generally due to the magnetic carriers of the surface rocks (up to tens of metres). Despite the limited data of ground surveys on Earth, a reasonable goal in terms of detectability for MOURA instrument is to be able to detect a variation of 1 % vol. concentration of pyrrhotite by the corresponding magnetic contrast (20 nT) apart from the daily variations corresponding to either the temperature swings and the solar wind incidence. Finally, to demonstrate experimentally the capability of the sensor, and for a fine re-calibration, we show a comparison of the corrected data registered by MOURA versus the official magnetic daily variation data provided by San Pablo de los Montes Geomagnetic Observatory (IAGA code: SPT) (Geomagnetic observatories, 2014).

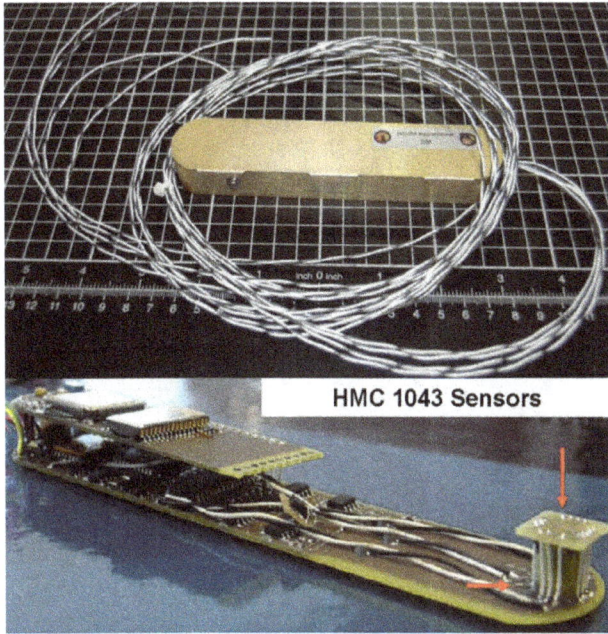

Figure 2. QM-MOURA box and detailed view of position of both HMC1043 sensors (red arrows).

2 Devices and equipment

In this section a brief description of the tested instrument and devices employed are presented.

2.1 Brief description of MOURA and tested parameters

MOURA is a vector magnetometer and gradiometer to measure the magnetic environment on the surface of Mars. It is based on AMR technology. The main characteristics of the instrument are summarized in Table 2.

The front-end is based on a flipping mode of the AMR, the SET/RESET flip recommended by the manufacturer (Honeywell Magnetic Sensors, 2014) in order to avoid cross-axis effects, increase repeatability by decreasing the thermal disorder of magnetic domains, and reduce hysteresis. Therefore, the measurement will consist in the subtraction of the two mirror states (after the SET and RESET pulses): Set-Reset mode (set pulse – V_{Set} acquisition – reset pulse – V_{Reset} acquisition – calculation of $V_{S/R} = (V_{Set} - V_{Reset})/2 \ldots$) with open loop conditioning of the AMR Wheatstone bridges, though either operations in set/reset (set/reset pulse – $V_{Set/Reset}$ – set/reset pulse – $V_{Set/Reset} \ldots$) or just one pulse based modes (set/reset pulse – $V_{Set/Reset} - V_{Set/Reset} \ldots$) are foreseen. In order to guarantee the correct flipping of the domains in the AMR, and therefore its repeatability, the pulses of SET and RESET are generated by the discharge of several capacitors charged to 12 V. The shape of the pulses is therefore that of the capacitors' discharge.

The noise is expected to be of the order of 1 nT. The offset coils are used for the calibration of the sensor gain and to double the dynamic range (to $\pm 130\,000$ nT max.) when the response of any axis is saturated. Due to mass and power constraints, the instrument is designed to operate in open loop (no feedback) and the thermal compensation is performed by calibration in contrast to other developments (Brown et al., 2012; Ripka et al., 2013; Díaz Michelena et al., 2015). The consequent cross-axis effects will be assumed.

The instrument has several temperature sensors based on platinum resistors (PT-1000 previously calibrated) for the compensation of the thermal drifts of the different elements. Of particular importance are the temperature sensors located on top of the two magnetometers (TMP1 and TMP2), which will be used for the compensation of the magnetic signals with temperature.

The instrument also comprises a tilt angle detector (a three-axis accelerometer ADXL327 by Analog Devices) for the correction of the inclination and northing. The accelerometer is selected amongst other devices because of its lower magnetic signature (magnetic moment lower than $1\,\mu\text{Am}^2$ when exposed to moderate fields (100 pT) contributing less than 0.5 nT in the position of the sensor).

The instrument has a physical envelope of $150 \times 30 \times 15$ mm^3 and a weight of 72 g.

For the present characterization we focus on the signals described in Table 3 (denoted as channels).

2.2 Equipment

All the calibration has been performed in the Space Magnetism Laboratory at INTA headquarters with the exception of the magnetic daily variations, which were registered in San Pablo de los Montes Observatory, Toledo, Spain.

Controlled magnetic fields are generated by a set of three pairs of high mechanical precision Helmholtz coils (HC), model Ferronato BH300-A. Each pair of coils (denoted as HC$_X$, HC$_Y$ and HC$_Z$) is calibrated by means of Bartington FG100 fluxgate (certified by Bartington, against the calibration references, in accordance with ISO10012: Mag-01 magnetometer, Mag Probe B, solenoid with current source and DC scaling solenoid, Table 4). The coil constants are: HC$_X = 524.38$ p TA^{-1}, HC$_Y = 542.15$ p TA^{-1} and HC$_Z = 525.60$ p TA^{-1}. The electric currents to generate the magnetic fields are supplied by a Keithley 6220 precision current source.

Non-orthogonalities in the HC are lower than $4''$; according to the documentation provided by the manufacturer (Honeywell Magnetic Sensors, 2014), orthogonality between x–y axes is better than $3.6''$ and it is checked experimentally that between the z axis and the XY plane the non-orthogonality is below $0.5°$.

For monitoring magnetic field pulses a fluxgate magnetometer FG-500 is used (Table 4).

Table 3. Channels involved in the study.

Sensor	Channel	Physical magnitude (units)
Sourcing voltage (V_{dd})	VREG	Voltage sourcing the magnetoresistive bridge (V)
Magnetic sensors axes	X1, X2, Y1, Y2, Z1, Z2	Magnetic field (nT)
Temperature sensors	TMP1, TMP2	Temperature (°C)

Table 4. Calibration of fluxgate magnetometers.

Parameter	Specified	FG-100			FG-500		
		X	Y	Z	X	Y	Z
Orthogonality error	(°) $\pm 0.1°$	$< 0.1'$	$< 0.1'$	$< 0.1'$	$< 0.1'$	$< 0.1'$	$< 0.1'$
Offset error in zero field	(nT) F. S.	0	−1.5	1	1.5	−10	−0.5
Scaling error	(%) @ 35 Hz, $\pm 0.5\%$	0.08	0.03	−0.08	0.07	0.15	0.20
Frequency response	(%) $\pm 5\%$	−0.06	0.06	0.08	1.00	1.17	1.05
Noise	(pT$_{RMS}$ $\sqrt{Hz^{-1}}$) @ 1 Hz	19.3	12.3	15.6	8.9	8.5	9.7

A thermal chamber (Binder MK53) is employed to set and control the temperature during the characterization tests. This chamber makes it possible to apply temperatures from −70 to 180 °C, and to circulate dry N_2 gas inside the chamber in order to control the humidity of the atmosphere. The N_2 flow is kept between 1 and 5 L min^{-1}. The measurement of the atmospheric humidity inside the chamber is performed by a Vaisala HMI31 humidity and temperature indicator, and always kept under 18 %. This is done to prevent water condensation in the low temperature range. No influence of humidity on our sensors' performance is observed.

For the characterization tests, the temperature register is performed by the included thermal chamber temperature sensors, those included in MOURA and two additional temperature sensors. These additional temperature sensors are two PT-1000 resistances calibrated by means of a SIKA TP 38165E, and connected to a data acquisition system (Agilent 3497A Automatic DAS, computer commanded).

For the calibration of the inclinometer a sine bar and gauge blocks with five values between 1.5 and 141.0 mm are used to generate the desired tilt angles around x and y axes (α and β angles). The rotations are obtained with one of the cylindrical plugs leaning on the gauge blocks and the other on the surface plate. The accuracy of the method is better than 10′.

The next section describes the step-by-step calibration. Section 3.1 deals with the calibration at a reference temperature. Section 3.2 deals with the characterization with temperature of the magnetic parameters. Figure 3 shows the details of the main setups for these measurements.

3 Methods and results

This section describes the procedure followed for MOURA calibration.

The results shown here correspond to the measurements taken with the flipping operation. Therefore, the response of the i axis (of the six measuring axes of the magnetometer) $B_{MOURA,i}$ corresponds to

$$\frac{(SET_i - RESET_i)}{2 \cdot VREG} = \cos\theta_i \cdot B_{REAL} \cdot GAIN_i - OFFSET_i, \quad (1)$$

where $i = X1, X2, Y1, Y2, Z1, Z2$, with SET_i and $RESET_i$ the registered magnetic signals from a given axis "i" after the application of a SET and RESET pulse, respectively. VREG is the registered voltage sourcing the magnetoresistive bridge. θ_i is the angle between the direction of the field and the measurement direction of "i" sensor. B_{REAL} is the external magnetic field modulus. $GAIN_i$ is the effective gain of the "i" sensor. $OFFSET_i$ is the offset of the "i" sensor.

The sensor always acquires both Set and Reset data and thus it is always possible to use the same calibration data for the data without flipping.

MOURA temperature sensors had been previously calibrated by means of a 38165E system by SIKA TP giving rise to a polynomial fit. However, this calibration is not used in the present work, but the calibration is performed with the direct readings of the temperature sensors no matter what the real temperature is.

To avoid influence of the voltage source fluctuations, the output values are always normalized with the bridge voltage, which is monitored (Eq. 1).

The parameters affected by temperature are:

1. gain;

2. offset.

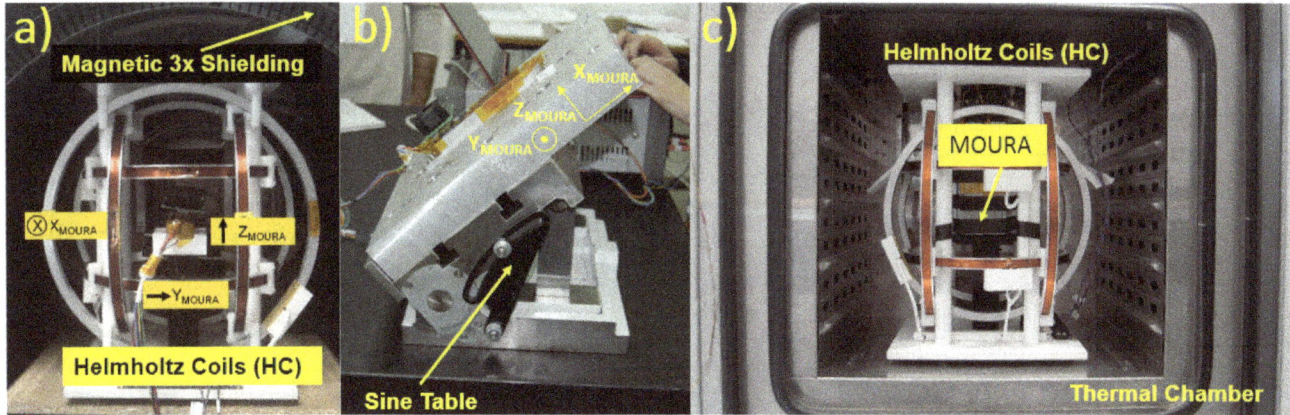

Figure 3. Different setups: (a) offset, gain and Euler angles characterization setup, (b) inclinometer characterization setup, (c) setup for gain characterization with temperature.

Therefore, MOURA response (for one of its axis) to a real magnetic field can be expressed as

$$B_{\text{MOURA}}(T)_i = \cos\theta_i \cdot B_{\text{REAL}} \cdot \text{GAIN}_i \cdot (1 - \Delta\text{GAIN}_i \cdot (T - T_{\text{G}}))$$
$$-\text{OFFSET}_i \cdot (1 - \Delta\text{OFFSET}_i) \cdot (T - T_{\text{OFFSET}}), \quad (2)$$

where ΔGAIN_i is the normalized GAIN temperature variation rate, ΔOFFSET_i is the normalized OFFSET temperature variation rate, T_{G} is the reference temperature for the gain, i.e. normalization temperature for GAIN_i, and T_{OFFSET} is the reference temperature for the offset, i.e. normalization temperature for OFFSET_i. Note that these temperatures do not need to be the same but they are subject to room temperature variations (seasonal).

From now on we will use M_i to refer to the "i" sensor measurement and leave B_{MOURA} for the total magnetic vector.

3.1 Room temperature characterization

This section describes the room temperature characterization of the offsets, gains and output field generated by the offset coils.

3.1.1 Characterization of MOURA magnetic offsets at room temperature

The offset characterization of the two magnetic sensors components was performed inside a three-layer magnetic shielded chamber of $2.5\,\text{m}^3$, previously characterized (at different points) by means of a three-axis fluxgate with minimum detectable fields of the order of $10\,\text{pT}$ (Fig. 3a).

Because the magnetic field inside the chamber ($\sim 0.1\,\text{nT}$) is lower than the minimum detectable field of the sensor ($\sim 1\,\text{nT}$), passive compensation of the Earth's magnetic field was considered to be sufficient.

After warming up and thermal stabilization at room temperature ($\text{TMP1} = 18.13 \pm 0.03\,^{\circ}\text{C}$ and $\text{TMP2} = 19.21$

$\pm 0.03\,^{\circ}\text{C}$, with variations $< 0.1\,^{\circ}\text{C}\,\text{min}^{-1}$), MOURA data were acquired constantly during the whole process.

Magnetic sensors outputs were analysed. Mean values and standard deviations are shown in Table 5.

3.1.2 Non-orthogonalities and Euler angle determination – gain characterization

In this section we describe the procedure and results for the determination of the geometrical directions of MOURA sensor axes.

The registered signal of MOURA "i" sensor, M_i, at room temperature can be written in vector expression as

$$M_i = (B_{\text{REAL}} \cdot u_i) \cdot \text{GAIN}_i - \text{OFFSET}_i,$$
$$B_{\text{real}} \cdot u_i = |B| \cdot |u_i| \cdot \cos\theta_i, \quad (3)$$
$$M_i = |B| \cdot \cos\theta_i \cdot \text{GAIN}_i - \text{OFFSET}_i,$$

with u_i being a unit vector in the measuring direction of "i" sensor.

The external field, B_{REAL}, is generated in the zero field chamber by the previously described set of HC. The sensor box is aligned with the axes of the HC. Hereinafter, MOURA basis will refer to the measuring directions of each three-axis sensor, and not to the box. Notice that the magnetic field generated by the HC in the HC basis will be denoted by B and the measurements of MOURA in MOURA basis will be denoted by M (Fig. 4).

If θ_i and GAIN_i are unknown it is not possible to distinguish between a misalignment and a scale factor.

To simplify the problem and to calculate the most accurate values of the GAIN and misalignment of sensors with the external system of reference (HC) some approximations can be applied:

1. The three axes of the magnetometers and the HC are taken as orthogonal due to the construction properties, described in Sect. 2.2.

Table 5. Gains and offset values as well as their temperature drifts.

SENSOR 1 axis	GAIN @ $T_G = \text{TMP1} = 25.9 \pm 0.2\,°\text{C}$	ΔGAIN $(°\text{C}^{-1})$ (referred to T_G)	OFFSET (nT) @ TMP1 $= 18.13 \pm 0.03\,°\text{C}$	ΔOFFSET $(°\text{C}^{-1})$ (referred to TMP1)
X1	0.910 ± 0.003	$(-0.00370 \pm 5 \times 10^{-5})$	764 ± 5	$(-0.0037 \pm 5 \times 10^{-5})$
Y1	0.902 ± 0.002	$(-0.00382 \pm 7 \times 10^{-5})$	-1130 ± 16	$(-0.00450 \pm 7 \times 10^{-5})$
Z1	0.832 ± 0.003	$(-0.00384 \pm 4 \times 10^{-5})$	1582 ± 8	$(-0.00352 \pm 4 \times 10^{-5})$
SENSOR 2 axis	GAIN @ $T_G = \text{TMP2} = 25.6 \pm 0.2\,°\text{C}$	ΔGAIN $(°\text{C}^{-1})$ (referred to T_G)	OFFSET (nT) @ TMP2 $= 19.21 \pm 0.03\,°\text{C}$	ΔOFFSET $(°\text{C}^{-1})$ (referred to TMP2)
X2	0.815 ± 0.003	$(-0.00591 \pm 5 \times 10^{-5})$	1107 ± 3	$(-0.00200 \pm 5 \times 10^{-5})$
Y2	0.807 ± 0.001	$(-0.00621 \pm 9 \times 10^{-5})$	-538 ± 5	$(-0.00794 \pm 9 \times 10^{-5})$
Z2	0.783 ± 0.002	$(-0.00616 \pm 6 \times 10^{-5})$	1427 ± 17	$(-0.0379 \pm 4 \times 10^{-4})$

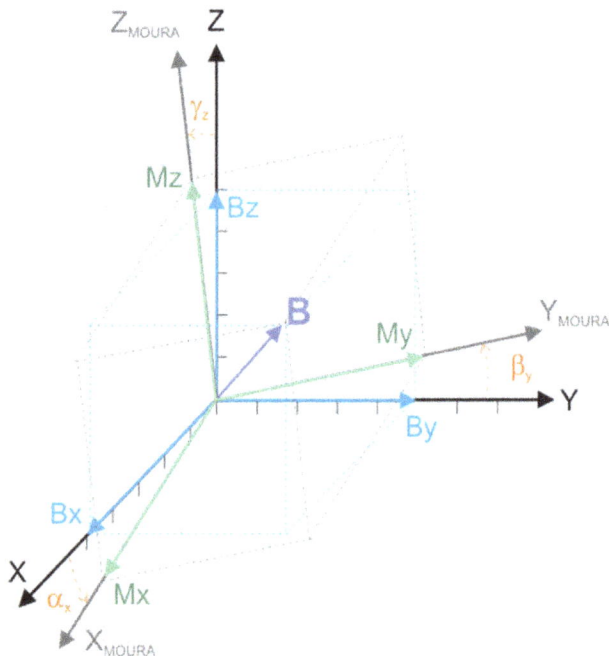

Figure 4. Approximation of system of reference of MOURA vs. external system of reference with the approximations taken into account.

2. The change of basis between HC and MOURA reference systems can be taken as a composition of small rotations around axes of the external reference system. Under this approximation it is not required the use of an unknown rotation matrix.

3. Gains of the different axes of the same sensor have similar values.

For each three-axis magnetometer (1 and 2), the angle between $\boldsymbol{B}_X$ and $\boldsymbol{M}_X$ is denoted as α_x, between $\boldsymbol{B}_Y$ and $\boldsymbol{M}_Y$ as β_y and between $\boldsymbol{B}_Z$ and $\boldsymbol{M}_Z$ as γ_z (Fig. 4). Notice that

these angles have two contributions: the already known non-orthogonality of the coils and the Euler angles referred to the HC system. These angles are to be determined for the two magnetometers composing the gradiometer.

In contrast with other highly precise calibration methods (Renaudin et al., 2010; Petrucha and Kaspar, 2009; Petrucha et al., 2009; Cai et al., 2010), in which a well calibrated and aligned three-axis goniometric platform is used, in the present case we have employed a method based on the calibration by means of the application of time varying circular magnetic fields (harmonic with a $\pi/2$ phase shift between components) around MOURA, in the planes of the HC system. As a result of these sine–cosine circular fields, an elliptical response, due to the expected misalignment and different gain of each sensor axis, will be detected by MOURA sensors.

MOURA was fixed in the centre and aligned with the set of HC (Fig. 3a), taking as a reference the geometrical shape of its box: for this measurement, a high-precision container was made ad hoc in order to fit rigidly the magnetometer, and a set of laser theodolites was used to align HC and the sides of the container. Doing so, we could set MOURA and the set of HC in co-axial position, with a calculated misalignment below $0.1'$.

The whole set was placed into the magnetic shielded chamber.

The calibration tests are performed in thermal equilibrium (thermal variations $< 0.2\,°\text{C}\,\text{min}^{-1}$). MOURA non-orthogonalities between i' and j' axes (MOURA reference system) are determined by comparison of orthogonalities between MOURA and HC system (i and j axes in HC reference system): $\Omega\text{MOURA}i'j'$, and $\Omega\text{HC}ij$. The comparison is performed by successive application of rotating magnetic fields in the different planes of the HC reference system (XY, ZX, YZ – Table 6) and the corresponding linear fit with MOURA synchronized readings (Fig. 5):

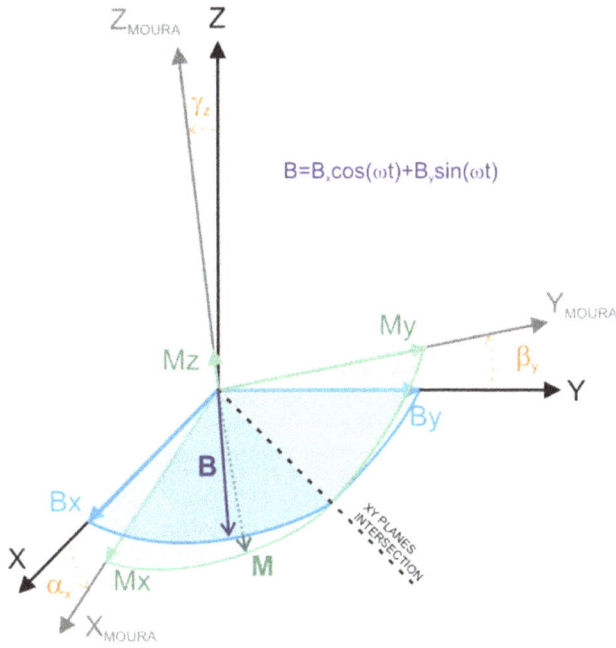

Figure 5. Sketch of the different shapes of magnetic field measured in external and MOURA reference systems.

Table 6. Applied electrical currents in the different planes.

Plane	Electrical current ($\omega = 1°\ \text{step}^{-1}$)	Sequence of steps
XY	$I_x(t) = 60\,\text{mA} \times \cos(\omega \times \text{step})$ $I_y(t) = 60\,\text{mA} \times \sin(\omega \times \text{step})$	From 1 to 360
ZX	$I_z(t) = 60\,\text{mA} \times \cos(\omega \times \text{step})$ $I_x(t) = 60\,\text{mA} \times \sin(\omega \times \text{step})$	From 361 to 721
YZ	$I_y(t) = 60\,\text{mA} \times \cos(\omega \times \text{step})$ $I_z(t) = 60\,\text{mA} \times \sin(\omega \times \text{step})$	From 722 to 1082

Table 7. Parameters δ and P of Eq. (4).

MOURA/HC planes	δ (°)	P
X1Y1/XY	0.64 ± 0.05	-0.997 ± 0.001
X1Z1/XZ	7.3 ± 0.2	-0.986 ± 0.004
Y1Z1/YZ	-0.42 ± 0.2	1.005 ± 0.004
X2Y2/XY	1.76 ± 0.05	-0.997 ± 0.001
X2Z2/XZ	-5.3 ± 0.1	0.973 ± 0.002
Y2Z2/YZ	1.12 ± 0.1	-1.006 ± 0.003

$$\Omega_{\text{HC}ij}(\text{step}) = \arctan\left(\frac{B_i}{B_j}\right);$$

$$\Omega_{\text{MOURA}i'j'}(\text{step}) = \arctan\left(\frac{B_{i'}}{B_{j'}}\right);$$

$$\Omega_{\text{HC}ij} = \delta + P \cdot \Omega_{\text{MOURA}i'j'}. \tag{4}$$

δ is the misalignment between ij and $i'j'$ HC and MOURA axes respectively. P is the slope of the linear fitting between $\Omega_{\text{HC}ij}$ and $\Omega_{\text{MOURA}i'j'}$.

Room temperature GAIN_x, GAIN_y, GAIN_z calculation is performed by comparison of the MOURA registered magnetic signals and successive reference positive and negative signals of intensity (P_i^+ and P_i^-) of 1 min duration applied in the different axes, using

$$B_{\text{MOURA}}(T)_i^+ - B_{\text{MOURA}}(T)_i^-$$
$$= \cos\theta_i \cdot \left(P_i^+ - P_i^-\right) \cdot \text{GAIN}_i \cdot (T_G). \tag{5}$$

Note that X1 and Y1 have opposite sense directions to X2 and Y2, respectively, for engineering purposes. This methodology is valid because it has been previously checked that the response of the sensor is linear with the magnetic field being the correlation coefficient between MOURA output and applied field higher than 0.9998 in all axes.

The results from the fittings are presented in Table 5.

Once the linear fitting and then the misalignment angles between planes were obtained, it was possible to approximate the misalignment of each axis by direct composition. Under this approximation the gains for each axis were obtained by direct calculus employing Eq. (5) and statistical

corrections of the measured magnetic moduli. The results are shown in Table 7.

With this correction, relative errors in the measurement of the field with the different axes are below 0.3 % except for the case of the Y1 sensor, which has an error of up to 0.9 %.

3.1.3 Characterization of output fields of the offset coils

The characterization of the offset coils constant (field vs. current) needs to be performed since the field vs. current provided by the manufacturer is subject to an error and these coils are used for the calibration of the sensors prior to use. Also these coils are used to extend the dynamic range of the magnetometer when it is saturated in the automated mode.

This characterization is performed in the same conditions as the gain characterization (using the same HC system in the zero-field chamber).

Decreasing and increasing field ramps are applied in 126 steps (between −45 655 and 45 665 nT). At room temperature, the field generated by the different offset coils is between 0.8617 and 0.9116 ± 0.0003 times that generated by the external field (Table 11). More details are given in Sect. 3.2.3, where the temperature calibration data are shown.

3.1.4 Inclinometer and gravimeter characterization

In order to be able to derive the horizontal and vertical components of the field and thus its orientation, it was required to characterize the response of the tilt angle detector, a three-axis accelerometer. This calibration is not direct since the ac-

Table 8. Tilt angles around $+X$ (α tilt angle) and experimental values (converted into g) for the first five steps.

α (°)	$\Delta\alpha$ (°)	ACC_X (g)	ACC_Y (g)	ACC_Z (g)	ACC (g)
4.9719	$< \pm 0.16$	0.2299	0.7839	−0.4297	0.92302
11.5369	$< \pm 0.16$	0.3043	0.7182	−0.4747	0.9131
19.4711	$< \pm 0.16$	0.3953	0.6261	−0.5170	0.9031
30.0000	$< \pm 0.16$	0.5080	0.4920	−0.5483	0.8948
41.8103	$< \pm 0.16$	0.6411	0.3583	−0.6114	0.9556

Table 9. Relative error between experimental 1 and theoretical values of α for different α.

α (°)	ACC_X (g)	ACC_Y (g)	ACC_Z (g)	ACC (g)
5	2.7 %	−1.5 %	0.0 %	−0.9 %
12	3.5 %	−3.5 %	−0.5 %	−2.0 %
19	4.6 %	−6.8 %	−1.2 %	−3.1 %
30	4.8 %	−12.3 %	−2.8 %	−3.9 %
42	7.8 %	−15.0 %	5.4 %	2.6 %

celerometer is placed on a PCB tilted 45° over the horizontal and rotated by an angle of −90° with respect to the MOURA box z axis, and by 45° with respect to the MOURA y axis. This rotation aims to linearize the accelerometer components response at zero tilt.

Also for field work it is interesting to characterize the gravity measurements in order to complement the magnetic measurements with gravity (low resolution) ones.

The following procedure designed by INTA and named "Procedure of Measurement Levels Calibration" consisted in the comparison between the real inclinations of the sine bar around x and y axes (α and β angles) with the measurements provided by the accelerometer axis (Table 8). The sequence of tilt is: zero tilt – maximum tilt – minimum tilt (negative) – zero tilt with five different values of tilt in absolute value (Fig. 3b).

The accelerometer and MOURA box system systems of reference are designed by {ACC} and {MOU}, respectively.

The theoretical change of basis matrix, **B**, between {ACC} and {MOU} and the inverse, $\mathbf{B}^{-1}$, with their respective errors, are

$$\mathbf{B} = \frac{1}{2}\begin{bmatrix} \sqrt{2} & 1 & 1 \\ -\sqrt{2} & 1 & 1 \\ 0 & -\sqrt{2} & \sqrt{2} \end{bmatrix}; \tag{6}$$

$$\Delta\mathbf{B} = \begin{bmatrix} 0.0371 & 0.0524 & 0.0524 \\ 0.0371 & 0.0524 & 0.0524 \\ 0 & 0.0371 & 0.0371 \end{bmatrix} \tag{7}$$

$$\mathbf{B}^{-1} = \frac{1}{2}\begin{bmatrix} \sqrt{2} & -\sqrt{2} & 0 \\ 1 & 1 & -\sqrt{2} \\ 1 & 1 & \sqrt{2} \end{bmatrix};$$

$$\Delta\mathbf{B}^{-1} = \begin{bmatrix} 0.0371 & 0.0371 & 0 \\ 0.0524 & 0.0524 & 0.0371 \\ 0.0524 & 0.0524 & 0.0371 \end{bmatrix}. \tag{8}$$

Table 9 displays the relative errors between the measured and theoretical values of α. The measured error allows us to calculate a slight extra rotation around the z axis (an angle of −1.7686°) due to the experimental positioning of the accelerometer PCB.

The new change of basis matrix corresponds to

$$\mathbf{B}_{\mathrm{exp}} = \mathbf{B}\mathbf{R}_\varepsilon = \frac{1}{2}\begin{bmatrix} \sqrt{2} & 1 & 1 \\ -\sqrt{2} & 1 & 1 \\ 0 & -\sqrt{2} & \sqrt{2} \end{bmatrix}\begin{bmatrix} 0.9995 & -0.0309 & 0 \\ 0.0309 & 0.9995 & 0 \\ 0 & 0 & 1 \end{bmatrix}. \tag{9}$$

Notice that this matrix will have to be characterized for each experimental setup, i.e. each model will have its own change of basis matrix.

Experimental values of α and β obtained subtracting the zero tilt measurement from those corresponding to the different inclinations can be adjusted easily to the real tilt values giving

$$\alpha_{\mathrm{EXP}}(°) = 1.0894 \cdot \alpha_{\mathrm{ref}}(°) + 0.7268° \tag{10}$$

$$\beta_{\mathrm{EXP}}(°) = 0.9779 \cdot \beta_{\mathrm{ref}}(°) - 1.7475°. \tag{11}$$

But these values have a slight difference from those derived from the accelerometer measurements in the MOURA box reference system:

$$\alpha'_{\mathrm{EXP}}(°) = \arctan\left(\frac{\mathrm{MOU\_Y}}{\mathrm{MOU\_Z}}\right) = \alpha_{\mathrm{EXP}}(°) - 7.2890° \tag{12}$$

$$\beta'_{\mathrm{EXP}}(°) = -\arctan\left(\frac{\mathrm{MOU\_X}}{\mathrm{MOU\_Z}}\right) = \beta_{\mathrm{EXP}}(°) + 21.4369°. \tag{13}$$

Consequently, the equations for the real inclinations around x and y axes from the accelerometer measurements in MOURA box reference system are

$$\alpha(°) = 0.9179 \cdot \arctan\left(\frac{\mathrm{MOU\_Y}}{\mathrm{MOU\_Z}}\right) + 6.0237° \tag{14}$$

$$\beta(°) = -1.0226 \cdot \arctan\left(\frac{\mathrm{MOU\_X}}{\mathrm{MOU\_Z}}\right) - 20.1347°. \tag{15}$$

These data permit the determination of the inclination with errors up to 3 % in α for $-30° < \alpha < +30°$, and of 6 % in $-30° < \beta < +30°$. A better adjustment with errors up to 1 % in α for $-40° < \alpha < +40°$, and of 3 % in $-40° < \beta < +40°$, can be obtained with the following polynomial fit (Fig. 6):

$$\alpha(°) = -0.0012 \cdot \arctan^2\left(\frac{\mathrm{MOU\_Y}}{\mathrm{MOU\_Z}}\right)$$
$$+ 0.8921 \cdot \arctan\left(\frac{\mathrm{MOU\_Y}}{\mathrm{MOU\_Z}}\right) + 6.5543° \tag{16}$$

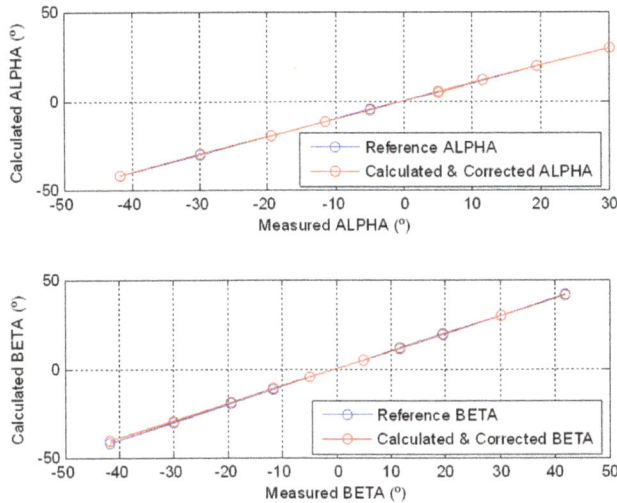

Figure 6. Corrected values of the tilt angles.

$$\beta(°) = 0.0038 \cdot \arctan^2\left(\frac{\text{MOU\_X}}{\text{MOU\_Z}}\right)$$
$$- 0.8656 \cdot \arctan\left(\frac{\text{MOU\_X}}{\text{MOU\_Z}}\right) - 20.3296°. \quad (17)$$

Also the gravimeter readings need to be calibrated since they are very dependent on the tilt angle measurements.

Without correction the errors in the gravity modulus are of 1 % for $\alpha = \pm 10°$, but of the order of 5 % in $\beta = \pm 10°$. By means of a quadratic correction with the tilt angles and considering that α and β are totally independent, the error is reduced to ± 0.0001 g (0.01 %) for $\alpha = \pm 40°$ and ± 0.004 g (0.4 %) for $\beta = \pm 40°$ ($g = 9.8$ m s^{-2}).

3.2 Temperature-dependent characterization

It is known that most of magnetic sensors have a temperature-dependent response, and therefore magnetic sensors need to undergo a temperature characterization when they are used in conditions of changing temperatures. Also the response of the conditioning electronics can change greatly when subject to huge temperature fluctuations.

The issue then is to determine the temperature of the device (core of the sensor, amplifier ...) whose response varies with the temperature. This is a very complicated problem since normally the only accessible part of the device is the package. In the steady state (thermal dynamic equilibrium), the temperature of the package can be assumed as an indicator of the core temperature, though this can have a non-negligible error when there are temperature gradients or/and fast temperature variations with time. Also the flipping of the AMR sensors implies temperature variations of the core. This situation makes it necessary to find an optimal working mode, which involves a trade-off between thermal changes, acquisition frequency and samples to average.

In this work we focus on a practical solution consisting in the characterization of the thermal behaviour in the steady state. Therefore, we focus on the two temperature channels, namely TMP1 and TMP2, corresponding to temperature sensors placed on top of the magnetic sensors. As has been introduced, these temperature sensors were not in close contact with the active sensing element of the magnetic sensors, i.e. magnetoresistive bridges, and therefore they are not expected to provide an accurate measurement of the magnetic sensors' instant temperatures. In addition, in this work it is considered that the orientation angles do not change by means of thermal expansion and contraction. This assumption responds to the fact that the calculated maximum angle deviation due to the worst case of an anisotropic thermal expansion is 0.015° approximately, of the order of the resolution limit of the sensor.

3.2.1 Offset characterization with temperature

The variation of the offset with temperature was formerly estimated with the daily fluctuation of the temperature outside the building (10–30 °C). It was observed that the variation of the offset with temperature was very similar to that of the gain. This test was performed inside a shielding chamber with a field stability better than 1.5 nT. Therefore the offset observed is only attributed to the variations of temperature inside the chamber, which are registered and are in good correlation with the offset values monitored. Consequently, for the extended range of temperature, both variations with temperature will be considered equivalent, i.e. ΔOFFSET $= \Delta$GAIN.

This assumption will be corrected with the long term records at San Pablo de los Montes Observatory with the local temperature data (Sect. 4).

3.2.2 Gain characterization with temperature – V_{REG} compensation

The extended expression of Eq. (1), the expression of the measured fields in the different axes, as a function of temperature is

$$M_i = (\mathbf{B}_{\text{REAL}} \cdot \mathbf{u}_i) \cdot \text{GAIN}_i (1 - \Delta\text{GAIN}(T - T_{\text{G}}))$$
$$- \text{OFFSET}_i (1 - \Delta\text{OFFSET}_i (T - T_{\text{OFFSET}}));$$
$$\mathbf{B}_{\text{REAL}} \cdot \mathbf{u}_i = |\mathbf{B}| \cdot |\mathbf{u}_i| \cdot \cos\theta_{\mathbf{B}_{\text{REAL}}M_i};$$
$$M_i = |\mathbf{B}| \cdot \cos\theta_{\mathbf{B}_{\text{REAL}}M_i} \cdot \text{GAIN}_i (1 - \Delta\text{GAIN}(T - T_{\text{G}}))$$
$$- \text{OFFSET}_i (1 - \Delta\text{OFFSET}_i (T - T_{\text{OFFSET}})). \quad (18)$$

Since the magnetic noise in the thermal chamber is higher than the precision of MOURA, it is not possible to control or monitor the magnetic field with the required precision and to discern absolute variations in the offset (OFFSET) and gain (GAIN) of MOURA. However, due to the linearity of the output with the field, it is possible to calculate the variation of the relative gain (ΔGAIN_i) by the measurement of the vari-

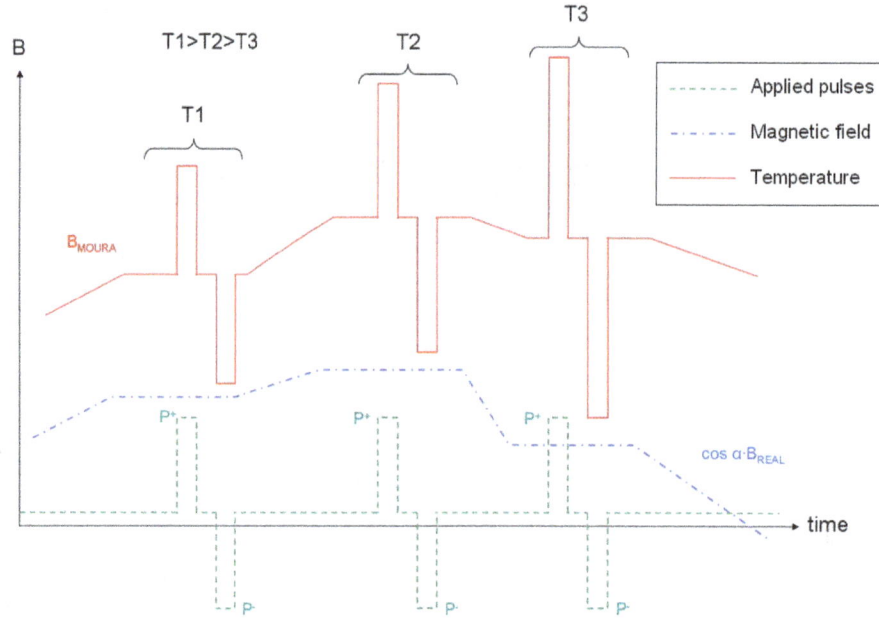

Figure 7. Sketch of applied pulses, real external magnetic field, registered signal by MOURA for different temperatures (T1, T2 and T3) as a function of time.

ation of the controlled amplitude between two applied magnetic pulses with the same modulus and direction but opposite polarization, assuming that the variation of the external magnetic field is much slower than the duration of the pulses (Fig. 3c). These pulses are denoted as P^+ and P^- (Fig. 7).

MOURA sensors' response for P^+ and P^- pulses will be denoted by $B_{\mathrm{MOURA}}(T)_i^+$ and $B_{\mathrm{MOURA}}(T)_i^-$, respectively:

$$B_{\mathrm{MOURA}}(T)_i^+$$
$$= \cos\theta_i \cdot P^+ \cdot \mathrm{GAIN}_i \cdot (1 - \Delta\mathrm{GAIN}_i \cdot (T - T_{\mathrm{G}}))$$
$$- \mathrm{OFFSET}_i \cdot (1 - \Delta\mathrm{OFFSET}_i) \cdot (T - T_{\mathrm{OFFSET}}) \, B_{\mathrm{MOURA}}(T)_i^-$$
$$= \cos\theta_i \cdot P^- \cdot \mathrm{GAIN}_i \cdot (1 - \Delta\mathrm{GAIN}_i \cdot (T - T_{\mathrm{G}}))$$
$$- \mathrm{OFFSET}_i \cdot (1 - \Delta\mathrm{OFFSET}_i) \cdot (T - T_{\mathrm{OFFSET}}). \quad (19)$$

Subtracting the pulses,

$$\frac{B_{\mathrm{MOURA}}(T)_i^+ - B_{\mathrm{MOURA}}(T)_i^-}{2}$$
$$= \cos\theta_i \cdot |2P| \cdot \mathrm{GAIN}_i \cdot (1 - \Delta\mathrm{GAIN}_i \cdot (T - T_{\mathrm{G}}))$$
$$= M_i^{\mathrm{A}}(T), \quad (20)$$

where M_i^{A} is the averaged amplitude of the magnetic pulses measured by the i axis of MOURA.

The magnetic field pulses are monitored by means of a three-axis fluxgate (FG-500). The accuracy of the magnetic field depends on the current in the circuit, which is controlled better than 1‰. Although it is not possible to determine analytically the absolute GAIN_i and θ_i, i.e. the metrics and orthogonal projection from the HC system to reference system of MOURA, it is possible to normalize the obtained signals,

using for the normalization the signal obtained at a reference temperature (T_{G}):

$$M_i^{\mathrm{A}}(T)/M_i^{\mathrm{A}}(T_{\mathrm{G}}) = \frac{\cos\theta_i \cdot |P| \cdot \mathrm{GAIN}_i \cdot (1 - \Delta\mathrm{GAIN}_i \cdot (T - T_{\mathrm{G}}))}{\cos\theta_i \cdot |P| \cdot \mathrm{GAIN}_i \cdot (1 - \Delta\mathrm{GAIN}_i \cdot (T_{\mathrm{G}} - T_{\mathrm{G}}))};$$
$$M_i^{\mathrm{A}}(T)/M_i^{\mathrm{A}}(T_{\mathrm{G}}) = (1 - \Delta\mathrm{GAIN}_i \cdot (T - T_{\mathrm{G}})). \quad (21)$$

Applying a linear fitting of the normalized signals as a function of $T - T_{\mathrm{G}}$, it is possible to calculate $\Delta\mathrm{GAIN}_i$ as a function of temperature.

The test is carried out at six temperature values in the range of temperatures in which field measurements were performed, using as first reference the thermal chamber temperature controller: -60, -30, 0, 15, 45, and $60\,^{\circ}\mathrm{C}$. The registered humidity inside the chamber is $< 18\,\%$ for the test (as explained in Sect. 2.2). The square magnetic pulses along the six semi-axes were applied by means of a Keithley precision source (10 mA current) supplied to the three pairs of HC simultaneously at thermal equilibrium $< 0.1\,^{\circ}\mathrm{C\,min^{-1}}$ (Table 9). The amplitude of each magnetic pulse was taken as the mean absolute value of each pulse applied along the same axis (positive and negative pulse). The registered magnetic field amplitudes were normalized to that obtained at room temperature: TMP1 $= 25.9 \pm 0.2\,^{\circ}\mathrm{C}$ and TMP2 $= 25.6 \pm 0.2\,^{\circ}\mathrm{C}$. Two additional temperature sensors (calibrated PT-1000, denoted as TT and TL) were placed on the top (TT) and on a side (TL) of MOURA in order to guarantee thermal equilibrium. Since the level of magnetic noise generated by the hardware of the thermal chamber (mainly rotor and pumps) makes it impossible to obtain suitable accurate data, the thermal chamber was switched off when the pulses were applied (Fig. 8).

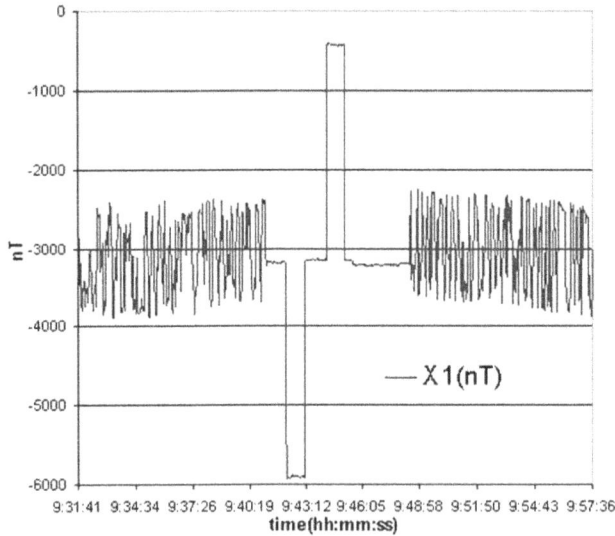

Figure 8. Detailed view of the register data by Sensor 1 x axis, minutes before and after the application of magnetic pulses (thermal chamber ON-OFF).

Table 10. Temperature registers and their temporal variation.

Measurement	TT (°C)	TL (°C)	TMP1 (°C)	TMP2 (°C)
1	59.4 ± 0.1	58.9 ± 0.2	50.4 ± 0.1	50.04 ± 0.04
2	32.6 ± 0.2	32.2 ± 0.1	25.9 ± 0.1	25.6 ± 0.1
3	5.4 ± 0.2	4.1 ± 0.1	-0.3 ± 0.2	-0.6 ± 0.2
4	16.6 ± 0.1	17.3 ± 0.1	11.8 ± 0.2	11.55 ± 0.2
5	44.8 ± 0.1	45.4 ± 0.1	38.4 ± 0.1	38.2 ± 0.1
6	58.4 ± 0.2	58.8 ± 0.2	50.6 ± 0.1	50.51 ± 0.01

Table 11. Offset coils constants at the different temperatures.

TMP1	Constant $(nT\,nT^{-1}) \pm 0.0003$		
(°C ± 0.05)	X1	Y1	Z1
16.41	0.8879	0.9116	0.8617
49.55	0.7743	0.8022	0.7644
26.92	0.8518	0.8767	0.8293
0.69	0.9406	0.9656	0.9086
11.42	0.9045	0.9290	0.8760
37.53	0.8160	0.8414	0.7961
50.37	0.7722	0.7988	0.7564

These variations of temperature affect the voltage sourcing of the magnetoresistive bridges. VREG has a variation with temperature of 0.1 %. The variation is recorded and will also be taken into account for the response correction.

The thermal chamber control is switched off for the measuring. The obtained values of amplitude for each axis were normalized by those obtained at TMP1 $= 25.9\,°C$ and TMP2 $= 25.6\,°C$. These normalized amplitudes were linearly fitted with the corresponding temperature ($T - 25.9\,°C$ for Sensor 1 data and $T - 25.6\,°C$ for Sensor 2 data) (Fig. 9).

The coefficients for the thermal drift correction of the magnetic data (ΔGAIN) for each axis are presented in Table 5. For example, the gain compensation with temperature in Eq. (17) for the x axis of Sensor 1 is

$$\Delta GAIN_{X1} \cdot (T - TG)$$
$$= \left(-0.00370 \pm 5 \times 10^{-5}\right) \, °C^{-1} \cdot \left[TMP1\,(°C) - 25.9°C\right].$$

3.2.3 Offset coils characterization with temperature

In this section the thermal variation of the offset coils constant ($nT\,nT^{-1}$ or $nT\,mA^{-1}$) is calibrated. For simplicity only Sensor 1 parameters are displayed.

This characterization is performed by means of a relative measurement of the constant variation with temperature and then referred to the reference temperature in a similar way as the gain characterization with temperature.

In this case, two ramps (Ramp I: decreasing from 45 665 to $-45\,665\,nT$, and Ramp II: increasing from $-45\,665$ to $45\,665\,nT$) of 126 steps (4.56 mA) have been applied with the offset coils.

Previously it has been checked that the current passing through the offset coils does not increase the temperature of the magnetoresistors and thus does not change the response.

In order to obtain the values of the offset coil constant variation with temperature ($nT\,nT^{-1}\,°C^{-1}$), the above-mentioned magnetic field ramps were applied at the five different temperatures in the same thermal chamber as the previous test.

In this case, MOURA temperature sensors were also employed to register the temperature variation during the test. The ramps were applied in thermal equilibrium ($< 0.1\,°C\,min^{-1}$) with the control of the chamber switched off. Table 10 shows the temperature readings at thermal equilibrium at the testing temperatures obtained with the different temperature sensors. The control of the humidity ($< 18\,\%$) was performed with N_2.

An example of the obtained data is presented in Fig. 10. It can be seen that the offset value is higher than that obtained in the magnetic shielded chamber due to the lack of a magnetic clean environment.

The obtained coil constants for Sensor 1 from the linear fits for each ramp at different temperatures, and averaging data corresponding to Ramp I and Ramp II, are presented in Table 11 and shown in Fig. 11 as a function of TMP1 normalized to the reference temperature.

The obtained gains for each temperature were linearly fitted versus the registered temperature by the corresponding TMP sensor (Fig. 11).

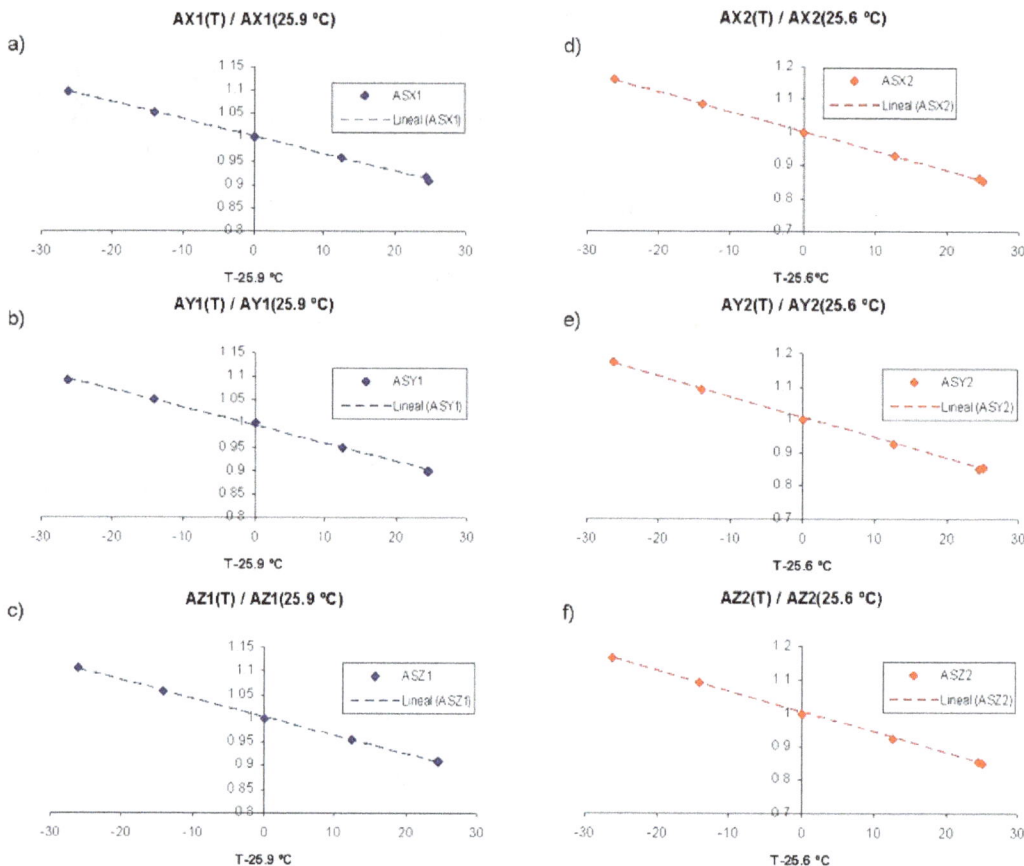

a) **AX1(T) / AX1(25.9 °C)**

b) **AY1(T) / AY1(25.9 °C)**

c) **AZ1(T) / AZ1(25.9 °C)**

d) **AX2(T) / AX2(25.6 °C)**

e) **AY2(T) / AY2(25.6 °C)**

f) **AZ2(T) / AZ2(25.6 °C)**

Figure 9. Normalized amplitude as a function of modified temperature for axis sensor **(a)** X1, **(b)** Y1, **(c)** Z1, **(d)** X2, **(e)** Y2, **(f)** Z2.

Table 12. Sensor 1 offset coils characterization with temperature.

ΔConstant axis	Value $((\text{nT nT}^{-1})\,\text{C}^{-1})$
ΔConstant$_{X1}$	$-0.088034 \pm 7 \times 10^{-6}$
ΔConstant$_{Y1}$	$-0.086506 \pm 1 \times 10^{-5}$
ΔConstant$_{Z1}$	$-0.078218 \pm 4 \times 10^{-5}$

The thermal variations of the offset coils constants are presented in Table 12. For example, the compensation with temperature parameter of X coil of Sensor 1 is

$$\Delta\text{Constant}_{X1}\left(\text{nT nT}^{-1}\right) \cdot (\text{TMP1} - T_{\text{ref}})\,(°\text{C})$$

$$= \left(-0.088034 \pm 7 \times 10^{-6}\right)°\text{C} - 1 \cdot (\text{TMP1} - 25.9)\,(°\text{C}).$$

4 Data comparison of MOURA and SPT reference magnetometers

This section describes on the one hand the calibration of the offset with temperature by means of comparison with

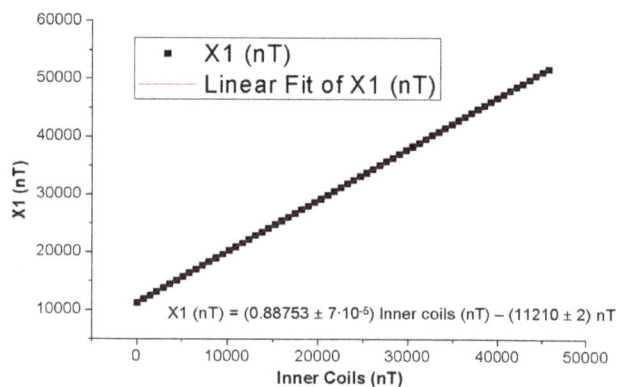

Figure 10. Data for Sensor 1 x axis "Ramp I" at 16.41 °C (TMP1).

another magnetometer (reference) at a temperature different from that of the laboratory, and on the other hand, a final measurement of a space weather event, as a demonstration of MOURA capabilities in terms of resolution. These measurements have been performed at the Geophysics Observatory of San Pablo-Toledo (SPT) (39.547° N, 4.349° W) in Spain, during late January and February 2013 (offset drift with temperature) and during June, July and August 2013

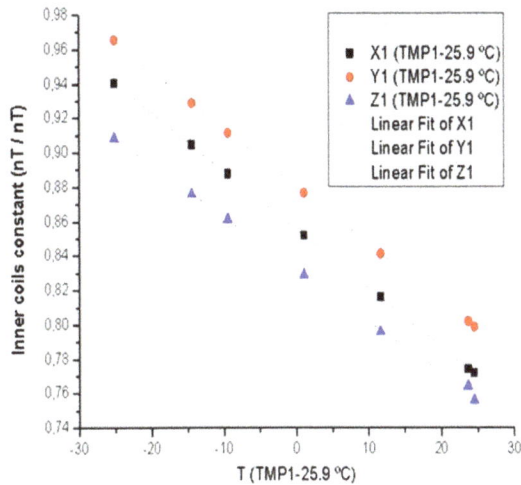

Figure 11. Linear fits of gain as a function of corrected temperature (TMP1-Tref).

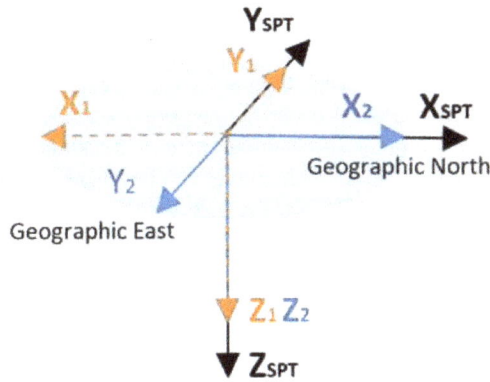

Figure 12. Relative axes of MOURA: X1, Y1, Z1 and X2, Y2, Z2 and SPT Observatory: XSPT, YSPT and ZSPT.

(geomagnetic storm). The comparison needs to be performed in situ for the large crustal magnetic anomalies variability in the peninsula and other factors like magnetic contamination (Martínez Catalán, 2012).

SPT belongs to INTERMAGNET (http://www.intermagnet.org), a global network of observatories (since 1997) and to the International Association of Geomagnetism and Aeronomy (IAGA) (available at http://www.iugg.org/IAGA). SPT has a fluxgate magnetometer FGE (Danish Meteorological Institute) and a fluxgate vector magnetometer Geomag M390. It is also equipped with Overhauser effect magnetometers GSM90 for calibration purposes. The instrumentation setup is completed by a dIdD Gemsystem equipment. Two declinometer–inclinometers (Zeiss 010B) with a fluxgate Bartington probe are used for absolute weekly observations. This suite of magnetometers offers raw data, which are further corrected by the observatory (contamination removal: instrumentation faults or man-made interferences, and daily basis filtering).

Table 13. Percentage of transmission errors during five consecutive days (21–25 February 2013).

Axis	Errors percentage (%)	
	Sensor 1	Sensor 2
X	0.20	0.0
Y	0.00	4.2
Z	0.35	4.2

In the present comparison partially treated and compensated available data will be used for the comparison. Final data are provided of the order of 1 year after the measurements.

For the test campaigns some auxiliary instrumentation was moved to SPT: MOURA instrument (with axes orientations as defined in Fig. 12), a voltage source with two output channels, a laptop, a 3G USB modem and a 82357B USB/GPIB interface (by Agilent Technologies).

Due to the distance between Toledo (test station) and Madrid (INTA headquarters), a 3G USB modem was used for remote control of the computer enabling all basic operations of MOURA. The complete setup with the elements described is shown in Fig. 13.

A first campaign between 21 and 25 February 2013 was used to refine the laboratory calibration.

During the acquisition, a percentage of erroneous data was detected (Table 13). This was attributed to transfer data errors during set and reset pulses or packing data errors. Retrieval software is able to detect and automatically suppress the errors.

The first campaign took place during quiet days. Figure 14 shows the variation of the different components of the magnetic field measured by MOURA sensors (Sensor 1 and Sensor 2) versus SPT for 21–24 February 2013. The typical terrestrial magnetic field daily variation can be seen, with a higher amplitude of the X component pointing to the north of the Earth during sunny hours, directly related with the exposure to the solar radiation during the day hours. MOURA data fit quite well with the reference data showing a daily variation of ± 35 nT with highest values at around 12:00 to 14:00 UTC on the X magnetic field component. A non-negligible offset deviation of the MOURA data can be seen. This is related to the variation of the offset with the temperature. Note that offset calibration performed in the laboratory takes place at 18 °C and the average temperature at SPT at the time of the campaign was 5 °C. This fact is used to correct the previous estimation performed in Sect. 3.2.1 according to the expression

$$\frac{\Delta \text{Offset}}{\Delta T}\,(\text{ppm}\,°C) = \frac{\frac{\text{Offset}\pm(\Delta\text{MOURA-SPT})-1}{\text{Offset}}}{T_{\text{Offset}} - T_{\text{SPT}}} \cdot 10^6. \quad (22)$$

The resulting values are given in Table 5. As well as in the drift of gain with temperature, the dispersion of the offset drift with temperature is very wide, which makes it necessary

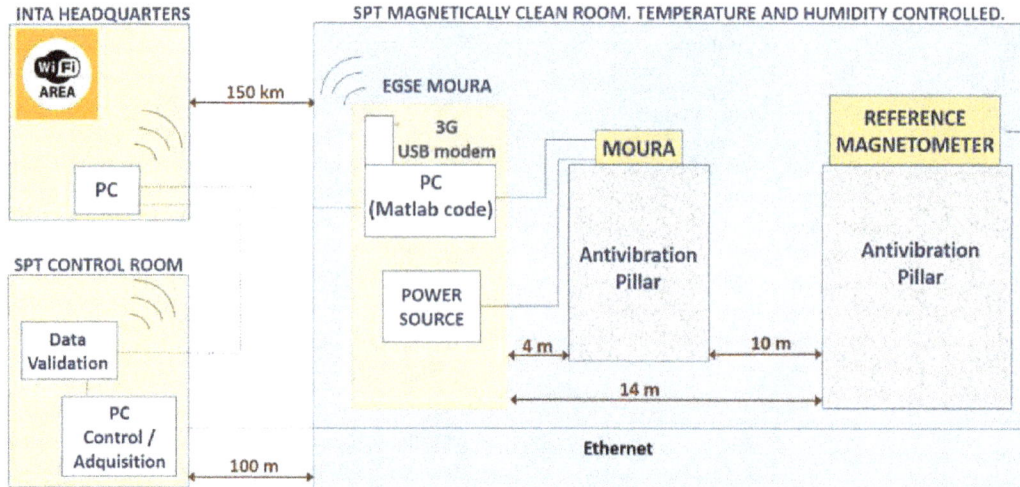

Figure 13. Setup of the measurements.

to screen the sensors to be used in the FM and filter the most suitable for the purpose.

After this last correction of the offset drift with temperature, and the corresponding modification of the retrieval software, a new campaign is performed with the double objective to validate the calibration and to demonstrate the suitability of the sensor to measure the space weather events. For this, the three months period from June to August 2013, with some solar activity, was selected. Figure 15 shows the data corresponding to the geomagnetic storm that occurred on 28 and 29 June: the horizontal component of MOURA sensors 1 and 2 as well as SPT reference data. Such event is characterized by a decrease of horizontal magnetic field component H, that is $H = (X^2 + Y^2)^{1/2}$, of about $100 - 200$ nT with respect to the initial level of H accompanied by irregular fluctuations of varying frequencies (periods from seconds to hours) and intensities (from nT to tens of nT). The results confirm that MOURA reproduces quite well the magnetic field variations measured by the official SPT magnetometer.

5 Discussion

MOURA magnetic instrument is based on an anisotropic magnetoresistive transducer with the purpose of significantly reducing weight in the instrument. Although these transducers do not present optimal magnetic properties and furthermore their response is very temperature dependent, after a careful calibration the instrument presents fairly good performance and fulfils the scientific goal.

In general, the magnetic parameters characterized are in agreement with the manufacturer data sheet. The non-orthogonalities between the in-plane components (X and Y) are negligible compared to our resolution, and the measured deviation between the z axis and the XY plane is lower than $1°$ as specified.

Sensitivities match very well the values of the data sheet, and offsets are lower than the maximum swing specified because the sensors have been screened to choose those with the lowest offsets at room temperature. Regarding the gain drifts with temperature the parameters measured are in accordance with the manufacturer data but there exists a wide dispersion of values as in the gain drift of Sensors 1 and 2. The observed offset drift with temperature is higher than the values specified by the manufacturer for Set and Reset operation. Also note the anomalous offset drift of the Z component of Sensor 2. Although the dispersion is attributed to the manufacturing process and is considered normal, it is an important factor, which needs to be taken into account in the selection of components for future missions.

The instrument has been tested in a real environment to measure a geomagnetic storm and the experimental data have been successfully contrasted with those of the reference magnetometer in an official geomagnetic observatory. These measurements demonstrate that the sensor is capable of following dynamically variations of the environmental magnetic field of the order of nT.

6 Conclusions and future work

A practical calibration of MOURA magnetometer and gradiometer has been performed to demonstrate its capability to fulfil the pursued scientific objectives on Mars: to measure the magnetic anomalies of the landing site and to observe the daily variation of the field and its perturbations with the solar activity. The calibration comprises the characterization of the offsets, gains, non-orthogonalities and Euler angles, as well as offset and gain drifts with temperature in a range from -60 to $60°C$, and the tilt angle detector characterization. The retrieval software includes the equations to derive

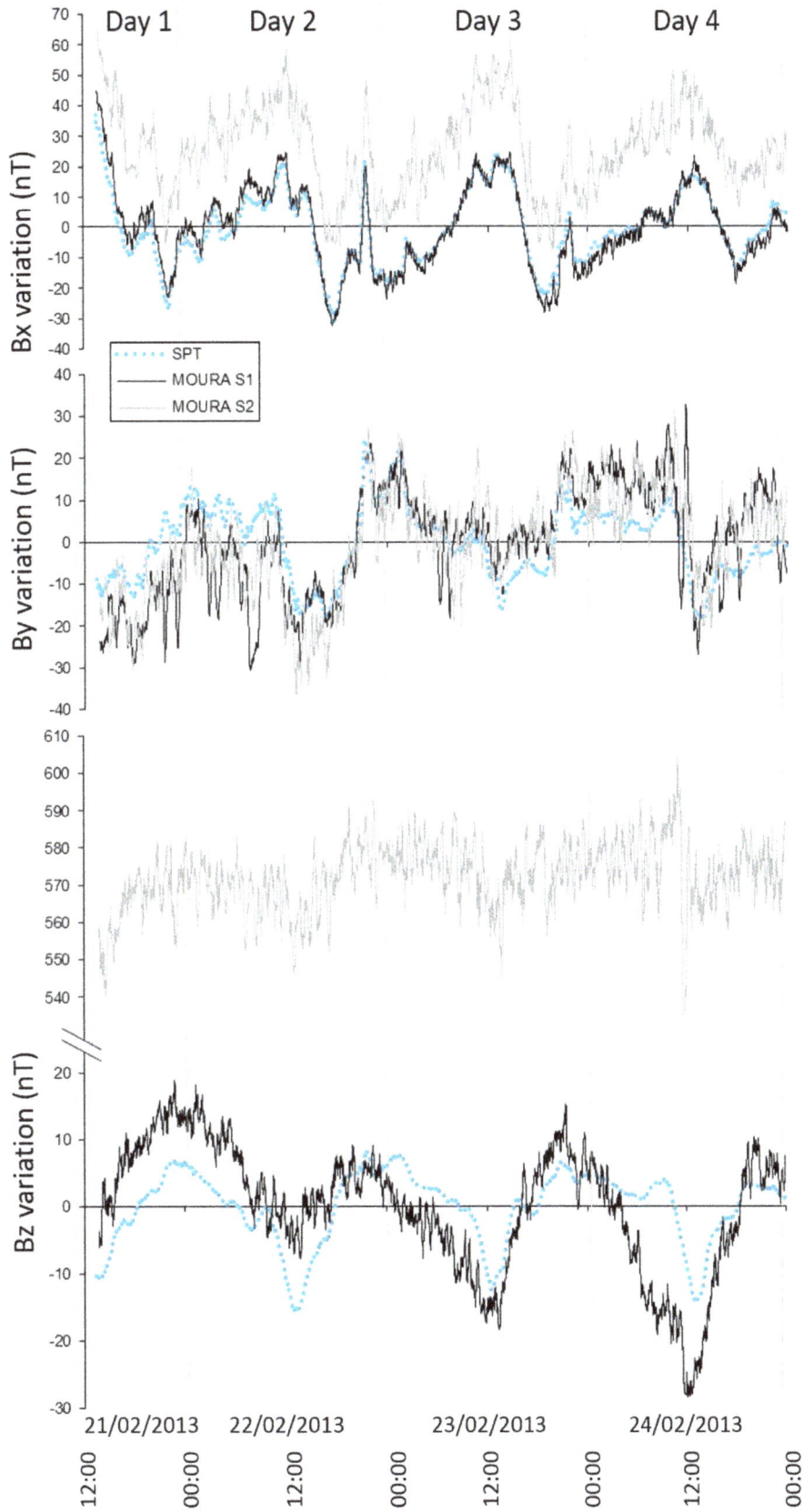

Figure 14. Comparison between measurements from SPT and MOURA, *x* axis (bottom panel), *y* axis (middle panel) and *z* axis (top panel).

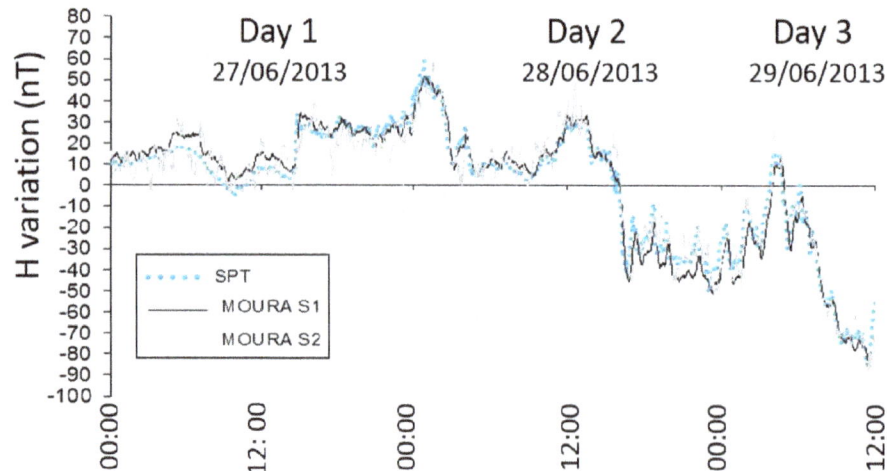

Figure 15. Horizontal component of the geomagnetic field measured with MOURA magnetometer and SPT reference magnetometers on 28–29 June 2013.

the magnetic field referred to the temperature-compensated Martian surface.

The offset drift with temperature has been characterized by means of measurements performed at a reference observatory, San Pablo de los Montes, Toledo.

Finally, a successful comparison of MOURA measurements with the reference magnetometer has been performed during a geomagnetic storm. The results are considered very useful: it is feasible to obtain scientific information on the magnetic environment with a 72 g compact magnetometer of < 0.5 W. The extended use of such instruments (net of landers/rover) could help the characterization of the unknown Martian magnetic scenario, highly improving the understanding of the remanence of the crust and possibly of the ancient magnetizing field.

In forthcoming works we will also report on our real and long-term prospections with MOURA in comparison with a scalar absolute magnetometer (Geometrics 858), and the data interpretation, to describe the potential of this miniaturized compact magnetometers for rovers and balloons.

Acknowledgements. This work was supported by the Spanish National Space Programme (DGI-MEC) project MEIGA-MET-NET under the Grant AYA2011-29967-C05-01. The authors wish to thank all the MetNet team for the support, especially Víctor de Manuel and Juan José Jiménez (INTA) and Víctor Apéstigue and Miguel González (ISDEFE), for their technical support, Miguel Ángel Herraiz (UCM) for his help in the discussion of the specifications, and Jose Manuel Tordesillas, Pablo Covisa and José Aguado (SPT) for great support during the campaigns and for allowing the use of the SPT facilities.

Edited by: W. Schmidt

References

Brown, P., Beek, T., Carr, C., O'Brien, H., Cupido, E., Oddy, T., and Horbury, T. S.: Magnetoresistive magnetometer for space science applications, Measure. Sci. Technol., 23, 025902, doi:10.1088/0957-0233/23/2/025902, 2012.

Cai, J., Andersen, N. L., and Malureanu, C.: In-Field Practical Calibration of Three-Axis Magnetometers, Proceedings of the 2010 International Technical Meeting of the Institute of Navigation, San Diego, CA, 2010.

Diaz Michelena, M., Cobos, P., and Aroca, C.: lock-in amplifiers for AMR sensors, Sensors Actuators A, 222, 149–159, 2015.

D. Michelena, M.: Small Magnetic Sensors for Space Application, Sensors, 4, 2271–2288, 2009.

D. Michelena, M., Arruego, I., Oter, J. M., and Guerrero, H.: COTS-Based Wireless Magnetic Sensor for Small Satellites, IEEE T. Aerospace Elect. Syst., 46, 542–557, 2010.

DTUsat: available at: http://www.dtusat.dtu.dk (last access: December 2010), 2014.

Fernández, A. B., Sanz, R., Covisa, P., Tordesillas, J. M., and Díaz-Michelena, M.: Testing the three axis magnetometer and gradiometer MOURA and data comparison on San Pablo de los Montes Observatory, vol. 15, EGU General Assembly 2013 Conference, Vienna, 2013.

Freitas, P. P., Ferreira, R., Cardoso, S., and Cardoso, F.: Magnetoresistive sensors, J. Phys.-Condens. Mat., 19, 165221, doi:10.1088/0953-8984/19/16/165221, 2007.

Geomagnetic observatories: available at: www.ign.esandwww.intermagnet.org (last access: October 2013), 2014.

Honeywell Magnetic Sensors: Morristown, NJ, USA, available at: http://www.magneticsensors.com/magnetic-sensor-products.phpHoneywell (last access: October 2013), 2014.

Martínez Catalán, J. R.: The Central Iberian arc, an orocline centered in the Iberian Massif and some implications for the Variscan belt, Int. J. Earth Sci., 101, 1299–1314, 2012.

MMPM – Mars MetNet Mission: available at: http://metnet.fmi.fi (last access: October 2013), 2014.

Petrucha, V. and Kaspar, P.: 'Calibration of a Triaxial Fluxgate Magnetometer and Accelerometer with an Automated Non-magnetic Calibration System, 2009 IEEE Sensors Conference, Christchurch, New Zealand, 1510–1513, 2009.

Petrucha, V., Kaspar, P., Ripka, P., and Merayo, J.: Automated System for the Calibration of Magnetometers, J. Appl. Phys., 105, 07E704-1–07E704-3, 2009.

Renaudin, V., Afzal, M. H., and Lachapelle, G.: Complete Triaxis Magnetometer Calibration in the Magnetic Domain, J. Sensors, 2010, 967245, doi:10.1155/2010/967245, 2010.

Ripka, P., Butta, M., and Platil, A: Temperature Stability of AMR Sensors, Sensor Lett., 11, 74–77, 2013.

Sanz, R., Cerdán, M. F., Wise, A., McHenry, M. E., and Díaz Michelena, M.: Temperature dependent Magnetization and Remanent Magnetization in Pseudo-binary x $(\mathrm{Fe_2TiO_4}) - (1-x)(\mathrm{Fe_3O_4})(0.30 < x < 1.00)$ Titanomagnetites, IEEE T. Magnet., 47, 4128–4131, 2011.

Sanz, R., Fernández, A. B., Dominguez, J. A., Martín, B., and D. Michelena, M.: Gamma Irradiation of Magnetoresistive Sensors for Planetary Exploration; Sensors, 12, 4447–4465, 2012.

Sordo-Ibáñez, S., Piñero-García, B., Muñoz-Díaz, M., Ragel-Morales, A., Ceballos-Cáceres, J., Carranza-González, L., Espejo-Meana, S., Arias-Drake, A., Ramos-Martos, J., Mora-Gutiérrez, J. M., and Lagos-Florido, M. A.: A Front-End ASIC for a 3D Magnetometer for Space Applications Based on Anisotropic Magnetoresistors, Conference Paper, European Conference on Magnetic Sensors and Actuators EMSA, 2014.

16

Innovations and applications of the VERA quality control

D. Mayer, A. Steiner, and R. Steinacker

Department of Meteorology and Geophysics, University of Vienna, Vienna, Austria

Correspondence to: D. Mayer (dieter.mayer@univie.ac.at)

Abstract. Quality control (QC) is seen today as an important scientific field to increase the value of observational data. Whereas most QC methods are linked to atmospheric modeling (being part of the data assimilation procedure), in this paper the focus is on the application of a model independent QC method based on data self consistency recently published: VERA-QC. A special challenge is the QC of data in complex terrain which requires special treatment in terms of data selection and data transformation. In this context, some special VERA-QC modules such as the consideration of significant elevation differences of adjacent stations or the consideration of transformed temperature values will be discussed. The system detects gross errors as well as biases and offers objective correction proposals (deviations) for each observation. The essential gross error detection is not only based on the statistical behavior of station specific deviations, but also on the rate of cost function reduction. Beside a two dimensional application, higher dimensionalities may also be chosen, for instance including the time coordinate. Applications and results are discussed for pressure, temperature as well as for precipitation data which needs, however, a very dense observation network. Real time application of VERA-QC allows the production of high quality fields of meteorological parameters, which can be used, e.g. for nowcasting as well as for model unbiased validation of prognostic models.

1 Introduction

Nowadays, it is well known that the output of a numerical weather prediction (NWP) is, to a great extent, dependent on the quality of the input data. This awareness of the need for a high data quality evolved from the very beginning of the NWP era when the quality control (QC) of observations was still considered a purely technical task (Gandin, 1988). Within the last decades, many QC methods have been developed and it can be stated that a new research area was born. Taking that into account and considering the progressive international exchange of meteorological observations, the WMO started to set up a global standard regarding the quality of measurements in 1980. Different guidelines and manuals have been published and are updated on a regular basis (WMO, 1993, 2003, 2007, 2008).

High data quality is also essential when analyzing the actual state of the atmosphere, which constitutes the basis for a reliable nowcasting (Häggmark et al., 2000) or for model validation. Furthermore, the evaluation of long-term climatological data with respect to climate change, or the evaluation of climate models is utmost dependent on high quality data. If measurements are biased or error affected, they may be interpreted wrongly as a climate trend or may conceal existing climate signals (Haimberger, 2007).

The many different QC methods can be distinguished according to their usage of prior knowledge (such as error statistics) or model information (such as first guess fields). Some QC procedures are directly embedded in the data assimilation process of NWPs, while others are designed to be stand-alone preprocessing tools. Depending on the availability of single point observations or spatially and temporally distributed measurements, the applicable QC methods range from simple limit and plausibility checks to more sophisticated tests for spatiotemporal consistency. Steinacker et al. (2011) gives a detailed overview of the nowadays common QC methods and summarizes the different error types.

Leading European operational numerical weather prediction centers put a lot of effort in their data assimilation systems that provide the initial conditions for the analyses and forecasts of atmospheric compounds (e.g. European

Centre for Medium-Range Weather Forecasts ECMWF, Rabier et al., 2000; UK's national weather service Met Office, Rawlins et al., 2007; French national meteorological service Météo-France, Fischer et al., 2005 or German meteorological service DWD, Wergen and Buchhold, 2002). The quality control of the used observations is part of the assimilation process (Andersson and Järvinen, 1999), but in most cases these observations have already been checked by preceding QCs carried out at the national institutions (e.g. Spengler, 2002). Smaller institutes usually do not run global models and therefore other methods are used to check the quality of their measurements. Salvati and Brambilla (2008) summarize the many different quality control procedures in Alpine meteorological services. Most of them can not afford the manpower to develop their own sophisticated QCs. This market niche has already been noticed and standalone QC routines can also be purchased (e.g. *QualiMET* used by DWD, Spengler, 2002 or the national meteorological and geophysical service of Austria ZAMG, Adler, 2009). Meanwhile, even open source QC-software has been developed and is distributed through the internet (e.g. *kvoss*, developed by the Norwegian Meteorological Institute (met.no) and the Swedish Meteorological and Hydrological Institute (SMHI)).

At the Department of Meteorology and Geophysics at the University of Vienna, an operational mesoscale analysis of basic meteorological parameters called VERA is carried out on an hourly basis. VERA is the abbreviation for Vienna Enhanced Resolution Analysis and describes a thin-plate spline interpolation method of irregularly distributed observations to a regular grid. A special feature of VERA is the possibility to combine measured values with measurement-independent a priori information called *Fingerprints* (Steinacker et al., 2006).

VERA analyses are used for teaching purposes at the department, for case studies, at field campaigns, for model validation or even for nowcasting at the Austrian Aeronautical Meteorology Service (Austro Control). In all these different fields of applications, the quality of the analyzed data is essential and a sophisticated preprocessing QC is required. Such a QC method has to be optionally independent of any prior knowledge, which is not available at field campaigns, and independent of model data, which would bias a model validation. Therefore, many existing advanced QC procedures (Bayesian QC, QC using optimum interpolation, variational QC) are not qualified for our purposes, and the performance of simpler spatial consistency checks (QC using spatial regression or inverse distance interpolation) was not convincing. An earlier QC procedure for VERA (Steinacker et al., 2000) was already based on the promising concept of minimizing the curvature of the analysis field. Nevertheless, its application was limited (e.g. to two dimensions, no cluster treatment, to a high degree depending on the station distribution) and hence, a new QC method had to be developed, named VERA-QC. The underlying mathematical concepts

of VERA-QC, the recognition of gross errors, and a special treatment of clustered stations have been outlined in a previous paper (Steinacker et al., 2011). The scope of this follow-up article is to discuss the applicability of VERA-QC, its operational applications and improvements, as well as to present some case studies.

In Sect. 2 we present the available data which VERA-QC is applied to operationally, as well as an associated *station selection algorithm* using the concept of the *minimum topography*. Section 3 is concerned with summarizing the basic principles and special features of VERA-QC. Additionally, we discuss the applicability of VERA-QC which depends on the parameter under consideration and scale of phenomenon, as well as on the density of the observation network. The section is concluded by the presentation of recent innovations. Different types of case studies are presented in Sect. 4 in order to demonstrate the efficiency of the different QC modules and the overall performance of VERA-QC itself. In Sect. 5 we give a summary of further fields of applications which made use of VERA-QC, and Sect. 6 contains the conclusions.

2 Data

Every hour, approximately 2000 meteorological surface observations all over Europe are operationally combined to an analysis by VERA. Prior to that, they are checked by the two dimensional VERA-QC. The available data are gathered from SYNOP (surface synoptic observations) and METAR (Aviation routine weather reports) bulletins, as well as from the dense automatic station network of the Austrian Meteorological Service, but originate also from the Regional Meteorological Service of Friuli Venezia Giulia, OSMER (Osservatorio Meteorologico Regionale). The used parameters are precipitation RR, the zonal and meridional wind components u and v, scalar wind speed $|\mathbf{V}|$, potential and equivalent potential temperature Θ and Θ_e and mean sea level (MSL) pressure p_{red}. Note that in a preprocessing step Θ, Θ_e and p_{red} are derived from the observed values for temperature, dew point and surface pressure. This step is necessary because different weather services may use different reduction formulas computing these parameters.

Supposing a two dimensional analysis of the actual ground observations (e.g. temperature measurements) in complex terrain, the predominant visible information is the height dependence. However, most applications require representative information regarding valleys and lowlands, and are not interested in the actual values at the mountain tops. Consequently, a station selection algorithm has been developed to exclude stations that are not representative for these regions. Therefore, the concept of the minimum topography is used.

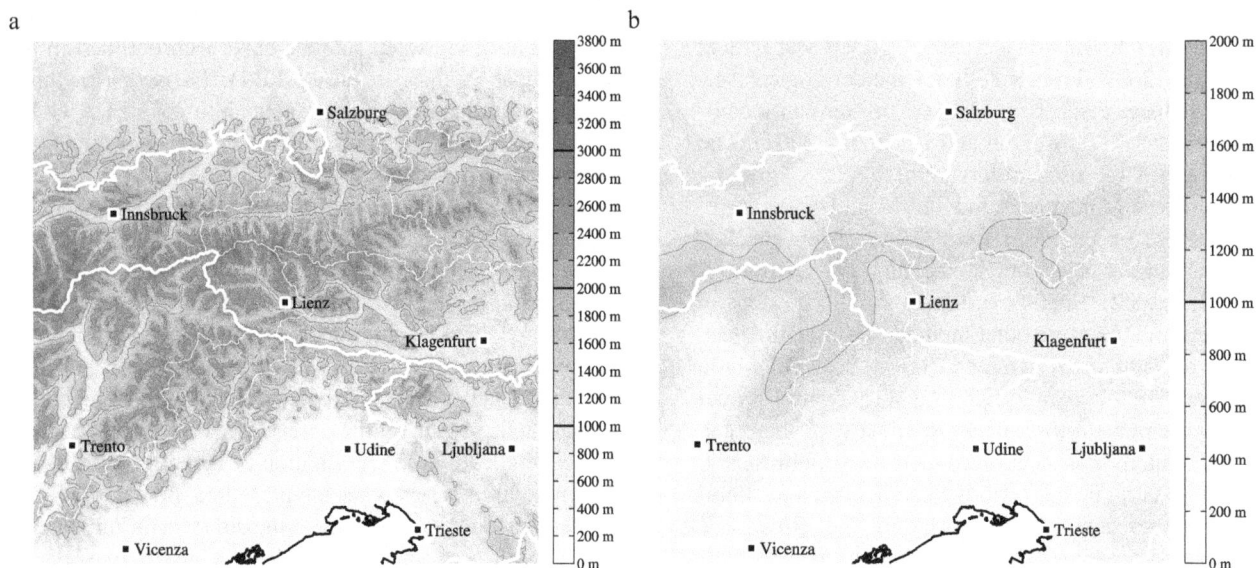

Fig. 1. Comparison of the real topography (**a**) and the idealized minimum topography (**b**) for the Central Alpine domain. The bold white lines illustrate national borders and thin white lines federal state borders. The black solid line in the south shows the coastline of the most northern part of the Mediterranean Sea. Major cities are marked by black squares with white frames. The shading refers to the absolute height above sea level.

2.1 Minimum topography

The minimum topography is derived from the real topography by smoothing summits and mountain ridges, but without lifting valleys, as conventional smoothing algorithms do. Plains are not affected by this modification, valleys and basins are widened, and the inclination of slopes is reduced, which may lead to the reduction of a mountain's elevation. Sufficiently narrow mountains between valleys disappear and, as a consequence, these valleys are connected to a broader artificial valley.

Mathematically, this procedure can be realized by replacing the real height at any point with the minimum height of all points within a certain (radial) distance from the considered point. This radius is typically set to a value of 10 km and the field obtained is smoothed in order to avoid rough structures.

In Fig. 1, the real topography (a) is compared to the idealized minimum topography (b) for a domain covering most of the Eastern Alps. Note that plains are not modified in (b) (e.g. the region between Vicenza and Udine), summits are smoothed (e.g. the main chain of the Alps), and only main valleys are preserved (e.g. Inn Valley with the city of Innsbruck, which is marked in the figure). For a more detailed example (Fig. 2), we consider a north–south cross section (held constant in east-west direction) through the mountains near Innsbruck. In addition to the height of the real topography (bold black solid line), the smoothed as well as the unsmoothed minimum topography is illustrated (bold gray solid line and thin gray solid line). The formerly mentioned properties of the minimum topography (smoothing summits, preserving main valleys,

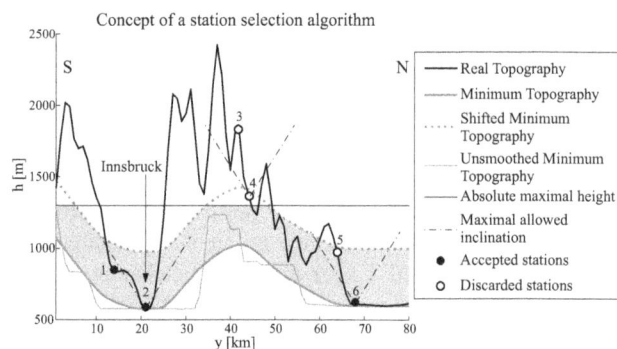

Fig. 2. Concept of the station selection algorithm on the basis of a north–south cross section through the mountains near Innsbruck (compare Fig. 1a). The reasons for accepting stations (compare stations 1, 2 and 6) or for discarding stations (compare stations 3, 4 and 5) are explained in the text. Note that the absolute maximal height was set to $h = 1300$ m (instead of $h = 1500$ m) and the ratio r to $r = 50$ m km^{-1} (instead of $r = 15$ m km^{-1}) in order to create an example containing all three criteria for excluding a station.

preserving main valleys, and leaving plains unmodified) are especially noticeable when comparing the real and the unsmoothed minimum topography.

This procedure is, for example, described in Bica et al. (2007). A similar approach is presented in Haiden (1998) where a comparable idealized topography is named *valley floor surface*.

2.2 Station selection algorithm

There are three criteria for an observation to be excluded from an analysis:

- The absolute height h of a station is too large. Computing reduced values (e.g. p_{red}, Θ, Θ_e), a considerable reduction error can be expected for high stations. Therefore, the station selection algorithm of VERA-QC excludes all pressure observations above $h = 1500$ m.

- The difference between the absolute height h of a station and the height of the smoothed minimum topography h_{MT} exceeds $\Delta h = h - h_{MT} = 400$ m. Note that the vertical distance between valleys in the real topography and those mapped on the minimum topography can exceed a few 100 m, depending on the inclination of the actual valley floor. This threshold is chosen in order to accept all stations along the valley floor, and, in contrast, to exclude stations that are located too high above the valley floor.

- If the ratio of the vertical and the horizontal distance ($r = \Delta z / \Delta x$) between two stations is too large, the higher lying station is excluded. The user defined threshold for the maximal allowed ratio r is defined as $r = 15$ m km^{-1}. The meaning of this criterion becomes obvious when considering exposed wind measurements on hills or temperature measurements above an inversion. These observations would add a micrometeorological error to the analysis because they are not resolvable by the available observational network and therefore are excluded by this third regulation.

Note that, for a station to be excluded, only one criterion has to be fulfilled. Figure 2 illustrates an example station distribution and schematically outlines the different criteria of the station selection algorithm. Accepted stations are symbolized by black circles and discarded stations by white circles. Table 1 shows which regulations are, and which are not fulfilled by each single station.

Naturally, this station selection algorithm can be skipped or altered depending on individual requirements. Operationally, the VERA-QC (including this station selection algorithm) is optimized to deliver the most representative input data for VERA analyses.

3 Methodology

This section summarizes the basic concepts of VERA-QC, it shows which parameters the spatial consistency check can be applied to and it also presents innovations. The latter help to improve the QC of operationally collected data as well as of data collected in complex terrain.

Table 1. Decisions of the station selection algorithm for the six stations in Fig. 2. The columns labeled with Crit 1, Crit 2, and Crit 3 refer to the three criteria listed in Sect. 2.2 (absolute height h, difference Δh between the absolute height h and the height of the minimum topography h_{MT}, and the ratio r of the vertical and the horizontal distance between two stations). Accepted stations are marked by the $\checkmark$ symbol and discarded stations by the $\times$ symbol.

Station	Crit 1	Crit 2	Crit 3	Total
1	$\checkmark$	$\checkmark$	$\checkmark$	$\checkmark$
2	$\checkmark$	$\checkmark$	$\checkmark$	$\checkmark$
3	$\times$	$\times$	$\times$	$\times$
4	$\times$	$\checkmark$	$\times$	$\times$
5	$\checkmark$	$\checkmark$	$\times$	$\times$
6	$\checkmark$	$\checkmark$	$\checkmark$	$\checkmark$

3.1 Basic principles of VERA-QC

Our previous article called *Data Quality Control Based on Self-Consistency* (Steinacker et al., 2011) was dedicated to explaining the mathematical background of VERA-QC in full detail. Subsequently, we give a short outline of these fundamentals.

After data passed simple limit checks, measurements of each parameter are examined for their spatial or, if requested, for their spatiotemporal self-consistency. For parameters Ψ featuring a high redundancy with regard to the available observational network (e.g. $\Psi = \Theta$ or p_{red}), we postulate that the error-free analysis field Ψ_a should be smooth. Mathematically, this can be expressed as an optimization problem minimizing the cost function $J(\Psi_a)$ which is defined as the sum of the squared curvatures $\mathcal{C}$ evaluated at the $n = 1, 2, ..., N(G)$ grid points:

$$J(\Psi_a) = \sum_n^{N(G)} (\mathcal{C}_{\Psi_a})_n^2 = \sum_n^{N(G)} \sum_{d_1=1}^{D} \sum_{d_2=1}^{D} \left(\frac{\partial^2 \Psi_a}{\partial d_1 \partial d_2} \right)_n^2 \rightarrow \min. \quad (1)$$

d_1 and d_2 denote the spatial or temporal coordinates of the considered D dimensional domain, which are used for computing the curvature.

Due to the fact that neither the analysis field Ψ_a nor its curvature are known, $\mathcal{C}(\Psi_a)$ has to be approximated by a first order Taylor series around the curvature of the observed field $\mathcal{C}(\Psi_o)$. Following this approach, we write the cost function J as

$$J = \sum_m \sum_s \sum_{d_1} \sum_{d_2} \left[\frac{\partial^2 \Psi_o}{\partial d_1 \partial d_2} + \sum_p \frac{\partial}{\partial \Psi_p} \left(\frac{\partial^2 \Psi_o}{\partial d_1 \partial d_2} \right) \Delta \Psi_p \right]_s^2. \quad (2)$$

Before explaining Eq. (2) in more detail, we have to define the concept of local neighborhoods. In contrast to some common consistency checks, which work with a constant radius of influence, VERA-QC uses the concept of natural neighbors: with the help of the Delaunay triangulation, a primary neighborhood ($p(m)$, enclosing the first neighbors of the regarded central station m and the central station itself) and a secondary neighborhood ($s(m)$, enclosing the neighbors of the primary stations, the primary stations as well as the regarded central station m) are defined for each station m. The advantage of this method is the possibility to adapt automatically to varying station densities within the whole considered domain.

These definitions in mind, Eq. (2) can be read as follows:

– The cost function J consists of as many terms as stations exist in the considered domain (summation over m).

– For each station, another summation over the corresponding secondary neighborhood (subscript s) is performed. The information for the Taylor series expansion related to the actually considered main station m is gathered from this local secondary neighborhood.

– Note the subscript p ($p \in s$), which denotes the primary neighborhood of the considered main station m. The values of these stations are allowed to vary; in other words, they are allowed to be erroneous.

– The differences $(\Psi_a - \Psi_o) = \Delta\Psi$ are the so-called deviations which present the only unknown variables in Eq. (2). By adding the deviations to the observation field, errors are corrected and the analysis field is obtained.

In order to compute these deviations, the cost function J from Eq. (2) is differentiated with respect to all Ψ_m, which are the values of the analysis field at the positions of the $m = 1, ..., N(\mathcal{M})$ stations in the considered domain. This leads to a system of linear equations that can be solved by matrix inversion.

3.2 Special features of VERA-QC

If the so calculated deviations are applied without further consideration, it could happen that an error propagates from one erroneous station to surrounding stations. The probability for such an error propagation is high in the case of gross errors, or if an erroneous station is located very close to others. In the following three subsections, we present a way to handle these problems.

3.2.1 Reduction of the cost function

Supposing an observation field with one central outlier, the VERA-QC would compute a high deviation for the error affected station, as well as smaller, but still significant deviations for the surrounding stations. In order to find out to what extent a deviation should be applied, its effect on the cost function J is considered. A weighting factor $\mathcal{W}_m$ is obtained by applying only the deviation in question and observing the

relative reduction of the analysis field's curvature. Weighting the deviations with $\mathcal{W}_m$ offers the advantage that deviations caused by the mentioned error propagation are reduced considerably.

3.2.2 Recognition of gross errors

These weights $\mathcal{W}_m$ are also used to recognize gross errors. If the cost function reduction exceeds a user defined threshold, and additionally, if the weighted deviation is large compared to all the other weighted deviations in the domain, the observation is assumed to present a gross error and therefore is discarded. After eliminating these gross errors, VERA-QC is repeated for the remaining stations.

Operationally, a cost function reduction of 80 % is chosen to indicate a potential gross error. Additionally, the absolute value of the involved weighted deviation has to exceed the median of all absolute deviations within the whole domain by a certain factor (in the operational setting 30 times).

3.2.3 Cluster treatment

The concept of weighting with the cost function reduction usually works very well. Nevertheless, there are still some problems if a station, located very close to others, is erroneous. Therefore, close stations are combined to a fictive cluster station. At first, VERA-QC is applied to this modified station distribution. The computed weighted deviation for such a fictive cluster station is assigned and applied to the single cluster members. Then, VERA-QC is executed for the second time. Finally, the proposed corrections from both iterations are combined to a final deviation.

3.3 Applicability of VERA-QC

The spatial consistency check of VERA-QC is only applicable to measurements featuring a high degree of redundancy. This property is fulfilled if the scale of the considered phenomenon is large compared to the mean distance between the observation points. Looking at the available surface observation network and the investigated parameters (described in Sect. 2), the requirement of a high redundancy is not satisfied by the parameter precipitation and especially not by convective precipitation.

A method to quantify this criterion is to compare the mean station distance $\bar{d}$ to the parameter specific decorrelation distance d_0. For this purpose we consider hourly pressure and precipitation observations from June 2011. In order to compute the parameter specific decorrelation distances, all time series of observations were compared to each other and the corresponding correlation coefficients were calculated. In a further step they can be described by an analytic autocorrelation function $\rho(d)$. The decorrelation distance d_0 is defined as the distance where the autocorrelation function reaches the value $\rho(d_0) = e^{-1} \approx 0.37$.

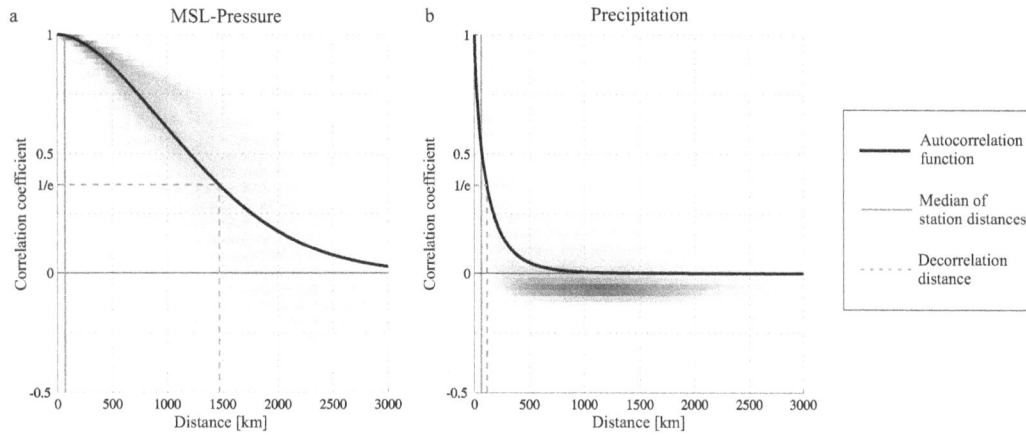

Fig. 3. Comparison of the autocorrelation for the parameters MSL-pressure (**a**) and precipitation (**b**) as a function of the station distances based on the hourly measurements of June 2011. The correlation is presented by the adjusted analytical autocorrelation function (black bold line). In addition to the decorrelation distance (gray dashed line), the median of all distances between neighboring stations is marked by a gray vertical line. Whereas in (**a**) the median of station distances is small compared to the decorrelation distance, in (**b**) they are of comparable magnitude.

Weber and Talkner (1993) give an overview of different autocorrelation functions of varying complexity. For our simple purpose, which is to compare the different spatial properties of precipitation and MSL-pressure, we chose the following exponential model as it is described, for example, in Gebremichael and Krajewski (2004):

$$\rho(d) = \exp\left[-\left(\frac{d}{d_0}\right)^{s_0}\right]. \tag{3}$$

The shape factor s_0 allows the autocorrelation function to differ from the common form of the exponential function. These two parameters d_0 and s_0 were adjusted with the help of a least square algorithm.

Figure 3 shows the autocorrelation as a function of the distance between all stations for the parameters MSL-pressure (a) and precipitation (b). Instead of illustrating all the single scattered points $(d, \rho(d))$, the density of this point cloud is indicated by different shades of gray. In addition, the point cloud is described analytically by Eq. (3) and is symbolized by the black bold line. The dashed gray line indicates the decorrelation distance, whereas the vertical gray line illustrates the median of all distances between neighboring stations.

Note that, for adequately resolving a phenomenon, the mean distance between neighboring stations $\bar{d}$ has to be much smaller than the decorrelation distance d_0 of the considered parameter. As the two graphs in Fig. 3 show, the parameter MSL-pressure fairly fulfills this criterion ($\bar{d} = 68$ km $\ll d_0 = 1475$ km), precipitation, however, does not ($\bar{d} = 62$ km $\approx d_0 = 113$ km).

It should be mentioned that the autocorrelation is not only a function of distance, but is also influenced by additional factors such as orography. Two (even more distant) stations at the same side of a main mountain ridge will most likely feature a higher autocorrelation than two stations separated by the mountain ridge (Lanzinger and Steinacker, 1990).

3.4 Innovations of VERA-QC

Since many observations in the operationally considered European domain are collected in complex terrain, special challenges regarding temperature measurements arise. Therefore, instead of considering potential temperature, the difference between Θ and the corresponding value of the standard atmosphere is controlled. Starting with the operational execution of VERA-QC, the deviations are stored continually and offer the possibility to compute a bias correction and are also used to improve the gross error recognition. In the following three subsections we will present these innovations.

3.4.1 Consideration of relative values

If the horizontal VERA-QC would be applied to temperature measurements, the temperatures of the few elevated stations would be increased significantly. In order to avoid this behavior, the potential temperature Θ is considered. Nevertheless, there still arises a problem. In general, the actual lapse rate is smaller than the dry-adiabatic lapse rate, but the computation of Θ bases on the assumption of a dry-adiabatic atmosphere. As a consequence, the 2-D VERA-QC systematically produces negative deviations for elevated stations as shown in Fig. 4. This behavior is more strongly pronounced in summer months due to the strong local heating in valleys (reduced air volume) and above elevated plains.

To overcome this problem, the relative value Θ_{rel}, computed as the difference between Θ and the corresponding value of the standard atmosphere at station height Θ^*, is calculated in a preprocessing step. This implicates the assumption that the actual lapse rate is approximated by the

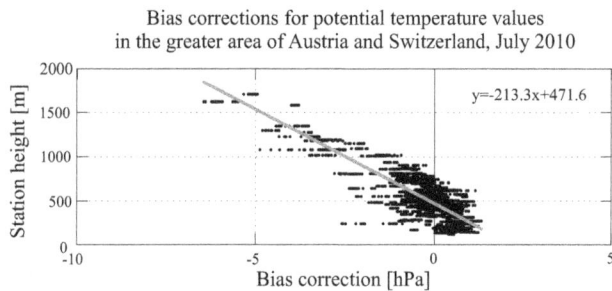

Fig. 4. Bias correction (black dots) for Θ as a function of the station height, evaluated for stations in the greater area of Austria and Switzerland in July 2010. The regression line (gray line) shows that the higher the stations, the more negative the bias corrections are. The absolute height limit of the station selection algorithm (operationally 1500 m) was raised to 1800 m for this case study.

moist-adiabatic lapse rate ($\Gamma_S = 0.0065 \, \text{K} \, \text{m}^{-1}$), which is in accordance with the mean atmospheric conditions. The modified variable Θ_{rel} has the advantage to be distributed more smoothly even in complex terrain and a 2-D VERA-QC can be applied to Θ_{rel} without producing additional systematic errors.

Figure 5 illustrates the improvements that can be attributed to the consideration of the relative value Θ_{rel}. The application of VERA-QC to Θ values yields the results presented in (a), whereas when applied to Θ_{rel} values (b), the systematic dependence on the height is reduced to a minimum.

For computing the equivalent potential temperature Θ_e, the lapse rate is already assumed to be the moist-adiabatic one. Thus, the described problem is not as pronounced for Θ_e as it is for Θ.

3.4.2 Bias correction

If a measurement of a single station is affected by a systematic error persisting in time (e.g. always yields in too high values), VERA-QC will compute deviations of the opposite sign (e.g. negative deviations) in order to correct the erroneous measurement. The operationally stored deviations for such a station are characterized by a median with absolute value significantly greater than zero. This shift in the deviations can be used to compute a bias correction.

The bias correction is defined as the median of the unweighted deviations $\Delta \Psi$ that have been collected within a parameter specific time interval. For the parameters p_{red}, u, v, and $|\mathbf{V}|$, this time interval spans the last 30 days (720 hourly observations). The deviations for the parameters Θ and Θ_e, as well as the parameters themselves, are subject to seasonal variations. Therefore, this time interval should span a whole year, but in order to save computational effort, the deviations of the last month and those of the month half a year ago are taken into account.

Figure 6 illustrates the chronological sequence of unweighted deviations for a station with bias affected pressure measurements. After 30 days, the first bias correction is computed and can be applied to the next day's measurements. Henceforward, at the end of each day, a new bias correction is computed. Note that the parameter specific time interval moves as well. As an example, the bias corrections for MSL-pressure, valid for 31 July 2010, are shown in Fig. 7. As apparent in this figure, the bias corrections are spatially uncorrelated. Note the high magnitudes of the corrections, which are mainly due to reduction errors, caused by erroneous station heights.

Operationally, the bias correction is carried out at the beginning of VERA-QC. As a result, the roughness of the observation field is already reduced before minimizing the curvature of the analysis field. The deviations of the bias corrected measurements are rather distributed around zero and are mainly caused by random errors. If a systematic error featured a high magnitude, it often occurred that VERA-QC without bias correction identified the affected measurement to be a gross error. Thanks to the bias correction, observations of such stations are retained and increase the amount of utilizable data.

3.4.3 Improvement of gross error detection

The stored deviations can also be used for computing variable station specific thresholds to detect gross errors. These limits enhance the previously described gross error recognition, which is based on the impact a deviation has on the curvature of the analysis field (correcting a gross error leads to a significantly lower curvature, cf. Sect. 3.2.2).

Every month, new thresholds are computed based on the last month's weighted deviations. For the definition of a lower limit (L_1) and an upper limit (L_2), the first and the third quartile (Q_1 and Q_3), as well as the interquartile range $\text{IQR} = Q_3 - Q_1$, are used:

$$L_1 = Q_1 - 3 \, \text{IQR}, \qquad L_2 = Q_3 + 3 \, \text{IQR}. \tag{4}$$

An observation is identified as a gross error if its value does not fall within the interval $[L_1, L_2]$ and if, additionally, its absolute value exceeds a user defined constant limit. Note that, for the gross error recognition, only one criterion (the one described in this section or the one in Sect. 3.2.2) has to be fulfilled.

Figure 8 illustrates this concept for the equivalent potential temperature values of an exemplary station for November 2010. The user defined constant limit is set to $\pm 4 \, °\text{C}$ (thin solid black lines) for this parameter. Both thresholds (black dash dotted lines) are used in December 2010 for the gross error recognition.

If only the individual time series of measurements would be used for a station specific gross error detection based on statistical thresholds, extreme weather related observations are likely to be rejected by mistake. Using deviation time series instead offers additional spatial information to identify unusual atmospheric phenomena.

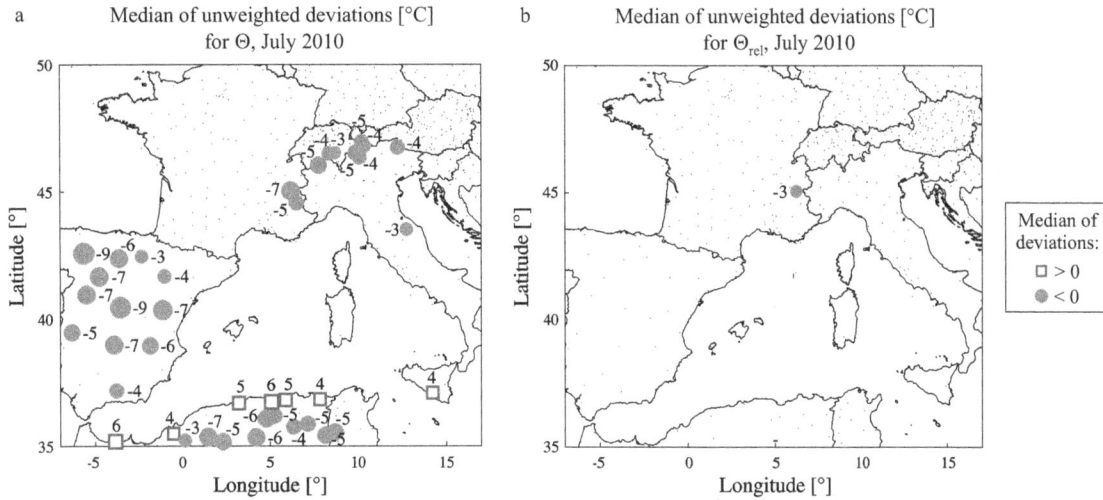

Fig. 5. Comparison of the median of unweighted deviations computed for Θ (**a**) and for the relative value $\Theta_{rel} = \Theta - \Theta^*$ (**b**) in July 2010 for a domain covering the Southwest Europe. Applying the VERA-QC to the potential temperature Θ (**a**), values of elevated stations are reduced systematically (circles), whereas values of neighboring coastal stations are increased (squares). If VERA-QC is applied to the difference between measurements and corresponding values of the standard atmosphere (Θ_{rel}), this height depending behavior is reduced significantly (**b**). For better visibility, only deviations of $|\Delta\Theta| \geq 3\,°C$ are illustrated.

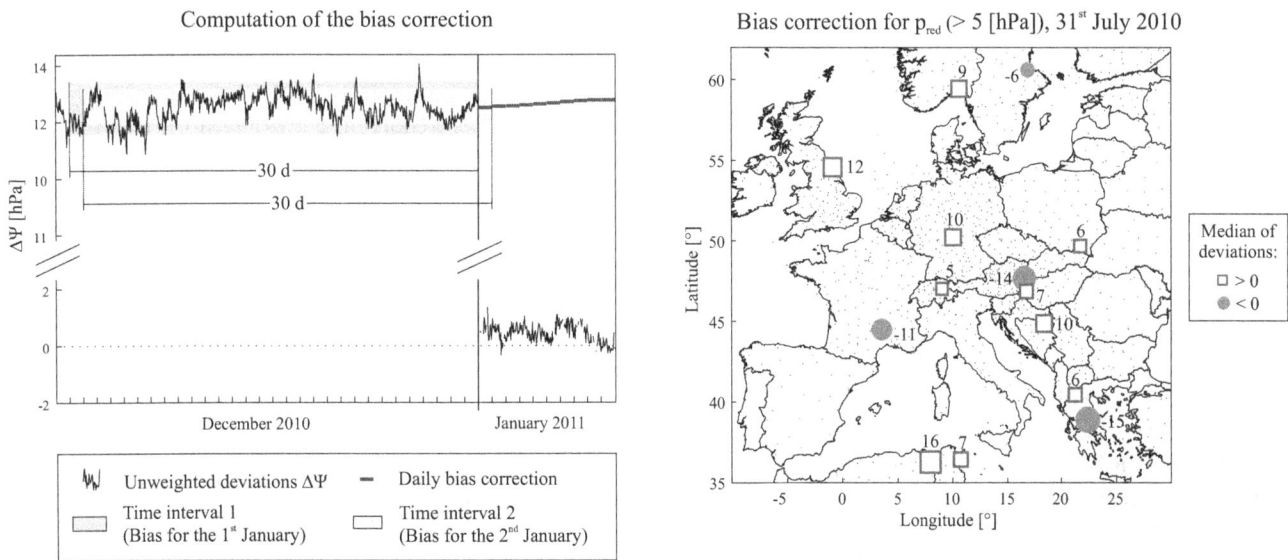

Fig. 6. Schematic illustration of computing a bias correction. The daily bias correction (dark solid horizontal bars starting at the beginning of January 2011) is defined as the median of the unweighted deviations (rough solid black line) within the user defined time interval (e.g. the last 30 days) and is valid for the next day. Note that the bias correction is allowed to vary and can automatically adapt to new situations (e.g. calibration of the instrument or changing station distributions).

Fig. 7. Example spatial distribution of the bias corrections $> 5\,hPa$ for MSL-pressure, valid for 31 July 2010 within the European domain. Circles symbolize negative bias corrections, and squares mark positive ones. In most cases, these can be explained by reduction errors (e.g. due to wrong station heights).

Within the two preceding chapters, which describe the station selection algorithm and the methodology of VERA-QC, many different threshold values have been presented. The process of optimizing the chosen threshold values can be considered as an iterative process taking quite some time.

Experienced meteorologists supervise and evaluate the results achieved by a certain set of thresholds. According to their suggestions, the thresholds are altered. This procedure is being repeated constantly with the goal to optimize the proportion of rejected observations and that of accepted non-representative observations for our available observational network (thus, for the available station network density). There are no objective configurations valid for an arbitrary

Computation of a station specific upper and lower limit
for the Gross Error recognition

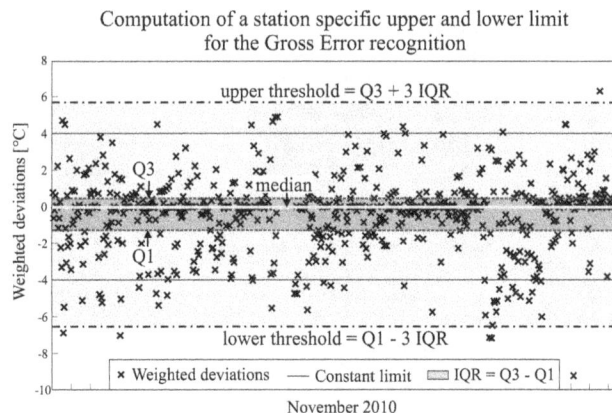

Fig. 8. Weighted deviations (black crosses) for Θ_e of an exemplary station for November 2010. These constitute the basis for the computation of a variable station specific lower (L_1) and an upper (L_2) limit (black dash dotted lines), which support the gross error recognition. Q_1 and Q_3 denote the first and third quartiles (dark gray dashed lines) and IQR the interquartile range ($Q_3 - Q_1$, dark gray area). Additionally to the variable limits, constant values (thin solid black lines) of $\pm 4\,^{\circ}\mathrm{C}$ have to be exceeded in order to recognize an observation as a gross error. In December 2010, all observations with weighted deviations outside the interval $[L_1, L_2]$ (light gray area) would be identified as gross errors.

observational network and the human component is essential at this point.

The setting of such thresholds is strongly constrained by the intended usage of observations and the relationship between hits, misses, false alarms and correct rejections has to be optimized individually. However, the intention of VERA-QC is to provide representative measurements regarding the available observation network. This means, VERA-QC only accepts atmospheric phenomena of a scale that is large compared to the mean station distance.

Concluding this methodology chapter, it should be emphasized that the VERA-QC's strictness is depending on the local station density. In areas of a denser station network, VERA-QC can – with a higher reliability – control, confirm, or criticize measurements that were caused by phenomena of a wider range of meteorological scales. The probability of gross error detection for an error of the same magnitude decreases with increasing median station distance. This can be explained by the fact that an error of a certain magnitude leads in a dense observation network to a higher curvature of the observation field than it does in areas of lower station density. In order to quantify this relationship, we investigated the dependency of the absolute value of a deviation (identified as a gross error) on the median of station distances (considering only primary neighbors). Expressing this relationship in terms of a linear regression, we found a dependence of approximately $+3.5\,\mathrm{K}/1000\,\mathrm{km}$ regarding potential temperature as an example (based on results from July 2011 to June 2012).

4 Results and examples

The following two sections present the changes and improvements visible in the VERA analysis if VERA-QC is used for preprocessing. First, monthly means of gridded hourly MSL-pressure values for July 2011 are considered. Second, the gradual improvements by applying different QC modules are investigated. In a third section, the 2-D and 3-D VERA-QC is applied to a high-resolution precipitation data set. Real problems that arise when working with raw data collected at a field campaign are addressed.

4.1 Effects of VERA-QC visible in mean values

In order to investigate how the bias correction (Sect. 3.4.2) copes with systematic errors, monthly mean values of gridded hourly MSL-pressure observations are considered. This strategy was chosen because random errors are eliminated by averaging and thus do not distort the improvements caused by the bias correction.

After the hourly measurements of July 2011 have been analyzed with the help of the analysis tool VERA, these hourly gridded data were averaged. First, this procedure was applied to the pure, uncorrected observations (Fig. 9a) and secondly, to the quality controlled data (Fig. 9b). VERA-QC included all presented innovations, especially the bias correction.

In July 2011, the predominant pressure distribution was characterized by two lows (Baltic Sea and Aegean Sea) and a high pressure system (Bay of Biscay). Averaging the analyzed uncorrected observations, the resulting pressure patterns are mostly covered by systematic errors. In Fig. 9a, these errors are noticeable as locally persistent, mesoscale pressure perturbations. Some high-magnitude errors increase the curvature of the analysis field significantly.

When applying VERA-QC without bias correction to these observations, most of the systematic errors are recognized as gross errors and therefore are discarded. As a result, one would obtain an averaged field comparable to the one in Fig. 9b. But the advantage of the bias correction becomes obvious when investigating the amount of recognized gross errors. Whereas 9193 measurements (0.99 %) within July 2011 have been recognized as gross errors by VERA-QC without bias correction, this number has been reduced to 3522 (0.38 %) by applying the bias correction. Especially in areas with only few observations, no measurement should be discarded unnecessarily. Figure 9b (QC with bias correction) illustrates the predominant pressure patterns of July 2011 without troublesome noise.

4.2 Step by step effects of VERA-QC

The VERA-QC tool allows to activate or deactivate many different QC-modules as they were described in Sect. 3: the usage of the weighting factor, the recognition of gross errors, the cluster treatment, the consideration of relative values (for

Monthly mean of the analyzed hourly pressure observations for July 2011

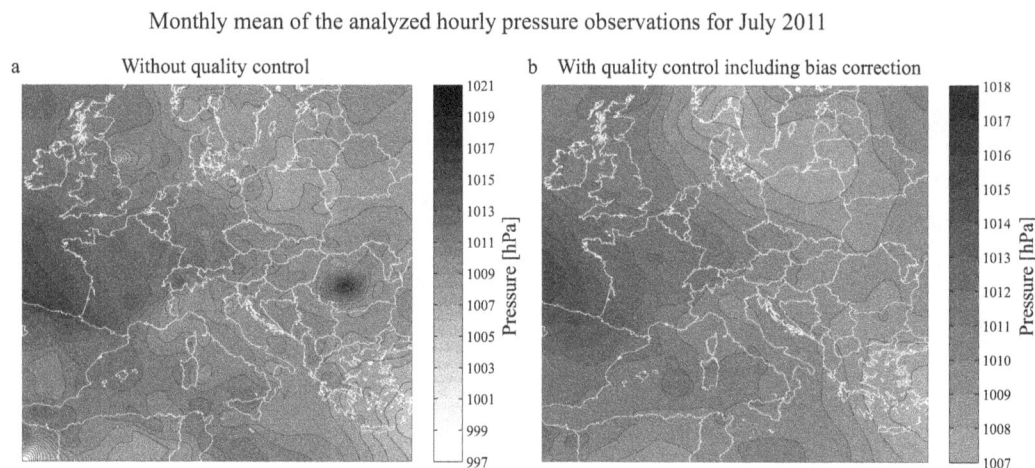

Fig. 9. Comparison of monthly mean values of the analyzed hourly MSL-pressure observations for July 2011 without QC (**a**) and with QC including bias correction (**b**). The hourly measurements have been analyzed with the help of the analysis tool VERA, the monthly mean values are based on the gridded data. Systematic errors, especially those with high magnitude (e.g. caused by reduction errors), have been corrected. As a result, the mean gridded values in (**b**) are distributed more smoothly.

potential temperature) and the bias correction. In Steinacker et al. (2011), we focused on presenting the effectiveness of the modules "usage of the weighting factors", "gross error recognition", and "cluster treatment". The "consideration of relative values" was the topic of Sect. 3.4.1 and in the following, the last module "bias correction" is investigated in more detail by means of a case study.

Conducting a field experiment or a case study, normally no prior knowledge to compute the bias correction is available and the corresponding bias correction module can not be executed. In order to present the quality achievable by VERA-QC with and without bias correction for one special case, Fig. 10 compares the original pressure observations of 5 July 2011, 06:00 UTC (a) to the quality controlled measurements after applying VERA-QC without (b) and with (c) activated bias correction. Note that both quality controlled fields (b) and (c) illustrate significant improvements compared to (a), and no essential difference between (b) and (c) is noticeable. Thus, the overall quality achievable for data collected at field campaigns or for case studies (in both cases no prior knowledge is available) is comparable to the quality achieved by a full QC run including bias correction. The advantage of a full VERA-QC, however, is that the amount of recognized gross errors and therefore rejected observations is reduced significantly (in the presented case: from 0.89 % to 0.38 %). Depending on the individual requirements, it can be important to keep as many observations as possible. Climate studies, for instance, would intend to retain as many uninterrupted time series as possible and thus would profit from a bias correction.

4.3 Usage of VERA-QC for field campaigns

In order to present the versatility of VERA-QC and how it could be used at field campaigns or for case studies, we intend to focus on different possible applications of VERA-QC to data, that were collected in 2007 at the field campaign COPS (Convective and Orographically-induced Precipitation Study, Wulfmeyer et al., 2011). We use measurements from the mesonet operated by the University of Vienna, which was located in the Black Forest region in the South West of Germany. Ninety-six automatic weather stations (type HOBO) were arranged to resemble a regular grid with a mean distance between stations of approximately one kilometer. The high spatial and temporal resolution of observations ($\Delta x \approx \Delta y \approx 1$ km, $\Delta t = 1$ min, but in the case of precipitation – due to the relatively low detection resolution $\Delta RR = 0.2$ mm – accumulated to $\Delta t = 10$ min for post-processing) allows to apply the 2-D or 3-D VERA-QC even to convective precipitation values as explained in Sect. 3.3. The considered time period (6 h from 18:00 to 24:00 UTC on 20 June 2007) was chosen because in this interval an intensive precipitation event occurred.

Depending on the particular requirements when using COPS data, an accurate QC has to fulfill certain specifications. One might be interested in keeping and therefore correcting as many observations as possible (minimal or no recognition of gross errors by the QC) or one might allow the QC to eliminate suspicious measurements. Thus, the optimal settings for a QC depend on the types of errors the user wants to discard from the data set by all means (e.g. random-, gross- or systematic errors). In the following subsections, we will present three case studies outlining how the settings of VERA-QC can be adapted to different requirements.

Comparison of different VERA-QC levels

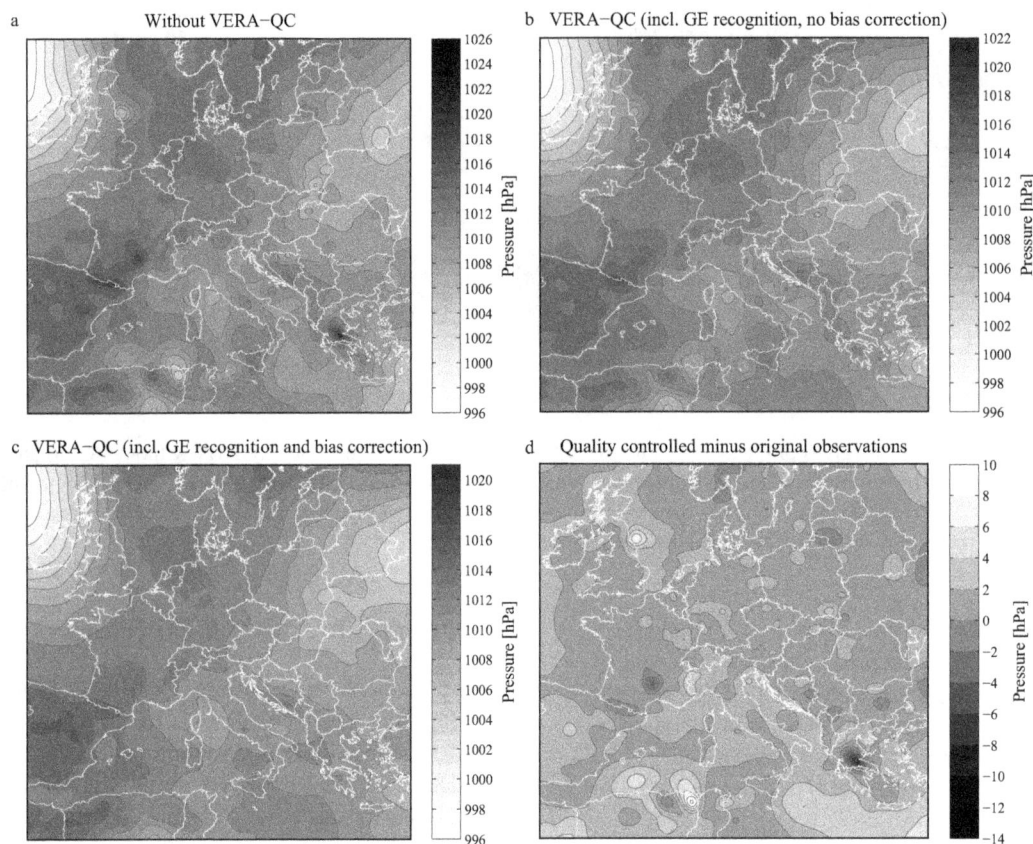

Fig. 10. Illustration of how observations are gradually improved by the VERA-QC. For all four figures, the values at the station positions in the European domain are interpolated to a regular grid by the analysis tool VERA. (**a**) Shows the original MSL-pressure values, computed from the 1567 observations on 5 July 2011, 06:00 UTC. (**b**) Presents the quality controlled observations, however no bias correction was applied. An amount of 0.89 % of all observations are identified as gross errors and thus are excluded from the analysis. (**c**) Shows the results after applying the complete QC (with bias correction). The amount of gross errors is reduced to 0.38 %. Finally, in (**d**) the differences between the quality controlled and original observations are illustrated.

Note that, because of the regular station distribution, clustering was deactivated. Furthermore, the cost function reduction limit for detecting gross errors was set to 20 % in all cases. Before controlling three dimensional (x, y, t) data with VERA-QC, the spatial and temporal resolutions of observations get adjusted to be comparable (accounting for the typical speed of propagation of the considered phenomenon). This is done by multiplying the temporal coordinates by a so-called anisotropic factor, formally defined as a velocity. Additionally, this factor can also be used to weight the spatial or temporal coherence individually.

4.3.1 QC of random tipping detected by rain gauges

A well known problem when working with rain gauges is the random (temporally singular) erroneous registration of the minimal resolvable reading (0.2 mm) although no precipitation occurs. This random tipping may be caused by

insects inside the gauge or by external forces shaking the instrument.

Figure 11 presents the observed 10 min precipitation values of 20 June 2007, 23:20 UTC. At this time two stations reported erroneous random tipping (light gray bars). Supposing the user does not allow to discard any observations, the gross error recognition module of VERA-QC was deactivated. Whereas the 2-D VERA-QC (x, y) is able to correct the two erroneous observation values to an extent of 90 %, the 3-D VERA-QC $(x, y, 18:00 \leq t \leq 24:00$ UTC) improves the percentage of correction to 98.5 %. The higher efficiency of a 3-D VERA-QC was already presented in Steinacker et al. (2011) by considering a comparable analytical example (see Fig. 9 of Steinacker et al., 2011).

Observations and corrected observations

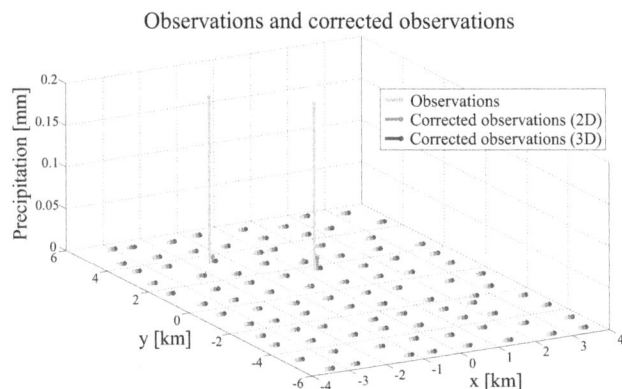

Fig. 11. Illustration of the station positions and observed 10 minutes precipitation values on 20 June 2007, 23:20 UTC, collected during the field campaign COPS (light gray bars). Note the two readings of 0.2 mm which are very likely due to erroneous random tipping or due to read out of memory, because no precipitation was observed at this time. The medium gray bars symbolize the observations corrected by a two dimensional (x, y) VERA-QC without gross error recognition, and the dark gray bars stand for those corrected by a three dimensional (x, y, t) VERA-QC without gross error recognition taking into account all measurements between 18:00 and 24:00 UTC.

4.3.2   QC of a 2-D high resolution precipitation field

When conducting measurements, it can always occur that one instrument fails. As in our case, we consider a rain gauge which registered constantly 0 mm independent of the occurrence of precipitation. Suppose doing a 2-D case study about a special precipitation event without having prior knowledge concerning previous or following measurements, one has to work, for example, with the information illustrated in Fig. 12 as light gray bars. At first sight, the observation marked by the black arrow seems to be erroneous. Applying the 2-D VERA-QC (x, y) leads to the corrected observations, shown as dark gray bars in Fig. 12. The considered erroneous measurement was successfully adapted to its surrounding observations, which all registered a heavy precipitation event, even though the erroneous station is located at the edge of the observation field.

4.3.3   QC of a systematic error without prior knowledge

Although an erroneous instrument may in fact be able to register precipitation, it may occur that a temporally persistent constant value is added to the observations. Such a case is illustrated in Fig. 13. The measurements of the erroneous rain gauge are marked by black dots, the observations of 12 neighboring stations are symbolized by gray asterisks. Their comparison indicates a bias of 2 mm in 10 min for the former mentioned erroneous observation. If a case study demands to achieve a high accuracy, one has to accept that observations are flagged as gross errors and therefore are

Observations and corrected observations

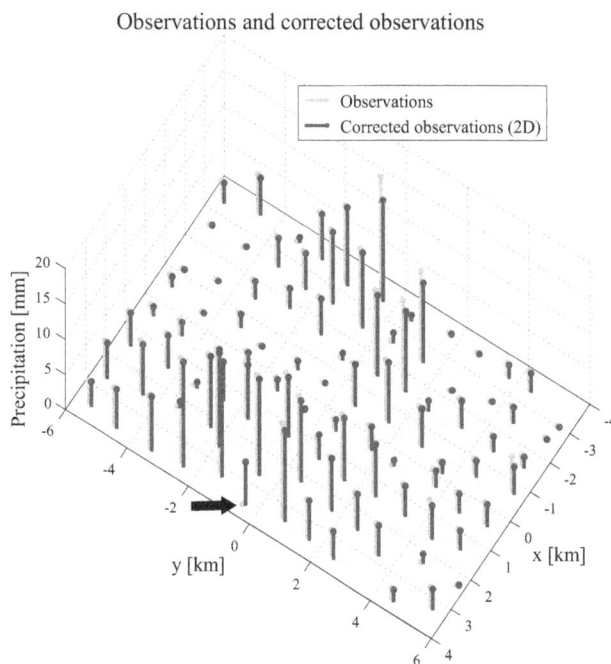

Fig. 12. Presentation of the observed 10 min precipitation values on 20 June 2007, 21:40 UTC as light gray bars. The dark gray bars mark the corrected observations after a two dimensional (x, y) VERA-QC (without gross error recognition) was applied. The black arrow points at a station that was located within a field of heavy precipitation but nevertheless was stuck at a reading of zero millimeters. This malfunction could be observed for the entire period of 20 June 2007.

excluded. For the considered time interval (spanning 19:30 to 24:00 UTC) a 3-D VERA-QC (x, y, t) with gross error recognition can be applied. In Fig. 13, the corrected observations are connected by the black and gray solid lines, respectively. Prior to and after the precipitation event centered at 22:00 UTC, the black solid line is interrupted, indicating that the biased observations were identified as gross errors and thus were excluded. Where no or only little precipitation was observed, the 3-D VERA-QC is able to correctly identify the biased measurement as a gross error. During the precipitation event, this erroneous observation is adjusted to the measurements in its surroundings. Although no bias correction was applied, the 3-D VERA-QC successfully handled this systematic error.

5   Applications

Quality controlled measurements constitute the basis for many different applications. Furthermore, also the computed deviations themselves offer valuable information. The following two sections overview different possibilities to make use of VERA-QC and its results.

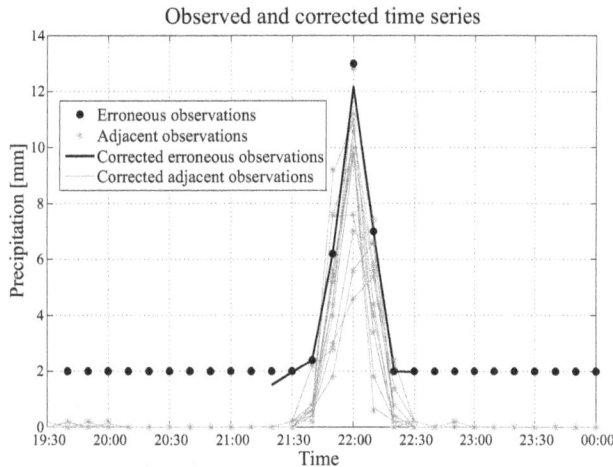

Fig. 13. Time series of observed (dots and asterisks) and corrected (solid lines) 10 min precipitation values on 20 June 2007. VERA-QC was carried out including all three dimensions (x, y, t). In this period, one station (symbolized by the black markers) constantly registered precipitation values of 2 mm even though no precipitation was observed in its surroundings (symbolized by the gray markers). Note that if the black line (presenting corrected values) is missing, the corresponding observation was identified to be a gross error and hence is excluded.

5.1 Usage of quality controlled measurements

At the department of Meteorology and Geophysics at the University of Vienna, several research projects have already been concerned with the application and improvement of VERA-QC. Many case studies have been conducted using different kinds of data sources, for example conventional GTS (Global Telecommunication Systems) data or even special observations taken at field experiments such as COPS. Another area of research within our department focuses on model verification. It is self-evident that high quality reference data are essential and VERA-QC was found to be able to provide these.

The most eminent application of VERA-QC is its usage as a preprocessing tool for the operational (hourly) analysis software VERA, which VERA-QC was named after. The so-called VERA-package (VERA-QC and VERA) delivers an exceptional performance and is therefore operationally used by the Austrian Aeronautical Meteorology Service (part of Austro Control) for nowcasting purposes. Currently, the VERA-package is also modified and prepared to be installed for operational use at the Swiss Federal Office of Meteorology and Climatology (MeteoSwiss).

5.2 Usage of deviations computed by VERA-QC

In Gorgas and Dorninger (2011), the idea of an analysis ensemble of deterministic, model-independent analyses is proposed. In the course of ensemble generation, the observations are manipulated using stochastic perturbations. One approach is to use the deviations computed by VERA-QC for determining the magnitude of these perturbations.

Investigating the temporal variations in the deviations helped Sperka and Steinacker (2011) to develop a method for creating a homogenized 3-hourly MSL-pressure analysis data set. After determining breaks in the deviations time series, the measurements within these breaks are corrected by the bias calculated from the deviations.

6 Conclusions and final remarks

As the mathematical concept of VERA-QC has already been described in Steinacker et al. (2011), this follow-up article is concerned with the operational application of the VERA-QC, innovations and the performance of the operational VERA-QC demonstrated with the aid of case studies.

First of all, the applicability of VERA-QC to different meteorological parameters was discussed and we showed that, depending on the density of the observational network, even precipitation can be controlled by VERA-QC as demonstrated in the results section. Accounting for special difficulties in complex terrain, a station selection algorithm that excludes high-elevation stations that are not comparable to those in valleys and lowlands was introduced.

As a consequence of the operational application of VERA-QC, we had the possibility to develop new QC-modules. The collected deviations are used to continually compute a bias correction and variable station-specific limits, which support the gross error recognition. Additionally, the consideration of relative values helps to minimize the deviation's dependency on height, a problem that occurs especially in complex terrain.

In order to point out the positive influence of VERA-QC on a subsequent analysis, two case studies focus on the improvements achievable by using a bias correction as well as an enhanced gross error recognition. The possible application of VERA-QC to data collected at the field campaign COPS is discussed in more detail, including the use of a 3-D (x, y, t) VERA-QC. In the concluding section, the many different fields of application of VERA-QC as well as of the computed deviations are summarized.

VERA-QC including the herein described innovations has been successfully carried out since January 2010. The proposed corrections and hence the performance of VERA-QC are continually being inspected by experienced meteorologists. This human quality control of VERA-QC ensures an optimal performance and recognizes potential for improvements. One suggestion for the operational use is to weight the observations with a priori knowledge about the trustworthiness of a station's measurement, for example, primarily trust the observations taken by a well maintained station at an airport. Another idea is to compute the curvature and its derivations directly by fitting a second order polynomial surface through a station and its surroundings instead of computing the curvature by differentiating the interpolated field at local grid points. This would avoid using an inverse distance interpolation method for computing the values at the local grid points and, additionally, there would be no need to use finite differences for the numerical approximation when computing the curvature.

Concluding, we would like to advert to our web site http://www.univie.ac.at/amk/veraflex/test/public/ showing the latest VERA analyses for many different domains and parameters. Note that, by keeping the left mouse button pressed at a station symbol, not only the measured and corrected values used for the VERA analysis are displayed but also the weighted and unweighted deviations as well as the bias corrections. Furthermore, we point out that VERA-QC is made available for scientific use.

Acknowledgements. Thanks are due to the Austrian Science Fund (Fonds zur Förderung der wissenschaftlichen Forschung, FWF; P19658) and to the Austrian Research Funding Association (Die Österreichische Forschungsförderungsgesellschaft, FFG; project 818110) for partial financial support of this work.

Edited by: M. Genzer

References

Adler, S.: Datenkontrolle der ZAMG mit dem Prüfprogramm QUALIMET von der Firma Ernst Basler & Partner, internal report, Vienna, Austria, 2009.

Andersson, E. and Järvinen, H.: Variational quality control. Q. J. Roy. Meteorol. Soc., 125, 697–722, 1999.

Bica, B., Steinacker, R., Lotteraner, C., and Suklitsch, M.: A new concept for high resolution temperature analysis over complex terrain, Theor. Appl. Climatol., 90, 173–183, 2007.

Fischer, C., Montmerle, T., Berre, L., Auger, L., and Stefanescu, S. E.: An overview of the variational assimilation in the ALADIN/France numerical weather-prediction system, Q. J. Roy. Meteorol. Soc., 131, 3477–3492, 2005.

Gandin, L. S.: Complex Quality Control of Meteorological Observations, Mon. Weather Rev., 116, 1137–1156, 1988.

Gebremichael, M. and Krajewski, W. F.: Assessment of the Statistical Characterization of Small-Scale Rainfall Variability from Radar: Analysis of TRMM Ground Validation Datasets, J. Appl. Meteorol., 43, 1180–1199, 2004.

Gorgas, T. and Dorninger, M.: Concepts for a pattern-oriented analysis ensemble based on observational uncertainties, Q. J. Roy. Meteorol. Soc., 138, 769–784, 2011.

Häggmark, L., Ivarsson, K., Gollvik, S., and Olofsson, P.: Mesan, an operational mesoscale analysis system, Tellus A, 52, 2–20, 2000.

Haiden, T.: Analytical aspects of mixed-layer growth in complex terrain. Preprints, Eighth Conference on Mountain Meteorology, Amer. Meteor. Soc., Flagstaff, Arizona, 368–372, 1998.

Haimberger, L.: Homogenization of Radiosonde Temperature Time Series Using Innovation Statistics, J. Climate, 20, 1377–1403, 2007.

Lanzinger, A. and Steinacker, R.: A Fine Mesh Analysis Scheme Designed for Mountainous Terrain, Meteorol. Atmos. Phys., 43, 213–219, 1990.

Norwegian Meterological Institute: kvalobs open source software project (kvqc2d, version: 1.3.1), an Open Source Software for the Quality Control of Geophysical Observations, http://kvalobs.wiki.met.no (last access: 4 October 2012), 2011.

Rabier, F., Järvinen, H., Klinker, E., Mahfouf, J.-F., and Simmons, A.: The ECMWF operational implementation of four-dimensional variational assimilation, I: Experimental results with simplified physics, Q. J. Roy. Meteorol. Soc., 126, 1143–1170, 2000

Rawlins, F., Ballard, S. P., Bovis, K. J., Clayton, A. M., Li, D., Inverarity, G. W., Lorenc, A. C., and Payne, T. J.: The Met Office global four-dimensional variational data assimilation scheme, Q. J. Roy. Meteorol. Soc., 133, 347–362, 2007.

Salvati, M. and Brambilla, E.: Data quality control procedures in Alpine meteorological services, Tech. rep., Regional Agency for Environmental Protection of Lombardia, Università degli Studi di Trento Dipartimento di Ingegneria Civile e Ambientale, 2008.

Spengler, R.: The new Quality Control- and Monitoring System of the Deutscher Wetterdienst, WMO Techn. Conf. on Meteorol. and Environm. Instrum. and Methods of Observ., Bratislava, 2002.

Sperka, S. and Steinacker, R.: A Quality-Control and Bias-Correction Method Developed for Irregularly Spaced Time Series of Observational Pressure Data, J. Atmos. Ocean. Tech., 28, 1317–1323, 2011.

Steinacker, R., Häberli, C., and Pöttschacher, W.: A Transparent Method for the Analysis and Quality Evaluation of Irregularly Distributed and Noisy Observational Data, Mon. Weather Rev., 128, 2303–2316, 2000.

Steinacker, R., Ratheiser, M., Bica, B., Chimani, B., Dorninger, M., Gepp, W., Lotteraner, C., Schneider, S., and Tschannett, S.: A Mesoscale Data Analysis and Downscaling Method over Complex Terrain, Mon. Weather Rev., 134, 2758–2771, 2006.

Steinacker, R., Mayer, D., and Steiner, A.: Data Quality Control Based on Self-Consistency, Mon. Weather Rev., 139, 3974–3991, 2011.

Weber, R. O. and Talkner, P.: Some Remarks on Spatial Correlation Function Models, Mon. Weather Rev., 121, 2611–2617, 1993.

Wergen, W. and Buchhold, M.: Datenassimilation für das Globalmodell GME, Promet, 27, 150–155, 2002.

WMO: Guide on the Global Data-processing System, World Meteorological Organization, WMO (Series), no. 305, 1993d Edn., http://www.wmo.int/e-catalog/detail_en.php?PUB_ID=380&SORT=N&q= (last access: 4 October 2012), 1993.

WMO: Manual on the Global Oberving System, World Meteorological Organization, WMO (Series), no. 544, 2003d Edn., www.wmo.int/pages/prog/www/OSY/Manual/WMO544.pdf (last access: 4 October 2012), 2003.

WMO: Guide to the Global Observing System, World Meteorological Organization, WMO (Series), no. 488, 2007th Edn., http://www.wmo.int/pages/prog/www/OSY/Manual/488_Guide_2007.pdf (last access: 4 October 2012), 2007.

WMO: Guide to Meteorological Instruments and Methods of Observation, World Meteorological Organization, WMO No. 8, 7th Edn., http://www.wmo.int/pages/prog/gcos/documents/gruanmanuals/CIMO/CIMO_Guide-7th_Edition-2008.pdf (last access: 4 October 2012), 2008.

Wulfmeyer, V., Behrendt, A., Kottmeier, C., Corsmeier, U., Barthlott, C., Craig, G., Hagen, M., Althausen, D., Aoshima, F., Arpagaus, M., Bauer, H., Bennett, L., Blyth, A., Brandau, C., Champollion, C., Crewell, S., Dick, G., Di Girolamo, P., Dorninger, M., Dufournet, Y., Eigenmann, R., Engelmann, R., Flamant, C., Foken, T., Gorgas, T., Grzeschik, M., Handwerker, J., Hauck, C., Höller, H., Junkermann, W., Kalthoff, N., Kiemle, C., Klink, S., König, M., Krauss, L., Long, C., Madonna, F., Mobbs, S., Neininger, B., Pal, S., Peters, G., Pigeon, G., Richard, E., Rotach, M., Russchenberg, H., Schwitalla, T., Smith, V., Steinacker, R., Trentmann, J., Turner, D., van Baelen, J., Vogt, S., Volkert, H., Weckwerth, T., Wernli, H., Wieser, A., and Wirth, M.: The Convective and Orographically-induced Precipitation Study (COPS): the scientific strategy, the field phase, and research highlights, Q. J. Roy. Meteorol. Soc., 137, 3–30, 2011.

Permissions

All chapters in this book were first published in GIMDS, by Copernicus Publications; hereby published with permission under the Creative Commons Attribution License or equivalent. Every chapter published in this book has been scrutinized by our experts. Their significance has been extensively debated. The topics covered herein carry significant findings which will fuel the growth of the discipline. They may even be implemented as practical applications or may be referred to as a beginning point for another development.

The contributors of this book come from diverse backgrounds, making this book a truly international effort. This book will bring forth new frontiers with its revolutionizing research information and detailed analysis of the nascent developments around the world.

We would like to thank all the contributing authors for lending their expertise to make the book truly unique. They have played a crucial role in the development of this book. Without their invaluable contributions this book wouldn't have been possible. They have made vital efforts to compile up to date information on the varied aspects of this subject to make this book a valuable addition to the collection of many professionals and students.

This book was conceptualized with the vision of imparting up-to-date information and advanced data in this field. To ensure the same, a matchless editorial board was set up. Every individual on the board went through rigorous rounds of assessment to prove their worth. After which they invested a large part of their time researching and compiling the most relevant data for our readers.

The editorial board has been involved in producing this book since its inception. They have spent rigorous hours researching and exploring the diverse topics which have resulted in the successful publishing of this book. They have passed on their knowledge of decades through this book. To expedite this challenging task, the publisher supported the team at every step. A small team of assistant editors was also appointed to further simplify the editing procedure and attain best results for the readers.

Apart from the editorial board, the designing team has also invested a significant amount of their time in understanding the subject and creating the most relevant covers. They scrutinized every image to scout for the most suitable representation of the subject and create an appropriate cover for the book.

The publishing team has been an ardent support to the editorial, designing and production team. Their endless efforts to recruit the best for this project, has resulted in the accomplishment of this book. They are a veteran in the field of academics and their pool of knowledge is as vast as their experience in printing. Their expertise and guidance has proved useful at every step. Their uncompromising quality standards have made this book an exceptional effort. Their encouragement from time to time has been an inspiration for everyone.

The publisher and the editorial board hope that this book will prove to be a valuable piece of knowledge for researchers, students, practitioners and scholars across the globe.

List of Contributors

B. U. E. Brändström
Swedish Institute of Space Physics, Kiruna, Sweden

C.-F. Enell
Sodankylä Geophysical Observatory, University of Oulu, Sodankylä, Finland

O. Widell
SSC, ESRANGE, Kiruna, Sweden

T. Hansson
SSC, ESRANGE, Kiruna, Sweden

D. Whiter
Finnish Meteorological Institute, Helsinki, Finland

S. Mäkinen
Finnish Meteorological Institute, Helsinki, Finland

D. Mikhaylova
Swedish Institute of Space Physics, Kiruna, Sweden

K. Axelsson
Swedish Institute of Space Physics, Kiruna, Sweden

F. Sigernes
The Kjell Henriksen Observatory, UNIS, Longyearbyen, Norway

N. Gulbrandsen
University of Tromsø, Tromsø, Norway

N. M. Schlatter
School of Electrical Engineering, Royal Institute of Technology, Stockholm, Sweden

A. G. Gjendem
Norwegian University of Science and Technology, Trondheim, Norway

L. Cai
Department of Physics, University of Oulu, Oulu, Finland

J. P. Reistad
University of Bergen, Bergen, Norway

M. Daae
Norwegian University of Science and Technology, Trondheim, Norway

T. D. Demissie
Norwegian University of Science and Technology, Trondheim, Norway

Y. L. Andalsvik
Department of Physics, University of Oslo, Oslo, Norway

O. Roberts
Aberystwyth University, Aberystwyth, UK

S. Poluyanov
Polar Geophysical Institute, Murmansk, Russia

S. Chernouss
Polar Geophysical Institute, Apatity, Russia

N. I. Kömle
Space Research Institute, Austrian Academy of Sciences, Graz, Austria

W. Macher
Space Research Institute, Austrian Academy of Sciences, Graz, Austria

G. Kargl
Space Research Institute, Austrian Academy of Sciences, Graz, Austria

M. S. Bentley
Space Research Institute, Austrian Academy of Sciences, Graz, Austria

L. M. Kistler
University of New Hampshire, Durham, NH, USA

C. G. Mouikis
University of New Hampshire, Durham, NH, USA

K. J. Genestreti
University of New Hampshire, Durham, NH, USA
Southwest Research Institute, San Antonio, TX, USA

A. Khokhlov
International Institute of Earthquake Prediction Theory and Mathematical Geophysics 79, b2, Warshavskoe shosse, 113556 Moscow, Russia
Geophysical Center of RAS, 3 Molodezhnaya St., 119296 Moscow, Russia
Institut de Physique du Globe de Paris, UMR7154, CNRS – 1 Rue Jussieu, 75005 Paris, France

J. L. Le Mouël
Institut de Physique du Globe de Paris, UMR7154, CNRS – 1 Rue Jussieu, 75005 Paris, France

M. Mandea
Centre National d'Etudes Spatiales, 2, Place Maurice Quentin, 75001 Paris, France

L. N. S. Alconcel
Department of Physics, The Blackett Laboratory, Imperial College London, Prince Consort Road, London, SW7 2BW, UK

P. Fox
Department of Physics, The Blackett Laboratory, Imperial College London, Prince Consort Road, London, SW7 2BW, UK

P. Brown
Department of Physics, The Blackett Laboratory, Imperial College London, Prince Consort Road, London, SW7 2BW, UK

T. M. Oddy
Department of Physics, The Blackett Laboratory, Imperial College London, Prince Consort Road, London, SW7 2BW, UK

E. L. Lucek
Department of Physics, The Blackett Laboratory, Imperial College London, Prince Consort Road, London, SW7 2BW, UK

C. M. Carr
Department of Physics, The Blackett Laboratory, Imperial College London, Prince Consort Road, London, SW7 2BW, UK

A. Blagau
Institute for Space Sciences, Bucharest, Romania

I. Dandouras
Institut de Recherche en Astrophysique et Planétologie, Université de Toulouse, Toulouse, France
CNRS, Institut de Recherche en Astrophysique et Planétologie, Toulouse, France

A. Barthe
Institut de Recherche en Astrophysique et Planétologie, Université de Toulouse, Toulouse, France
CNRS, Institut de Recherche en Astrophysique et Planétologie, Toulouse, France
AKKA Technologies, Toulouse, France

S. Brunato
Institut de Recherche en Astrophysique et Planétologie, Université de Toulouse, Toulouse, France
CNRS, Institut de Recherche en Astrophysique et Planétologie, Toulouse, France
Noveltis, Toulouse, France

G. Facskó
Laboratoire de Physique et Chimie de l'Environnement et de l'Espace, Orléans, France
Geodetic and Geophysical Institute, Research Centre for Astronomy and Earth Sciences, HAS, Sopron, Hungary
Finnish Meteorological Institute, Helsinki, Finland

V. Constantinescu
Institute for Space Sciences, Bucharest, Romania

R. Nakamura
Space Research Institute, Austrian Academy of Sciences, 8042 Graz, Austria

F. Plaschke
Space Research Institute, Austrian Academy of Sciences, 8042 Graz, Austria

R. Teubenbacher
Space Research Institute, Austrian Academy of Sciences, 8042 Graz, Austria

L. Giner
Graz University of Technology, 8010 Graz, Austria

W. Baumjohann
Space Research Institute, Austrian Academy of Sciences, 8042 Graz, Austria

W. Magnes
Space Research Institute, Austrian Academy of Sciences, 8042 Graz, Austria

M. Steller
Space Research Institute, Austrian Academy of Sciences, 8042 Graz, Austria

R. B. Torbert
University of New Hampshire, Durham, NH 03824, USA

H. Vaith
University of New Hampshire, Durham, NH 03824, USA

M. Chutter
University of New Hampshire, Durham, NH 03824, USA

K.-H. Fornaçon
Institut für Geophysik und extraterrestrische Physik, Technische Universität Braunschweig, 38106 Braunschweig, Germany

K.-H. Glassmeier
Institut für Geophysik und extraterrestrische Physik, Technische Universität Braunschweig, 38106 Braunschweig, Germany

C. Carr
Blackett Laboratory, Imperial College London, London, UK *now at: Materials Center Leoben Forschung GmbH, Leoben, Austria

M. van de Kamp
Finnish Meteorological Institute, P.O. Box 503, 00101 Helsinki, Finland

Y. V. Khotyaintsev
Swedish Institute of Space Physics, Uppsala, Sweden

P.-A. Lindqvist
Royal Institute of Technology, Stockholm, Sweden *now at: Department of Physics and Astronomy, University of Calgary, Calgary, Canada

C. M. Cully
Swedish Institute of Space Physics, Uppsala, Sweden

A. I. Eriksson
Swedish Institute of Space Physics, Uppsala, Sweden

M. André
Swedish Institute of Space Physics, Uppsala, Sweden

N. Doss
Mullard Space Science Laboratory, University College London, Dorking, UK

A. N. Fazakerley
Mullard Space Science Laboratory, University College London, Dorking, UK

B. Mihaljčić
Mullard Space Science Laboratory, University College London, Dorking, UK

A. D. Lahiff
Mullard Space Science Laboratory, University College London, Dorking, UK
Rutherford Appleton Laboratory, Oxford, UK

R. J. Wilson
Mullard Space Science Laboratory, University College London, Dorking, UK
University of Colorado Boulder, Colorado, USA

D. Kataria
Mullard Space Science Laboratory, University College London, Dorking, UK

I. Rozum
Mullard Space Science Laboratory, University College London, Dorking, UK
European Centre for Medium-Range Weather Forecasts, Reading, UK

G. Watson
Mullard Space Science Laboratory, University College London, Dorking, UK

Y. Bogdanova
Mullard Space Science Laboratory, University College London, Dorking, UK
Rutherford Appleton Laboratory, Oxford, UK

T. A. Boden
Oak Ridge National Laboratory, Carbon Dioxide Information Analysis Center, Oak Ridge, TN 37831-6290, USA

M. Krassovski
Oak Ridge National Laboratory, Carbon Dioxide Information Analysis Center, Oak Ridge, TN 37831-6290, USA

B. Yang
Oak Ridge National Laboratory, Carbon Dioxide Information Analysis Center, Oak Ridge, TN 37831-6290, USA

F. Sigernes
University Centre in Svalbard, Longyearbyen, Norway

S. E. Holmen
University Centre in Svalbard, Longyearbyen, Norway

D. Biles
Magnetosphere Ionosphere Research Lab, University of New Hampshire, USA

H. Bjørklund
University Centre in Svalbard, Longyearbyen, Norway

X. Chen
University Centre in Svalbard, Longyearbyen, Norway

M. Dyrland
University Centre in Svalbard, Longyearbyen, Norway

D. A. Lorentzen
University Centre in Svalbard, Longyearbyen, Norway

L. Baddeley
University Centre in Svalbard, Longyearbyen, Norway

T. Trondsen
Keo Scientific Ltd, Calgary, Alberta, Canada

U. Brändström
Swedish Institute of Space Physics, Kiruna, Sweden

E. Trondsen
Department of Physics, University of Oslo, Oslo, Norway

B. Lybekk
Department of Physics, University of Oslo, Oslo, Norway

J. Moen
Department of Physics, University of Oslo, Oslo, Norway

S. Chernouss
Polar Geophysical Institute, Murmansk Region, Apatity, Russia

C. S. Deehr
Geophysical Institute, University of Alaska, Fairbanks, USA
Birkeland Centre for Space Science, University of Bergen, Bergen, Norway

C. R. Clauer
Center for Space Science Engineering and Research, Virginia Polytechnic Institute and State University, Blacksburg, VA, USA
Bradley Department of Electrical and Computer Engineering, Virginia Polytechnic Institute and State University, Blacksburg, VA, USA

H. Kim
Center for Space Science Engineering and Research, Virginia Polytechnic Institute and State University, Blacksburg, VA, USA
Bradley Department of Electrical and Computer Engineering, Virginia Polytechnic Institute and State University, Blacksburg, VA, USA

K. Deshpande
Center for Space Science Engineering and Research, Virginia Polytechnic Institute and State University, Blacksburg, VA, USA
Bradley Department of Electrical and Computer Engineering, Virginia Polytechnic Institute and State University, Blacksburg, VA, USA

Z. Xu
Center for Space Science Engineering and Research, Virginia Polytechnic Institute and State University, Blacksburg, VA, USA
Bradley Department of Electrical and Computer Engineering, Virginia Polytechnic Institute and State University, Blacksburg, VA, USA

D. Weimer
Center for Space Science Engineering and Research, Virginia Polytechnic Institute and State University, Blacksburg, VA, USA
Bradley Department of Electrical and Computer Engineering, Virginia Polytechnic Institute and State University, Blacksburg, VA, USA

S. Musko
Space Physics Research Laboratory, University of Michigan, Ann Arbor, MI, USA

G. Crowley
Atmospheric & Space Technology Research Associates, Boulder, CO, USA

C. Fish
Space Dynamics Laboratory, Utah State University, Logan, Utah, USA

R. Nealy
Bradley Department of Electrical and Computer Engineering, Virginia Polytechnic Institute and State University, Blacksburg, VA, USA

T. E. Humphreys
Department of Aerospace Engineering and Engineering Mechanics, The University of Texas at Austin, Austin, TX, USA

J. A. Bhatti
Department of Aerospace Engineering and Engineering Mechanics, The University of Texas at Austin, Austin, TX, USA

A. J. Ridley
Space Physics Research Laboratory, University of Michigan, Ann Arbor, MI, USA

P. Robert
Laboratoire de Physique des Plasmas, CNRS, Palaiseau, France

N. Cornilleau-Wehrlin
Laboratoire de Physique des Plasmas, CNRS, Palaiseau, France
LESIA-Observatoire de Paris, CNRS, Meudon, France

R. Piberne
Laboratoire de Physique des Plasmas, CNRS, Palaiseau, France

Y. de Conchy
LESIA-Observatoire de Paris, CNRS, Meudon, France

C. Lacombe
LESIA-Observatoire de Paris, CNRS, Meudon, France

V. Bouzid
Laboratoire de Physique des Plasmas, CNRS, Palaiseau, France

B. Grison
Institute of Atmospheric Physics, Prague, Czech Republic

D. Alison
Laboratoire de Physique des Plasmas, CNRS, Palaiseau, France

P. Canu
Laboratoire de Physique des Plasmas, CNRS, Palaiseau, France

M. Díaz-Michelena
National Institute of Aerospace Technology, Ctra. de Ajalvir, km 4, 28850 Torrejón de Ardoz, Madrid, Spain

R. Sanz
National Institute of Aerospace Technology, Ctra. de Ajalvir, km 4, 28850 Torrejón de Ardoz, Madrid, Spain

M. F. Cerdán
National Institute of Aerospace Technology, Ctra. de Ajalvir, km 4, 28850 Torrejón de Ardoz, Madrid, Spain
Departamento de Física de la Tierra, Astronomía y Astrofísica I, Universidad Complutense de Madrid, Pza. de las Ciencias, s/n, 28040 Madrid, Spain
CNR-IMM MATIS at Dipartimento di Fisica e Astronomia, Università di Catania, Via S. Sofia 64, 95123 Catania, Italy

A. B. Fernández
National Institute of Aerospace Technology, Ctra. de Ajalvir, km 4, 28850 Torrejón de Ardoz, Madrid, Spain

D. Mayer
Department of Meteorology and Geophysics, University of Vienna, Vienna, Austria

A. Steiner
Department of Meteorology and Geophysics, University of Vienna, Vienna, Austria

R. Steinacker
Department of Meteorology and Geophysics, University of Vienna, Vienna, Austria